T0203085

Measurement Models for Psychological Attributes

Chapman & Hall/CRC Statistics in the Social and Behavioral Sciences

Series Editors

Jeff Gill
Washington University, USA

Steven Heeringa
University of Michigan, USA

Wim J. van der Linden
University of Twente
Enschede, the Netherlands

Tom Snijders
Oxford University, UK
University of Groningen, NL

For more information about this series,
please visit: https://www.crcpress.com/go/ssbs

Measurement Models for Psychological Attributes

Klaas Sijtsma
L. Andries van der Ark

CRC Press
Taylor & Francis Group
Boca Raton London New York

CRC Press is an imprint of the
Taylor & Francis Group, an **informa** business

A CHAPMAN & HALL BOOK

First published in 2020
by CRC Press
Taylor & Francis Group
6000 Broken Sound Parkway NW, Suite 300
Boca Raton, FL 33487-2742

and by CRC Press
2 Park Square, Milton Park, Abingdon, Oxon, OX14 4RN

© 2021 by Taylor & Francis Group, LLC

CRC Press is an imprint of Taylor & Francis Group, an Informa business
No claim to original U.S. Government works

International Standard Book Number-13: 978-1-4398-8134-7 (Hardback)
International Standard Book Number-13: 978-0-367-42452-7 (Paperback)
International Standard Book Number-13: 978-0-4291-1244-7 (eBook)

Library of Congress Cataloging-in-Publication Data

Library of Congress Control Number: 2020942992

Visit the Taylor & Francis Web site at
http://www.taylorandfrancis.com

and the CRC Press Web site at
http://www.crcpress.com

Contents

Acknowledgement

Scientists were already undertaking measurement efforts of psychological attributes in the nineteenth century, but at that time psychological measurement looked more like physical measurement, for example, taking the pressure exercised by the squeeze of the hand as the expression of mental energy. Measurement of psychological attributes, such as intelligence and personality traits, originated in the beginning of the twentieth century. From our perspective, measurement thus became better recognizable as psychological, but this also introduced reliance on human behavior instead of physical properties. Even though human behavior probably is much more informative about psychological phenomena than physical properties, we say that reliance on human behavior is problematic, because like the weather, economic processes, and societal developments, human behavior shows patterns that seem volatile and rarely seem to reveal underlying causes and processes. One is interested in a person's intelligence but can only consider her dexterity to rotate geometric shapes mentally to a known position and her ability to find her way through a maze to reach a target object. When we assess a person's neuroticism, we ask her questions about how she responds in typical situations and hope her answers are informative about neuroticism. These are not bad ideas, but their operationalization results in data that are noisy and contain many signals the origins of which we do not know. Whether we are really measuring intelligence or neuroticism requires inference and the availability of well-founded theories about the attributes of interest. It is here that measurement efforts are most vulnerable, because such theories are rare, weak, or simply absent.

On the contrary, the development of statistical measurement models has seen an enormous growth over the past century, and it is intriguing that we now know so much more about formal requirements of psychological measurement than about the attributes we intend to measure. The gap this has created between psychometrics and psychology or, if one wishes, the backlog of psychology with respect to psychometrics is both fascinating and alarming. However, one may also argue that the lead psychometrics seems to have is only appearance without reference to reality, precisely because we know so little about many attributes, their structure, and the cognitive and affective processes on which they rest. So, how could psychometric models be correct in describing real-life attributes other than by coincidence? One might argue that the models rather posit desiderata for measurement of which no one knows precisely whether they are consistent with most or even some psychological attributes.

Several of our colleagues have noticed the gap between psychometrics and psychology and expressed their concern. What is necessary is that psychology focuses more on theory development with respect to the attributes psychologists measure and find pivotal in their research. Likewise, psychometrics probably must do much of the same thing, that is, contribute to the development of attribute theory and then provide the correct model for constructing useful scales to measure these attributes. Much interesting work has been done, for example, in intelligence, but the road probably is long, a situation not much different from the long history of measurement in physics. It seems to us, given where we are now, that we should give up on the idea that psychological measurement is quantitative, providing interval and ratio scales. Such claims require strong theory about attributes and natural laws that connect different measurements in a multiplicative way, but psychology has neither strong theory nor natural laws, at least, not now. The best that psychological measurement can attain, we believe, is the ordering or the classification of people with respect to an attribute. In this book, we explain that this is what psychological measurement has to offer. And it is not bad at all!

The book is useful for graduate and PhD students, researchers in the areas of social science, psychology, and health, and for fellow psychometricians. Some basic knowledge of psychometrics and statistics helps the reader to work her way through the chapters. Knowledge of at least one of the human sciences helps the reader to appreciate the problematic nature of non-physical measurement. The book may be of help in academic courses on measurement and as a guide for practical instrument construction. The last part of Chapter 1 gives an overview of the measurement models that we discuss in the book. In a nutshell, they are classical test theory and factor analysis (Chapter 2), nonparametric item response theory and Mokken scale analysis (Chapter 3), parametric item response theory and structural extensions (Chapter 4), latent class models and cognitive diagnostic models (Chapter 5), and pairwise comparison, unfolding, response time, and network models (Chapter 6).

John Kimmel of CRC press was endlessly patient with us after the first author had to interrupt writing when a scientific fraud case in his School required him in 2011 to become dean for what turned out to be a long period. At least this provided him with a sabbatical leave six years later that very much helped in finally shaping the book. The department of Psychological Methods of the University of Amsterdam, in particular, its chair, Han van der Maas, kindly provided hospitality to the first author and offered a stimulating environment in which writing the book was a treat. The second author moved from Tilburg University to the department of Child Development and Education of the University of Amsterdam in 2014. Our close proximity during the year of writing, no more than a pedestrian bridge linking two university buildings, surely helped the process of writing along.

Unlike many other books on psychometrics, the book that lies before you gives you a rather complete overview of the methods available, but we must admit that we also left out several small-scale but valuable efforts, only because the book would become too long and the distinction between ripe and unripe methods might fade. We also left out all of the monumental measurement theory that Krantz, Luce, Suppes, and Tversky (1971) and other authors proposed, because we expected this theory would be so much different from what modern psychometrics has to say that we feared it would be almost unrecognizable for most readers. We distinguished accepted psychometric methods that are used "frequently" and assigned them separate chapters, from psychometric methods that are used "not as frequently", which we collected in Chapter 6. We emphasize that these rarer methods may have great psychometric potential and practical importance; they only are relatively unknown and less frequently used, but this may change eventually and for some methods sooner than foreseeable. Throughout the book, we emphasize the dominance that psychological theory about attributes must have over psychometric theory; measurement without theory is either art or technology, measurement based on theory is science.

The book profited from the first author's visits in 2018 to Jimmy de la Torre at the University of Hong Kong and Matthew Johnson at Columbia University, New York. Inspiring discussions during the time of writing with many colleagues and receiving help from others greatly contributed to shaping the thought processes that produced the chapters. We explicitly thank Anton Béguin, Riet van Bork, Denny Borsboom, Elise Crompvoets, Jimmy de la Torre, Wilco Emons, Sacha Epskamp, Arne Evers, Cees Glas, Willem Heiser, Bas Hemker, Matthew Johnson, Maarten Marsman, Dylan Molenaar, James Ramsay, Jim Roberts, Tom Snijders, Jesper Tijmstra, Han van der Maas, and Matthias von Davier. We also thank the many colleagues we hope do not mind we forgot to thank them here. Several colleagues provided their comments on earlier drafts of the book chapters. In alphabetical order, they are Daniel Bolt, Samantha Bouwmeester, Elise Crompvoets, Jules Ellis, Sacha Epskamp, Jimmy de la Torre, Cees Glas, Bas Hemker, Matthew Johnson, Maarten Marsman, Dylan Molenaar, Jeroen Vermunt, and Matthias von Davier. Their attempts to keep us from making mistakes deserve our deepest gratitude. We are thankful to Samantha Bouwmeester, Johan Denollet, Leo van Maanen, and Gina Rossi for making their data available to us. Samantha Bouwmeester provided the artwork in Figure 1.1, and Lars de Boer provided the artwork in Figure 2.1.

Our former and present PhD students contributed considerably to the development of modern psychometrics, but we could not include all of their contributions in this book. This only means that we needed to focus and that some of their work, however much appreciated, did not fall within the boundaries we set for the book. We gratefully acknowledge the influence that our students had on psychometrics and our ideas about the field. The Department of Methodology and Statistics of Tilburg University proved to be a reliable home base providing inspiring research projects and discussions for two decades. We will not ignore the fact that the book profited from the endless discussions we had among each other for some 20 years, and the

numerous papers that resulted from these discussions. We are grateful to
Christine Selvan at Deanta Global for proofreading and typesetting the book
manuscript. We thank John Kimmel of CRC Press not only for his patience
but also for his wise and experienced guidance throughout the process of
writing and editing. The data sets and the statistical software we used in the
examples can be found in the Open Science Framework: https://osf.io/e9jrz/.
Anything incorrect that remains in the text is entirely our responsibility.

Klaas Sijtsma and L. Andries van der Ark

Authors

Klaas Sijtsma is professor of Methods of Psychological Research at the Tilburg School of Social and Behavioral Sciences, Tilburg University, the Netherlands. His research specializes in psychometrics, in particular, all issues related to the measurement of psychological attributes by means of tests and questionnaires. He is a past President of the Psychometric Society, an editorial board member for several journals, and has authored two other books on measurement.

L. Andries van der Ark is professor of Psychometrics at the Research Institute of Child Development and Education, Faculty of Social and Behavioral Sciences, University of Amsterdam, the Netherlands. His primary research interests include reliability analysis and nonparametric item response theory. The authors have published over 40 papers together on measurement in the social and behavioral sciences.

Glossary of Notation and Acronyms

Notation

Notes:

a. Below, notation passes in the approximate order in which it is encountered when reading the book cover to cover.

b. Notation incidentally used is mostly left out to avoid the already long list from becoming even longer; exceptions are made depending on the importance of the topic discussed.

c. Arithmetic and logical operators are not included in the list.

d. Some notation can have varying meaning throughout the book, but the meaning will be clear from the context in which the notation is used; below, occasionally, chapter reference is added for notation carrying a particular meaning.

e. Finally, with many models to discuss, and for each model, many methods, techniques, and quantities, it is impossible not to run out of notation and in different contexts, use the same notation with different meanings. We have done our best to use common notation for quantities that appear throughout the text (e.g., X_j always stands for the score on item j), and for each topic, we have unambiguously defined the notation needed.

j, k, l	item indices
J	number of items in a test or questionnaire
X_j	score on item j
x_j	realization of X_j
X	unweighted sum of J item scores, sum score, test score
M_j+1	number of different integer item scores x_j, $x_j=0, ..., M_j$
N	size of sample of observations
$\mathbf{X}$	data matrix having N rows and J columns, containing item scores x
$R_{(j)}$	rest score, sum score on $J-1$ items not including item j
r	realization of rest score $R_{(j)}$, $r=0, ..., (J-1)M$
n_r	sample frequency of people having rest score r (Chap. 3)

n_{jr}	sample frequency of people having both a score of 1 on item j and rest score r (Chap. 3)
N_{jx}	random variable of people having score 1 on binary scored item j and sum score x on test; $x = 1, \ldots, J - 1$, $x = 0$ and $x = J$ are uninformative (Chap. 4)
n_{jx}	realization of N_{jx} (Chap. 4)
n_x	sample frequency of people having sum score x (Chap. 4)
$n_\mathbf{x}$	sample frequency of people having item-score vector $\mathbf{x} = (x_1, \ldots, x_J)$; $n_\mathbf{x} = n_{x_1 x_2 \ldots x_J}$ (Chap. 5)
$e_\mathbf{x}$	expected frequency having item-score vector $\mathbf{x}$ based on latent class model (Chap. 5)
u	index for item-score vectors $\mathbf{x}, u, 1, \ldots, M^J$ (Chap. 5)
$n_{\mathbf{x}_u}$	sample frequency of people having item-score vector $\mathbf{x}$ indexed u (Chap. 5)
i, v, w	person indices
a, b, c, d	scalars; indices for test scores or for items (Chap. 3)
S_j	sum score on item j across all N persons (Chaps. 3, 4)
$S_{j(i)}$	sum score on item j across $N - 1$ persons except person i (Chap. 3)
Y	random variable
Z	standard score
μ_Y	population mean of random variable Y
$\bar{Y}$	sample mean
σ_Y	population standard deviation
S_Y	sample standard deviation
σ_Y^2	population variance
σ_{XY}	population covariance between random variables X and Y
ρ_{XY}	population product-moment correlation
r_{XY}	sample product-moment correlation
T	true score, expectation of a person's sum score or another measure
E	random measurement error
X'	sum score on test parallel to test with sum score X
$\rho_{XX'}$	reliability of sum score X
r	index for independent replications of test (note: not a sample correlation)
α	preceded by "coefficient" stands for reliability lower bound (Chap. 2)
$\lambda_3(\alpha)$	coefficient α (Chap. 2)
Σ_X	covariance matrix for item scores
λ_m	reliability lower bound λ_m, $m = 1, \ldots, 6$ (Chap. 2)
U_m	upper bound to the sum of J item error-variances, $m = 1, \ldots, 6$ (Chap. 2)

t_j	short-hand for $\sigma^2_{T_j}$, the variance of the true score on item j (Chap. 2)
μ_r	r-th reliability lower bound from lower-bound series μ (Chap. 2)
ε	residual of a linear model, such as the regression model
σ_E	standard error of measurement, standard deviation of random measurement error
σ_ϵ	standard error of estimate, ϵ is residual in linear regression of T on X (note: $\epsilon \equiv \varepsilon$)
R^2	squared multiple correlation, proportion of variance in linear regression model
Q	number of item subsets, number of principal components, number of latent variables or factors; quantiles Q, with realization q_i, $i = 1, \ldots, N$ (Chap. 3)
q	index for item subsets, number of principal components, latent variables or factors, $q = 1, \ldots, Q$
$\mathbf{R}_X$	correlation matrix for item scores
ξ_q	qth latent variable or factor (Chaps. 2, 4)
b_j	intercept of factor model for item j (Chap. 2)
a_{jq}	loading of item j on latent variable/factor q (Chap. 2)
δ_j	residual of factor model for item j (Chap. 2)
U_j	unique attribute in factor model for item j (Chap. 2)
h^2_j	communality of item j, proportion of variance factors explain of X_j (Chap. 2)
d^2_j	proportion of unexplained variance; $d^2_j = 1 - h^2_j$ (Chap. 2)
ω_t	reliability coefficient based on multi-factor model (Chap. 2)
θ	latent variable in IRT models
$G(\theta)$	cumulative distribution function for θ
$\boldsymbol{\theta}$	vector of latent variables, $\boldsymbol{\theta} = (\theta_1, \ldots, \theta_Q)$
$P(X_j \geq x \mid \theta)$	item-step response function; probability of having at least a score of x on item j, given θ
$P(X_j = 1 \mid \theta)$	item response function for dichotomous items (scored 0,1)
$P_j(\theta)$	item response function, $P_j(\theta) \equiv P(X_j = 1 \mid \theta)$
$\mathbf{X}_{(j)}$	vector containing $J - 1$ item-score variables except X_j
$\mathbf{x}_{(j)}$	vector containing $J - 1$ item scores except x_j
δ_j	location or difficulty parameter in a logistic IRT model (Chaps. 3, 4)
π_j	population proportion having a 1 score on item j
π_{jk}	joint population proportion having 1 scores on both items j and k
H_{jk}	scalability coefficient for items j and k
H_j	scalability coefficient for item j

H	scalability coefficient for the whole scale
c	user-specified scalability lower bound for scalability coefficient, $0 < c \le 1$ (Chap. 3)
β_{YZ}	regression coefficient of variable Y on variable Z (note: Not a standard score)
$\rho_{XY.Z}$	partial correlation between variables X and Y corrected for variable Z
G	number of scales (Chap. 3), number of exclusive and exhaustive groups of people (Chap. 4)
g	index for scales, $g = 1, \ldots, G$ (Chap. 3), or groups of people (Chap. 4)
J_g	number of items in scale g
b_j	location or difficulty parameter in the normal-ogive model
a_j	slope or discrimination parameter in the normal-ogive model
c_j	lower asymptote parameter in the normal-ogive model
Ω_j	hypothetical continuous score on item j (Chap. 4)
ω_j	threshold value for continuous item score Ω_j
α_j	slope or discrimination parameter in a logistic IRT model
γ_j	lower asymptote parameter in a logistic IRT model
ϵ_j	item parameter, $\epsilon_j = \exp(\delta_j)$
$L(\mathbf{X} \mid \theta, \delta)$	likelihood of the data given the model parameters of the Rasch model
$I_j(\theta)$	Fisher's information function for item j
$I(\theta)$	Fisher's information function for a set of items
$\sigma_{\hat{\theta} \mid \theta}$	standard error of maximum likelihood estimate of latent variable; $\sigma_{\hat{\theta} \mid \theta} = I(\theta)^{-1/2}$
δ_{jx}	location parameter of the ISRF for item score x_j in adjacent category model (Chap. 4)
π_x	probability of having a score x on item j, given θ; $\pi_x = P(X_j = x \mid \theta)$ (Chap. 4)
τ_x	threshold parameter for item score x in the rating scale model (Chap. 4)
λ_{jx}	location parameter of ISRF for item score x in cumulative probability model (Chap. 4)
β_{jx}	location parameter of ISRF for item score x in continuation ratio model (Chap. 4)
η_f	basic parameter in the linear logistic test model, $f = 1, \ldots, F$ (Chap. 4)
w_{jf}	weight of basic parameter η_f for item j (Chap. 4)
w	value of discrete latent variable in latent class model (Chap. 5), $w = 1, \ldots, W$
$P(\theta = w)$	probability of random person being in latent class w, class size
$P(X_j = x \mid \theta = w)$	probability of having score x when in latent class w

$u^j_{x_j}$	log-linear-model effect of item-score x_j on log of probability; similar effects pertain to latent variable and combination of item and latent variable (Chap. 5)
$\boldsymbol{\alpha}_i$	vector of binary attribute parameters α_{ik}, $k = 1, \ldots, K$, for person i; $\alpha_i = (\alpha_{i1}, \ldots, \alpha_{iK})$
K	number of latent attributes needed to solve a set of items
$\mathbf{Q}$	matrix of order J (items) by K (attributes)
q_{jk}	binary elements of matrix $\mathbf{Q}$
ξ_i	vector of latent binary item scores, $\xi_i = (\xi_{i1}, \ldots, \xi_{iJ})$, in DINA model
s_j	slipping parameter, probability of failing the item when the latent item score equals 1
g_j	guessing parameter, probability of succeeding the item when the latent item score is 0
η_{ij}	latent binary item score in DINO model
β_{jx}	latent item parameters in GDM and MGDM
U	latent class variable with class scores for individuals, u_i (Chap. 5)
K^*_j	number of attributes needed to solve item j; $K^*_j \leq K$
m	index for different attribute vectors (Chap. 5)
λ_{jk}	latent item parameters in logit version of G-DINA model, k indexes latent attributes α_k
δ_{jk}	latent item parameters in identity-link version of G-DINA model
u_{ij}	response of person i to item j based on discriminal process; $u_{ij} = u_j$ (Chap. 6)
u_{ijk}	response of person i comparing items j and k; $u_{ijk} = u_{jk} = u_j - u_k$ (Chap. 6)
$p_{k>j}$	proportion of repetitions in which item k is ranked higher than item j (Chap. 6)
X_{ijk}	binary variable; value 1 means person i prefers item j to item k, value 0 means i prefers k to j
s_{jk}	sample frequency preferring item j to item k
T_{ij}	random variable for the time it takes person i to solve item j (Chap. 6)
t_{ij}	realization of T_{ij}
D_{ij}	binary random variable for completion of item j by person i (Chap. 6)
d_{ij}	realization of D_{ij}, person i completed item j ($d_{ij} = 1$) or not ($d_{ij} = 0$)
τ^*_i	speed of person i's labor in lognormal model; also, $\tau_i = \ln \tau^*_i$ (Chap. 6)
β^*_j	amount of labor the item requires in lognormal model; also, $\beta_j = \ln \beta^*_j$

ς_j	item discrimination parameter for response time in lognormal model
ν	item drift rate parameter, quality of collected information in diffusion model (Chap. 6)
α	boundary separation parameter in diffusion model
z	starting point of information collection process in diffusion model
T_{er}	time needed for encoding and responding in diffusion model
τ_j	item threshold parameter in Ising model (Chap. 6)
ω_{jk}	network parameters for interactions between nodes j and k in Ising model

Acronyms in alphabetical order

ACM	Adjacent category model
CDM	Cognitive diagnostic model
CFA	Confirmatory factor analysis
CPM	Cumulative probability model
CRM	Continuation ratio model
CTT	Classical test theory
DINA	Deterministic input, noisy "AND" gate
DINO	Deterministic input, noisy "OR" gate
EFA	Exploratory factor analysis
G-DINA	Generalized deterministic input, noisy "AND" gate
GDM	General diagnostic model
GLB	Greatest lower bound to the reliability
IRF	Item response function
IRT	Item response theory
ISRF	Item-step response function
LCDM	Log-linear cognitive diagnostic model
LCM	Latent class model
MGDM	Mixture distribution version of the GDM
MFM	Multi-factor model
OFM	One-factor model
PCA	Principal component analysis
PEA	Per Element Accuracy
RMSEA	Root mean squared error of approximation
RRUM	Reduced reparametrized unified model
TLI	Tucker–Lewis Index

1

Measurement in the Social, Behavioral, and Health Sciences

In the nonphysical sciences, measurement has always been problematic, and it has become increasingly evident to nearly everyone concerned that we must devise theories somewhat different from those that have worked in physics.

—David H. Krantz, R. Duncan Luce, Patrick Suppes, and Amos Tversky

Introduction

Without adequate measurement of psychological attributes, such as people's intelligence or personality traits, many research areas would not be capable of producing valid knowledge. Measurement is paramount to scientific research that relates attributes to one another in an effort to test hypotheses about human activity and find support for theories or reject those theories. For example, one may study the relationship between a child's intelligence and her parental support of cognitive activity, such as the degree to which her parents stimulate her to read and game. Without adequate measurement, intelligence and parental support can only be observed imperfectly and subjectively. To improve the assessment of attributes, one needs valid and objective measurement instruments, constructed using accepted methodological procedures and psychometric measurement models. Without valid and objective measurement, the recording of intelligence and parental support would be arbitrary to a high degree, depending on the behavior people accidentally exhibit when observed and how the observer interprets the behavior. The information collected this way would be incomplete and not representative of the target attributes, and the knowledge obtained from the research would be unconvincing. The measurement of psychological attributes is central in this monograph, and in this introduction, we comment on its complex and problematic nature.

Methodological Procedures and Psychometric Measurement Models

One needs methodological procedures and psychometric measurement models to construct measurement instruments. Methodological procedures provide researchers with sets of rules they can use to manufacture measurement instruments, similar to a manual that instructs one to assemble a cupboard from a box of parts, screws, and nuts. These instruments are used to measure psychological attributes, and when administered to people, the instruments yield data that can be used in two ways. First, in the phase of instrument construction, the researcher uses the preliminary instrument that is under construction and still liable to modification to collect data in a representative sample of people from the population in which the instrument will later be used for measuring the attribute. The sample data are used for research that is intended to help make decisions about the composition of the final instrument and that we explain shortly. Second, once the instrument is assembled in its final form and ready for use, it can be employed in scientific research to study the relation between attributes or to measure individual people's attributes. Depending on the purpose of the measurement, measurement values of individuals can be used, for example, to advise the measured individual about her job aptitude or to advise the organization who had an individual assessed about that individual's suitability for a job. In a clinical context, measurement values can be used to advise both patient and practitioner about the necessity of treatment of a pathology. This is the instrument's use for diagnostic purposes. In this monograph, we do not focus on methodological procedures for instrument construction nor do we concern ourselves with the use of established measurement instruments that are used to diagnose individuals. Instead, our attention is focused on the phase of instrument construction when the researcher has collected a representative sample of data and wishes to construct a scale. This is where psychometric measurement models enter the stage.

The researcher has a wide array of psychometric measurement models available that they can use to analyze the data and assess whether the preliminary measurement instrument elicits responses from people that contain information that can be used to locate them on a scale for the attribute. Measurement models help one to extract information from the data from which one can infer whether a scale has been attained or how the preliminary instrument can be modified to realize a scale by means of the modified instrument. In past decades, influenced by quickly growing computing power, the arsenal of psychometric measurement models has increased rapidly and their usability for data analysis has grown impressively. We discuss the best-known of these measurement models and explain how they can be used to assess whether a measurement instrument enables a scale for the measurement of the target attribute.

Relation of Measurement Model to Attribute Scale

Psychometric measurement models are based on formal assumptions that together imply the properties of the scale on which an attribute is measured. The assumptions concern the degree to which the measurement structure is simple, which refers to the number of potential scales; the internal structure of the instrument; and the relations of the internal structure with one or more scales. For simplicity, assuming one scale, depending on the measurement model, the scale can have different properties. For example, almost all item-response theory (IRT) models (Chapters 3 and 4) imply an ordinal scale on which people's locations inform one about their relative position with respect to other people. Individuals may be ordered with respect to their levels of intelligence or their degree of providing parental support. On the other hand, latent class models (LCMs; Chapter 5) may imply unordered categories representing, for example, strategies for solving particular cognitive problems or qualitatively different support-profiles informing one about parents' style of providing support to their offspring. Some of the measurement models are highly complex, and this complexity translates to the *expected* structure of the data collected by means of the measurement instruments. The expected data structure can be compared with the structure of the *real* data, and this comparison provides the key to assessing whether the instrument enables a scale or how to modify the preliminary instrument to make this possible. It often happens that the model expectations about the data and the real data show inconsistencies. What does this mean?

Model–data inconsistency, also called model–data misfit, means that the model does not provide an adequate representation of the data and that one cannot trust the model and its features, such as scale properties and indices of scale quality, to represent meaningful empirical phenomena. Realizing that the model represents an ideal and the data represent reality, one may readily grasp what misfit is. Misfit is a concept that also appears in other, everyday contexts, often representing an unpleasant situation in which a person or a situation is not accepted by other people. We use the term in a neutral way, only indicating that two entities are different or do not go together well. For example, a business suit may not fit one's physique, because the suit is too tight or too wide, too long or too short, and hence fails as an approximate description of one's bodily dimensions. Approximate, because like models, suits never fit precisely. Some misfitting suits only fail a little and do not make the wearer look bad, in which case the misfit may be acceptable, but when the misfit is gross, parameter values such as arm length and waist become meaningless, calling for another, better-fitting suit.

In social, behavioral, and health measurement of psychological attributes, model–data misfit often is a serious event. When two entities show a misfit, either can be seen as the cause of the misfit. This means that one may have used the wrong model to fit the data. Replacing a misfitting model by another, hopefully, better-fitting model should be pursued in any serious measurement effort. However, we will argue that finding a fitting model will remain a problem, because the data collected in these research areas usually are messy, rendering the data seriously inconsistent with the neat expected-data structure models impose. The reason data collected by means of psychological measurement instruments are messy, is that, in addition to the target attribute, many unwanted sources influence the measurement process and its outcome. Later in this chapter, we discuss three main causes of unwanted influences on data. First, theories about attributes often are not well developed or are even absent, which renders the attribute's operationalization in the phase of instrument construction inaccurate. Second, measured persons show a tendency to respond to being the subject of an assessment procedure, known as participant reactivity (Furr & Bacharach, 2013), which influences the way they respond to the items and increases the complexity of the data. Third, any measurement procedure in the social, behavioral, and health sciences suffers from the influence of many random and systematic error sources beyond the researcher's control and not attributable to imperfect theory and participant reactivity. These and possibly other measurement confounders affect the composition of the data, producing often-complex data sets signaling much noise and many weak signals, some of which are wanted and others unwanted, while it proves difficult to disentangle which is which.

Developing Attribute Theory Is Important

We contend there exists a serious discrepancy between several measurement models' far-reaching implications with respect to the data structure on the one hand, and the typical messy, erratic, and error-ridden data structures one finds in research on the other hand. More than participant reactivity and inability to control for multiple error sources, we think the main problem with the measurement of social, behavioral, and health attributes is that the theory of the target attributes is often poorly or incompletely developed. This results in imperfect operationalizations that one cannot expect to produce data consistent with the expected data structures measurement models imply. Studying and developing measurement procedures and measurement models are important activities, but at least as much time and effort should be devoted to constructing and testing attribute theories providing guidance to instrument construction and meaningful data collection. Without a well-founded theory of what one should measure, constructing a measurement instrument for the attribute is an educated guess at best.

We discuss measurement models that do not expect too much of often-messy data and that focus on the data features that provide just enough information to measure people's attributes. These are models that enable the use of simple sum scores, such as counts of the correct answers to a set of items or counts of the rating-scale scores on a set of items, for ordering people on a scale, and models that use patterns of item scores to identify subgroups that are in a particular state. Most models fall into one of these two categories, but several other models impose more structure on the data, unavoidably inviting greater misfit than less-demanding models already do. Several of such more-demanding models were designed to study particular features of sets of attributes, which we think is useful, in particular, when the theory of the target attribute predicts such features to happen in real psychological processes and to be perceptible in corresponding responses people provide when measured, and finally to be reflected in real data. We emphasize that it is not our aim to criticize such ambitious models nor do we intend to blame the data for not fitting the expectations of these or less restrictive models. Rather, by discussing the models and emphasizing how to investigate their fit to the data, we call attention to the importance of considering the origin of the data and the overriding importance of developing well-founded theory for the attributes one intends to measure in the presence of so many confounding influences on measurement.

Measurement Instruments

Measurement instruments are called *tests* when the assessment of an achievement is of interest, and inventories or, more generally, *questionnaires* when the assessment of typical behavior is of interest. Achievement refers to maximum performance, for example, when students are tested for their arithmetic ability or applicants for their intelligence, and one has to solve as many problems correctly as possible, relying on one's abilities. Maximum performance is of interest, for example, in educational assessment and in job applications assessing job aptitude. In general, it is in the tested person's best interest to perform maximally to pass the exam or to attain the job. Typical behavior refers to how one behaves and reacts in general or in particular situations and thus does not require a maximum performance but rather an exposition of one's way of doing. Assessment of typical behavior usually is based on the answers people provide to questions about how they typically are or react, or by a judgment of the person provided by others, such as siblings, parents, or friends, or teachers in an educational context, therapists in a clinical context, and recruitment officers in a job selection or work assessment context. Typical behavior is of interest, for example, when a person's personality is assessed or her attitudes or opinions are inventoried.

Psychological measurement instruments traditionally consist of a set of problems or a set of questions or statements. A generic label for problems, questions, and statements is *items*. Typically, people who are measured respond to the items, and one assumes that the responses contain relevant information about the target attribute. Examples of classes of psychological attributes are abilities (e.g., arithmetic ability, verbal comprehension, analytical reasoning, and spatial orientation), knowledge (e.g., geography, history, anatomy, and vocabulary), skills (e.g., typing, attention, concentration, and motor skills), personality traits (e.g., openness to experience, conscientiousness, extraversion, agreeableness, and neuroticism, together known as the Big Five personality traits), and attitudes (e.g., toward euthanasia, immigrants, democracy, and one's mother). We use the collective name of *attributes* for any ability, knowledge domain, skill, personality, and attitude. Sociologists using questionnaires measure attributes such as political efficacy and religiosity, health researchers measure quality of life and pain experience, marketing researchers measure acquaintance with brands and satisfaction with services, and there is much more.

The appearance of psychological measurement instruments varies, from a set of arithmetic problems and questions, either printed or exhibited on a computer screen, to animated games or filmed behavioral sequences, both presented on a screen. Another possibility is that the tested person has to engage in real behavior, as in a typing test or as a member of a group of people who have to perform particular tasks in which raters assess the tested person's input, creativity, social skills, and leadership. Irrespective of the instrument's appearance, we assume that the tested person's responses provide information relevant to the measurement of the target attribute. In the next step, the responses a person provides to a set of items are transformed to numerical item scores.

Measurement Models

Let X_j denote the random variable for the score on item j. Variable X_j has realizations x, so that $X_j = x$. Items are indexed $j = 1, ..., J$. Two types of item scores are dominant in social, behavioral, and health measurement. The first type is *dichotomous* scores. Dichotomous item scores often represent a positive (score: $x = 1$) and a negative (score: $x = 0$) response with respect to the attribute. Examples are correct (score: 1) and incorrect (score: 0) solutions of arithmetic problems, and agreeing (score: 1) and disagreeing (score: 0) with an introversion statement such as "I like to be alone most of the time". If the statement had been "I like to be among people most of the time", scoring would have been 0 for agreeing (indicating a relatively low position on an introversion scale) and 1 for disagreeing (indicating a higher position). The second type is *polytomous* scores. Such scores represent the degree to which one has solved a free-response arithmetic problem correctly or the degree to which one endorses, for example, a particular introversion statement on a typical rating scale; that is, higher scores indicate higher scale positions. Scores are integers $0, 1, ..., M$, where $M + 1$ represents the number of different integer item scores. The number of different scores usually is equal across items, but if not, M_j indicates the number of possible scores for item j. The resulting data matrix contains dichotomous or polytomous item scores from N persons on J items and is denoted by $\mathbf{X}_{N \times J}$ or simply $\mathbf{X}$.

Because of their prevalence, we focus this monograph primarily on measurement models for dichotomous and polytomous item scores that suggest a higher scale value as the item score is higher. Measurement models for other data types exist but are used less frequently than the data we discussed. In Chapter 6, we briefly discuss models for pairwise comparison data (Bradley & Terry, 1952; Thurstone, 1927a), proximity data (e.g., Andrich, 1988a; Johnson, 2006), response times in conjunction with dichotomous incorrect/correct scores (Tuerlinckx, Molenaar, & Van der Maas, 2016; Van der Linden, 2007, 2009), and an approach that considers attributes as networks of symptoms (Borsboom & Cramer, 2013; Epskamp, Borsboom, & Fried, 2018). One may consult Coombs (1964), who provided a classical foundational treatment of the data types typical in psychological research.

Whether the quantification step transforming qualitative responses to items into numerical item scores makes sense has to be established by means of a measurement model that is based on particular assumptions about scales. Typical assumptions are that measurement errors are uncorrelated (classical test theory, Chapter 2); together, several factors linearly produce the item score (factor analysis, Chapter 2); one latent variable represents the attribute (IRT, Chapters 3 and 4); and different patterns of item scores identify distinct subgroups of people (LCMs, Chapter 5). A measurement model is identical to a set of assumptions that together restrict the structure of the data. This means that a particular measurement model predicts the structure of the data, and will be consistent with some real-data sets but inconsistent with other real-data sets. Only when the measurement model is consistent with a real-data set does it make sense to transfer the features of the model to the particular test or questionnaire that generated the data when administered to a group of people. For example, if an IRT model with one latent variable fits a particular real-data set well, one latent variable explains the structure of the real data and we assume that the test or the questionnaire measures the attribute on one scale. Similarly, if an LCM with three latent classes fits the data well, we assume that a three-subgroup structure explains the data and that the test or the questionnaire can be used to assign people to these subgroups based on their item scores. Measurement models differ in the degree to which they produce a mismatch with the messy data typical of much research in the social, behavioral, and health sciences. Diagnostic misfit statistics may help to identify reasons for misfit and choose a better-fitting model.

Classical test theory (CTT) not only is older than the other models, but it is different in that its assumptions do not restrict the data in a falsifiable way. This means that one cannot make decisions on whether the model fits a particular data set. CTT's contribution primarily lies in its focus on the repeatability of test performance, expressed by the definition of the reliability of measurement and the many methods available to estimate reliability. Researchers appreciated the concept of measurement reliability but realized they also needed a methodology to distinguish sets of items that were suited to define a scale and item sets that were unsuited for this purpose. Principal component analysis and factor analysis became popular methods to investigate whether the data could identify suitable item sets as well as items that were unsuited to contribute to the definition of a scale. The advent of structural equation modeling opened the possibility to formulate alternative versions of CTT that allowed the statistical assessment of model–data fit (e.g., Bollen, 1989; Jöreskog, 1971).

Assessing a measurement model's fit to data is a complex process. A considerable part of this monograph deals with methods for investigating the fit of measurement models to data. Here, we illustrate the misfit of the assumption of unidimensional measurement. Ignoring the mathematical complexity of this assumption, we notice that a questionnaire intended to measure introversion ideally only measures this attribute and nothing else. Hence, we assume that unwanted influences did not confound the measurement process. One may compare this to measuring the outside air temperature; then, one assumes that the measurement value read from the thermometer only reflects temperature but no other aspects of the weather, such as air pressure, humidity, and wind velocity. If measurement values of temperature also reflected these influences, one would greatly mistrust their meaning, but with psychological measurement, confounding of measurement values with unwanted influences is the rule rather than the exception. For example, arithmetic problems often require language skills as well, and it proves difficult to construct arithmetic items that avoid language skills' confounding influence on measurement.

Returning to introversion as an example, if introversion is a trait or a process that unequivocally causes people to respond at a particular level in situations that elicit introvert reactions, we may hypothesize that this mechanism causes the data to have a particular structure that is consistent with the structure a measurement model formalizes in mathematical terms. The one-factor analysis model (Bollen, 1989) and unidimensional IRT models (e.g., Van der Linden, 2016a) mathematically define this structure by means of one *factor* and one *latent variable*, respectively. If a one-factor model or a unidimensional item-response model fit the data well, by implication one unobservable or latent variable one estimated from the data summarizes the information contained in the data.

Measurement models include parameters representing people's performance on the items and parameters representing item properties such as the items' difficulty and items' potential to distinguish low and high scale values. If the measurement model approximately fits the data, one displays estimates of these parameters on the dimension that explained the data structure, thus defining a scale of measurement. One has to check whether unidimensionality and each of the other assumptions hold in the data to establish whether one can assume a scale as defined by the measurement model to be meaningful.

Scales of Measurement

Many measurement models follow the logic that a higher score on an item represents a higher position on a scale for the attribute. Consequently, counting the number of correct solutions on J arithmetic items or adding the polytomous scores on J rating-scale introversion items would be the next logical step. Addition of item scores produces a test score or a sum score, defined as

$$X = \sum_{j=1}^{J} X_j. \tag{1.1}$$

We use the terminology of sum score throughout this monograph when referring to Equation (1.1). Indeed, the history of psychological measurement shows that the use of sum scores or simple linear transformations thereof, such as standard scores and IQ-scores, and non-linear transformations, such as percentiles, has been and still is common practice (Chapter 2). For transformations of sum scores, we use the generic term of test scores.

Principal component analysis and factor analysis, both in combination with CTT (Chapter 2), are the most frequently used approaches to constructing scales for attributes measured by means of particular instruments. In factor models, person scale values are *factor scores*, but researchers rarely use factor scores for person measurement and readily replace them with sum scores on the subsets of items identified by their high loadings on a particular factor. These item subsets define separate scales. Models based on IRT (Chapters 3 and 4) are the second most frequently used models for scale construction. Nonparametric IRT models (Chapter 3) use the sum score, and parametric IRT models (Chapter 4) use *latent-variable scores* to measure people. The LCM (Chapter 5) uses the pattern of scores on the J items in the test to classify people, usually in a limited number of classes. Within classes, there is no ordering of people, and classes may or may not be ordered. Thus, people's class membership represents either a nominal or an ordinal scale.

Interestingly, as we noticed, the assumptions that define CTT do not restrict the structure of the data; in fact, they render CTT a tautological model, which does not logically imply a scale for the sum score or any other score. Instead, CTT is a theory of random measurement error, and its reliability coefficient, which is one of the most frequently reported quantities in the social, behavioral, and health sciences, quantifies the degree to which a set of sum scores from a group of people can be repeated when the same people are re-tested, using the same test under the same administration conditions. However, repeatability does not imply the existence of a scale for measurement; that is, repeatability only shows that scores can be repeated, not that they lie on a scale. If one uses a set of sum scores as if it were a scale, but without a model that justifies this, measurement takes place *by fiat*—the researcher decides she has a scale—rather than *by implication*—the model implies the researcher has a scale.

Sum scores are simple and appeal to intuition, and it would be convenient if one could provide a solid justification for their use in measurement. In fact, the class of nonparametric IRT models provides the justification for the use of the sum score that its greatest advocate, which is CTT, cannot provide. Nonparametric IRT is based on assumptions that imply the use of sum score X to order people on the latent-variable scale of IRT. Thus, measurement according to nonparametric IRT is ordinal. Chapter 3 demonstrates that this result is restricted to tests and questionnaires consisting of dichotomous items. For polytomous-item measurement instruments, the ordering result does not hold precisely but the scale of X is a robust ordinal estimator of people's ordering on the latent-variable scale, meaning that deviations found in practical measurement are small and within reasonable error bounds. In addition, compared to dichotomous-item instruments, for instruments consisting of polytomous items, nonparametric IRT logically implies a weaker ordering property for pairs of groups defined by subsets of disjoint and exhaustive (and within groups, adjacent) sum scores. It is interesting to realize that these ordinal scale results also hold for almost all parametric IRT models, but this result has found little resonance among parametric IRT theorists, who tend to focus on latent variables.

Parametric IRT uses the latent-variable scale that results from maximum likelihood estimation or Bayesian estimation methods (e.g., Baker & Kim, 2004). This produces real person-scale values that are usually normed between, say, −3 and 3, and that one assumes follow a standard normal distribution. For the simple one-parameter logistic model or Rasch model, the sum score X [Equation (1.1)] is a sufficient statistic for the latent person parameter, so that in this model the sum score and the latent person parameters on the latent-variable scale have a simple, albeit non-linear relation. In other parametric IRT models, the pattern of J item scores is used to estimate person values on the latent-variable scale, and the simple relation with sum score X seems to be lost. Fortunately, this is not true. From nonparametric IRT it follows that, for dichotomous items, sum score X orders people on the latent-variable scale. For polytomous items, sum score X is a robust ordinal estimator and a logically justified ordinal estimator of pairs of subsets of sum scores defining low and high-scoring groups. As far as we know, test practitioners and test agencies tend not to use latent-variable values to measure persons but prefer simpler scores that have better intuitive appeal. Sum scores are well-justified candidates, but other scores that are transformations of sum scores, such as Z-scores, IQ-scores, and percentile scores, are also used (Chapter 2). These simpler scores facilitate communication with tested persons, such as students, their teachers, and their parents, applicants for a job, or patients awaiting treatment.

Causes of Messy Data

Three causes of messy data mentioned in this chapter's introduction are the following. First, the absence of a well-founded theory about the target attribute or the presence of a partly founded, incomplete theory unavoidably renders the attribute's operationalization inaccurate. This inaccurate operationalization invites many sources influencing people's responses to items and producing data, more complex than when a well-founded theory had been available. Second, the tendency measured persons show to respond to being the subject of an assessment procedure, known as participant reactivity, influences the way people respond to the items and increases the complexity of the data, because person-specific deviations from the attribute's influence play a role. Moreover, the effects of participant reactivity probably vary across persons, further complicating data structures. Third, any measurement procedure in the social, behavioral, and health sciences suffers from the influence of many error sources beyond the researcher's control. We further discuss each of the three causes in this section.

First, a well-founded theory about the target attribute provides the researcher with guidance for the operationalization of the attribute into a number of measurement prescriptions. Absence of a well-founded theory withholds such guidance, leaving the researcher to rely on vaguer information about the attribute, her experience, habit and tradition in the research area, and educated guesses. Absence of theory introduces subjectivity, meaning that the composition of a test or a questionnaire is arbitrary to some degree at least and that measurement is off target to an unknown degree. The data will reflect this and have a structure inconsistent with the (partly) unknown structure of the attribute. Examples of attributes that are not or hardly based on empirically well-founded theory are creativity, emotional intelligence, mindfulness, and self-efficacy, and competencies that are popular but ill-defined and often based on intuition in modern work and organizational psychology. In addition, many attributes are supported to some degree by empirically tested hypotheses but are too abstract to imply cognitive and behavioral phenomena that can be separated from other influences on the measurement outcome. A positive view of the problem of inconsistent data is that the statistical analysis outcomes may contribute to the development of the theory that should be at the basis of the instrument construction by suggesting at least some associations that may be worth pursuing, possibly inspiring more targeted follow-up research. This perspective inspires the metaphor of the Baron Von Münchhausen who pulled himself and his horse out of a swamp by his hair and has proven to be a fruitful but labor-intensive approach, requiring perseverance from the researcher and a science conception allowing slow progress rather than quick and publishable results.

Second, people who are tested show a tendency to respond to being the subject of an assessment procedure, which is known as participant reactivity. For example, a person responding to questions asking that person about her introversion level may be inclined to reflect on her own personality and let her thoughts affect the answers she gives in person-dependent, hence, unpredictable ways. One possibility is that she composes an image that shows more coherence than is realistic, and another possibility is that she gives answers that express how she conceives her ideal self, independent of the truth. Other possibilities of participant reactivity in testing situations are someone very nervous, producing negatively biased results on a high-stakes arithmetic test, clearly inconsistent with her best interests, and someone consciously manipulating a personality inventory assessing neuroticism, thus producing answers that are socially desirable, estimating that answering truthfully may lower her chances of getting the job. The classical text by Rosenthal and Rosnow (1969) discussed various psychological mechanisms underlying the reactivity people show when they participate in psychological research, and we have little reason not to expect similar effects in testing thus introducing unwanted effects in the data (e.g., Rosenthal, 1969).

Two well-known causes of undesirable influences on research outcomes are demand characteristics and experimenter expectancy. Demand characteristics refer to features of the test administration context that inspire tested persons to accommodate their responses to the context they perceive. For example, the formulation of the items in an opinion poll about immigration policy may inspire some respondents to either soften or harden their initial viewpoints. Experimenter expectancy refers to signals the test practitioner may send unconsciously to the tested persons. For example, some tested persons may interpret an oral instruction to relax to mean that the test is not as important as they thought it was, resulting in a worse test performance. These are just a few of many influences that threaten to confound psychological measurement. The influences witness that, other than in the natural sciences, in the human sciences the measurement object responds when measured and thus invalidates the measurement outcome to some, usually unknown degree. As an aside, we notice that measurement of an electrical current, the concentration of a substance in a chemical compound, and animals' blood pressure are not without problems either, and a recurring issue also relevant with these measurement procedures is how to reduce or eliminate influences that confound the measurement procedure. Possible confounders are disturbances of the magnetic field, pollution of compounds, and distracting events, respectively. With human beings, the additional problem thus is their awareness of the measurement procedure and their capability of responding to the situation and even manipulating it to some degree.

Third, there are also unpredictable features of the test administration situation randomly influencing responses to items. Examples are a noisy environment affecting persons' concentration or, alternatively, a pleasant room with coffee and tea available and no distractions whatsoever, or the tested person's temporary mental state, because, for example, she slept badly the night before being tested or, alternatively, she slept better than average and felt relaxed when she took the exam. CTT focuses on precisely this third category of influences and assumes they exert a random effect on test performance, quantified as random measurement error. Standardization of the test administration conditions helps to reduce random measurement error and may have some positive effect on people's responsiveness to being assessed.

The three causes of messy data together constitute a call for reflection on a difficult topic: Do messy data typical of the social, behavioral, and health sciences justify the fine-grained analysis that advanced measurement models facilitate? Given that any data set contains numerous signals, often of unknown origin, what do the analysis outcomes signify? We contend that the improvement of social, behavioral, and health measurement can only happen when better theories are developed about what it is one measures. We anticipate relatively little improvement from trying to reduce people's responsiveness and further standardization of the administration conditions. The true Achilles' heel of measurement is weak attribute theory foundational to measurement. Meanwhile, we acknowledge that several practical problems call for their study, and possible solutions, if only preliminary, cannot wait for better data to become available.

A Scale for Transitive Reasoning

Transitive reasoning provides a fine example from cognitive developmental psychology of the construction of a measurement instrument based on theory about the target attribute, using multiple measurement models to accomplish the goal (Bouwmeester & Sijtsma, 2004). Transitive reasoning entails the following elements, which we discuss by means of an example of a transitive reasoning task. Let Y be, say, the length of a stick, and let three sticks be identified as A, B, and C. Figure 1.1 shows two boxes in its left-hand panel, the first box containing sticks A and B and the second box containing sticks B and C. In the example, the sticks have different length and different color. The two boxes appear on a screen. The first box opens showing the entire sticks A and B, and the experimenter asks the child to identify the longer of the two sticks by its color. Then, the box closes and the second box opens, showing the entire sticks B and C, and the experimenter repeats the question. The second box closes and, finally, the last box (right-hand panel) is shown but this box stays closed, the two sticks A and C only sticking out

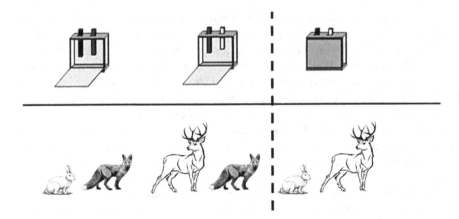

FIGURE 1.1
Examples of transitive reasoning tasks for length (upper panel) and age (lower panel), showing two premises (left) and one inference task (right). Artwork by Samantha Bouwmeester.

but not revealing which is longer, and the experimenter asks the child which of the sticks is longer, A or C (actually, the experimenter mentions their colors). Transitive reasoning entails that if one knows that $Y_A > Y_B$ and $Y_B > Y_C$, then one is able to infer from these two premises the conclusion that $Y_A > Y_C$. Other objects (e.g., animals) and properties (e.g., age) are possible (Figure 1.1), in which case, unlike the length of sticks, the property is invisible.

Several theories about transitive reasoning exist that lead to different and even contradictory operationalizations and different measurement instruments. We focus on three of these theories. We discuss only their essentials and refer to Bouwmeester, Vermunt, and Sijtsma (2007) for more discussion and ample references. The three theories are the developmental theory of Piaget (Piaget, Inhelder, & Zseminska, 1948), information-processing theory (e.g., Trabasso, 1977), and fuzzy trace theory (Brainerd & Kingma, 1984). Each theory assumes different cognitive processes governing transitive reasoning, and each theory differentially predicts how transitive-reasoning tasks' difficulty depends on presentation mode and task characteristics. Presentation mode refers to simultaneous or successive presentation of the premises. The examples in Figure 1.1 concern successive presentation. Examples of task characteristics are the use of premises involving inequalities (as in the examples in Figure 1.1), equalities, or both, and use of physical content (the property the subject has to assess can be seen; length) or verbal content (experimenter tells subject age relation between physically visible animals). Tasks may also vary with respect to the number of objects, implying varying numbers of premises and, possibly, inferences. Theories differ also with respect to the responses they consider informative of transitive reasoning ability, such as verbal explanations of a judgment, response times, and performance on particular task formats.

Bouwmeester and Sijtsma (2004) investigated (1) whether one or two qualitatively different abilities underlay transitive reasoning; (2) whether data based on verbal explanations of judgments (called strategy scores) produced better scales than judgment data (i.e., incorrect/correct scores, also called product scores) alone; and (3) whether task characteristics or combinations of task characteristics affected task difficulty. The authors used a fully crossed factorial design to construct 16 items: Presentation mode (simultaneous, successive), task format ($Y_A > Y_B > Y_C$; $Y_A = Y_B = Y_C = Y_D$; $Y_A > Y_B > Y_C > Y_D > Y_E$; $Y_A = Y_B > Y_C = Y_D$), and type of content [physical (sticks, length), verbal (animals, age)]. A sample of 615 children ranging in age from 6 to 13 years, coming from different schools and SES background, took the computerized test version. Responses were coded as incorrect (0) or correct (1), producing product data, and according to incorrect (0) or correct (1) use of information about solution strategy the children reported verbally after having finished a task, producing strategy data.

Using three different methodologies from nonparametric IRT, based on the analysis of the product scores the authors had to reject the hypothesis of unidimensionality, meaning that one explanatory ability underlying transitive reasoning appeared unlikely. In particular, the items having an equality format stood out from the other items, and physical-content items were dimensionally different from verbal-content items, both results suggesting multiple active abilities. The authors concluded that at least three abilities underlay the realization of product scores and explained the result by the fact that a correct answer may be the result of abilities, skills, and use of tricks that are unrelated to transitive reasoning. Items having different presentation modes were not dimensionally distinct. Analysis of the strategy scores produced different conclusions about dimensionality, and equality items again seemed to stand out. The deviations from unidimensionality were rather small, so that the use of a unidimensional IRT-model for analyzing the strategy scores led to the conclusion that children could be ordered on a scale spanned by 15 items. Here, a 1-score results from an explanation for the reasoning the child gave that the experimenter could tie more tightly to the logic of transitive reasoning, thus producing a score that has better validity than a bare judgment score. A regression analysis showed that equality items positively influenced item easiness and that simultaneous presentation was easier than successive presentation.

Bouwmeester and Sijtsma (2004) fed back their results to the three theories for transitive reasoning, and Bouwmeester et al. (2007) used latent variable models to test in data collected for the purpose, which of the three theories about transitive reasoning received the most support from the data. They focused on fuzzy trace theory, which posits that children cognitively process incoming information in verbatim traces, thus storing the premises precisely, and in gist traces, only storing the essence as fuzzy, reduced, pattern-like gist. In typical transitivity tasks, one may distinguish tasks called memory test-pairs, which test a child's memory of the premise information in premise test-pairs, such as $Y_A > Y_B$ and $Y_B > Y_C$ in the three-object example we gave (Figure 1.1), and tasks called transitivity test-pairs, here, $Y_A > Y_C$, which test whether the child is able to draw the correct inference from the premises. According to fuzzy trace theory, children usually have access to both kinds of traces, verbatim and fuzzy, but Brainerd and Kingma (1984) found that children tend to use the gist traces to solve both the memory and transitivity test-pairs. In non-cognitive parlance, children will use shortcuts to solve transitive reasoning tasks whenever possible. Bouwmeester et al. (2007) noticed that the results ignored possible individual differences between children when they respond to particular task characteristics and set out to test an individual differences model based on fuzzy trace theory, using a multilevel LCM for this purpose.

The authors tested a three-level model that predicted task performance as probabilities for the way in which continuous verbatim and fuzzy abilities (level 1) guided combinations of different ordered, discrete levels of verbatim and gist traces (level 2) to produce performance on three kinds of tasks (level 3). The tasks differed with respect to presentation mode, that is, the order in which the premise information and the memory and transitivity tasks were presented, which was either following an order that was increasing or decreasing or an order that was scrambled. The measurement model was an LCM (Chapter 5), which the authors used to identify classes of children using particular traces and to estimate within classes the probabilities of successfully solving memory and transitivity test-pairs. A multilevel model tied the LCM for the responses children gave to the memory and transitivity task-pairs at level 3 to the higher level-2 ordered discrete latent verbatim and gist traces. The partial credit model, which is an IRT model (Chapter 4), tied the probability of using a particular, discrete trace to the guiding continuous latent ability at the highest level 1.

The fuzzy trace model fitted the data better than the two competing theories, a result that held up upon cross-validation. The estimated response probabilities were compared to the response probabilities predicted based on fuzzy trace theory, and their degree of consistency determined the degree to which the model supported the fuzzy trace theory of transitive reasoning. Largely, support for fuzzy trace theory and not for developmental theory and information-processing theory was found. This means that the authors provided evidence of two abilities and that the estimated probability structures gave support to the interpretation of the abilities as verbatim and gist abilities. These results can be useful for the future construction of scales for transitive reasoning.

Cycle of Instrument Construction

We use a simple scheme (Figure 1.2; Sijtsma, 2012) for explaining the process of instrument construction. The scheme has a cyclic character, acknowledging the possibility for continuous improvement (e.g., De Groot, 1961).

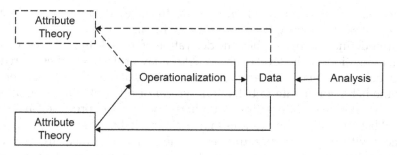

FIGURE 1.2
Scheme for the construction of psychological measurement instruments. Note: Based on Figure 1 in Sijtsma (2012).

A measurement instrument assesses people's position on a scale representing the target attribute. Thus, the attribute presents the point of departure for the construction of the instrument. Different theories can support the same attribute, and although this may seem somewhat odd, it is possible that different theories provide competing views on the same attribute, and the evidence collected in favor of one or another theory may not be convincing enough to make a decision regarding which theory is the best. In physics, the alternative wave and particle interpretations of light provide a well-known example, and both theories explain particular phenomena well albeit based on different perspectives. In psychology, the circumstance of competing theories is common, but we will see that the cause of the diversity is often that theories lack the level of articulation needed to make precise and testable predictions about behavior and hence, to distinguish different theories with respect to their success of prediction. The example of transitive reasoning provided a convincing exception.

Setting up a theory about a psychological attribute, for example, intelligence, entails several part components. First, for intelligence, one needs to delineate the set of cognitive processes that are relevant to the attribute. The cognitive processes may refer to, for example, inductive reasoning and spatial orientation, but in case of the relatively complex attribute of intelligence, several other cognitive processes, such as those relevant to verbal abilities, are involved. With other attributes, such as personality traits, processes related to motivation and environmental adaptation rather than cognition may be involved. Second, in the case of intelligence one needs to explain how the cognitive processes relate to observable phenomena. Observable phenomena, often behaviors, allow us to recognize the presence of particular levels of intelligence, whereas the causal cognitive processes are inaccessible to observation and assessment. Observable behaviors may be the solutions people provide to inductive reasoning and spatial orientation tasks, and the fact that solutions are correct or incorrect, or partially correct, provides evidence of different levels of intelligence. Third, one needs to explain how the cognitive processes and the intelligent behaviors relate to sets of different behaviors, such as school performance, or different attributes, for example, coping styles in response to environmental threats.

A well-defined theory enables the derivation of testable hypotheses, which entail predictions at the level of observable behavior. The hypotheses predict which behaviors relate to one another as correlates of the attribute of interest, how the behaviors relate to behaviors typical of other attributes, and how the manipulation of the cognitive (or other) processes can produce particular typical behaviors. The hypotheses may include predictions about group differences with respect to measurement values, change of measurement values over time, and the internal structure of the measurement instrument (Cronbach & Meehl, 1955), but also the investigation of the processes that were activated by the items and led to the observable responses (Borsboom, Mellenbergh, & Van Heerden, 2004; Embretson & Gorin, 2001). The hypotheses are tested using empirical data, and the outcome of the test may support or reject the theory or suggest how to modify the theory of the attribute.

Figure 1.2 is based on two assumptions. First, there is always an attribute theory, however primitive, that underlies the construction of the instrument. If a theory were absent, the attribute would coincide with the operations the researcher has identified to measure it; that is, in the absence of a theory, the items used together define the attribute (e.g., Boring, 1923). This viewpoint is considered unacceptable today, and intelligence theorists and researchers assume that intelligence does not coincide with what the test measures (but see Van der Maas, Kan, & Borsboom, 2014, for a provocative counter-viewpoint). Likewise, in physics, the thermometer does not coincide with the concept of temperature in the sense that temperature coincides with the expansion of a mercury column, but the measurement procedure uses the expansion properties of substances as they relate to the movement of particles and the energy this produces. That is, the measurement of temperature is based on theory about the target attribute, which may even inspire the construction of differently devised thermometers. Back to measurement of psychological attributes, a test can represent only a more or less successful attempt to provide good estimates of the attribute. When the psychological attribute theory is well developed and tested, it should be possible to derive precise operations producing a valid measurement instrument, but in the absence of such a theory, it is impossible to know well what the test scores represent.

Second, we assume that an arbitrary albeit consistent rule for assigning numbers to objects with respect to an attribute only produces numbers but not measurement values (but see Stevens, 1946, for a different viewpoint). For example, many tests consist of a number of problems, and intuition, habit, or copycat behavior may posit the count of the number of correct answers as a measure of the attribute the set of items allegedly assesses. However, without further precautions, counting the number of correct solutions produces a count and nothing else. This conception of measurement or better, of what measurement is not, excludes *operationism* (Michell, 1999, pp. 169–177) from this discussion.

Next, we discuss the four steps of instrument construction (Figure 1.2).

Step 1: *Identification of the Attribute and Its Theory*
First, if one intends to construct a measurement instrument for attribute A, as a starting point one needs to identify the theory of interest or rivaling theories that one compares so that a choice can be made. In the intelligence domain, as we discussed, at least three rivaling theories explain the attribute of transitive reasoning (Piaget, 1947), and Bouwmeester et al. (2007) succeeded in identifying among these three theories the theory receiving most empirical support and thus best suited for constructing a measurement instrument for transitive reasoning (Bouwmeester & Sijtsma, 2004). This is an example of how we think one should construct measurement instruments. However, in reality, theories that are well articulated and supported by research are rare, and many measurement attempts show a larger degree of arbitrariness and a lesser degree of success than the example of transitive reasoning. Hence, for most other attributes and their rivaling theories, we face a situation that is more complex. For example, crucial experiments, that is, experiments convincing psychologists of the superiority of one of the several competing intelligence theories, are absent (Van der Maas et al., 2006). What happens instead is that different intelligence theories inspire different intelligence tests, which in fact do exist, whereas research results supporting the choice of a foundational theory are unavailable. The solid box in Figure 1.2 stands for the set of relatively well-developed attribute theories, steering instrument construction.

The dashed box in Figure 1.2 collects the psychological attributes based on theories that are more abstract and vaguely defined. Examples of such attributes are creativity, emotional intelligence, mindfulness, and self-efficacy, and many personality traits (e.g., competencies) and *ad hoc* attitudes. The theories supporting these and other attributes from this set use abstract words and notions and do not or only vaguely relate to concrete behavior. Only a few relationships with other attributes or variables, such as gender and age, are specified, and if relationships are quantified, this mostly concerns whether the relationships are positive or negative. Because of the incipience of their theory, rules and principles guiding the researchers toward the relevant behaviors are absent or uncertain, and valid instrument construction is almost impossible or at least problematic. The construction process is subjective, often based on personal experience and preference, and particular instruments may become standard not because of their theoretical foundation but because the instrument at least provides some straws in the wind.

Step 2: *Operationalization of the Attribute*
Operationalization involves the specification of the concrete, tangible operations needed to measure the attribute. The attribute theory should preferably guide this specification by providing a logically consistent analysis of how to translate the theory into a set of measurement prescriptions. Obviously, theories that are well founded can guide the operationalization well (Figure 1.2, solid line), but theories still in their infancy cannot provide such guidance,

leaving the researcher on her own, relying to a high degree on her subjectivity (Figure 1.2, dashed line). Two remarks are in order.

First, one needs to delineate the domain of behaviors typical of the attribute. The presence of these behaviors in persons suggests a high level with respect to the attribute, and the absence suggests a low level. If one thinks of behaviors as symptoms, many behaviors are nonspecific of particular attributes. For example, an item in an introversion inventory may be "I like to be alone most of the time", but someone agreeing may also provide evidence of a perfectly natural desire for occasional privacy, may be a professional musician needing to practice a lot alone, or someone who has recently lived in a noisy apartment and is ready for some solitude. Thus, wanting to be alone now and then is not necessarily a symptom of introversion, even though introvert people typically will want to be alone more often than non-introvert people. Even a preference for being alone to being among other people may not be unique to introvert people but may also be found with other people, such as those who, more than others, value their privacy or those who prefer individual sports to team sports. Likewise, the measurement of arithmetic ability among 12-year olds is not served by problems like "$15 + 8 = ...$", because they are much too simple, and perhaps addition is not even the most valid operation to test. A good theory surely helps to overcome intuition to make the right choices.

Second, one needs to define the domain of items from which one presents a selected subset to people, and the items elicit responses that provide information about the attribute. For maximum-performance testing, items can be arithmetic problems that need to be solved ("write down the steps leading to the answer"), maze tasks ("show the way through a maze"), building-block tasks ("build a particular example construction"), computer games ("guide a ship safely into the harbor"), and mental rotation of abstract geometrical shapes ("choose the correct outcome from four options"). In the examples, each task elicits responses that are incorrect or correct. Alternatives for incorrect/correct response recording could reflect information about the solution strategy (e.g., which approach to solving the problem was used, which steps were taken correctly, how many steps were correct) the child reported based on thinking-aloud protocols (Bouwmeester & Sijtsma, 2007), and response times in addition to incorrect/correct responses provide another source of information (Van der Linden, 2006; also, see Chapter 6). For a typical-behavior inventory, items often are statements describing a behavior typical of the trait, and people rate the degree to which they engage in the typical behavior on an ordered scale but also to which degree they perceive other people, such as students, friends, family, parents, and siblings, to engage in those behaviors. Kagan (2005) questioned the degree to which people are capable of assessing their or other people's typical behavior and noticed that personality testing relies almost solely on "the semantic structures activated when participants answer questionnaires" (ibid., p. 126) and challenged the validity of such data. Instead, he proposed to consider using data stemming from motor activity, and distress to unfamiliar visual, auditory, and olfactory stimuli.

Technological innovations, such as smartwatches, enable the continuous collection of data from individuals, such as biomedical data that are recorded without the person being aware of it and lifestyle and personality data that require the person to answer a couple of questions on her smartphone at several occasions. These and other data types have attracted psychometrics' attention (Brandmaier, Prindle, McArdle, & Lindenberger, 2016; Cheung & Jak, 2016; Klinkenberg, Straatemeier, & Van der Maas, 2011) but challenge the quality of administration conditions and data. In the context of massive data sets, alleged possibilities elicit both unlimited enthusiasm (Domingos, 2015) and skepticism (Hand, 2013). We concentrate on the dominant data types being the typical maximum-performance and self- or other-description data and occasionally include other data types as well (e.g., Chapter 6).

Step 3: *Quantification of Item Responses and Psychometric Analysis*
The solutions people provide to most maximum-performance items are qualitative, for example, whether stick A is longer than stick C in a transitive reasoning task, how to move through a maze from entrance to exit in an intelligence test, and which of four options of a multiple-choice item they choose they think is correct. Typical-behavior statements require people to rate a box on a rating scale, and a trained observer may check the degree to which applicants participate in a discussion, whereas other types of items require different types of responses. All examples we discussed have in common that the responses are qualitative, hence non-numerical and thus extremely difficult to process by means of statistical procedures, let alone display on scales.

To be accessible to statistical manipulation, one transforms the qualitative reactions to quantitative item scores (Figure 1.2, Data box), following the rule that higher item scores reflect higher levels on a hypothetical scale for the attribute. For example, a higher score on a rating scale of an item hypothetically measuring an aspect of the trait of introversion is assumed to reflect a higher introversion level, but whether the item really measures an aspect of introversion and the quantification makes sense depends on the validity of the supporting theory and the operationalization. One can use psychometric models to check whether the quantification stands the test of the data (see Figure 1.2, Analysis box). A result inconsistent with the quantification could be that one finds that for a particular item the relation of the item score with the latent variable the model uses to represent the scale for the attribute is non-monotone (Chapter 3), thus providing evidence against the hypothesis that the item measures the target trait.

Step 4: *Feedback to the Theory of the Attribute*
The psychometric analysis can have two outcomes. The model fits the data, in which case the data support (but not prove!) the psychometric model.

This outcome enables the construction of a scale. The other outcome, which is more common, is that model and data show misfit. This outcome triggers the search for the cause of the misfit. Many possibilities for misfit exist. For example, the attribute theory may be incorrect, and one can use the data-analysis outcome to modify the theory. This can happen only when the theory was well developed and well tested. For example, if the theory predicts that children at a particular developmental level should be able to solve a particular logical problem if they also master a part problem that precedes the more involved problem, and the data analysis shows this prediction to be consistently incorrect, the researcher may decide to modify this part of the theory. Modification may involve research directed at testing this part of the theory without the intention of constructing a measurement instrument. The case where theory is not perfect but able to provide guidance to the instrument construction process is shown in Figure 1.2, where a dashed line runs between the data analyses box and the theory box.

On the other hand, the theory may be correct but the operationalization may have been unfortunate. For example, because of the formulation of the items, verbal skills have unintendedly influenced the solution process underlying analytical reasoning. This necessitates modifying the operationalization, using different items to assess analytical reasoning but without letting individual differences caused by verbal skills get in the way of the analytical-reasoning achievement. That is, people having the same analytical-reasoning level should have the same test score, and the need for verbal skills, if unavoidable, should be suppressed to a low level, so that they do not cause individual differences, allowing the person having the best verbal skills to obtain the highest alleged "analytical-reasoning" test score. We contend that it is impossible to purify measurement of psychological attributes; hence, one should neutralize unwanted influences as much as possible but without expecting they can be nullified. Structural equation modeling may help to separate wanted from unwanted sources of variance in test scores (Green & Yang, 2009), of course after having constructed the items and collected the data.

If the attribute theory is rather abstract and remote from actual behavior expressed in concrete situations, the theory necessarily results in an arbitrary operationalization, and the psychometric analysis predominantly provides information about the test or the questionnaire but much less about the attribute theory (Figure 1.2, dashed line). This is the situation that approximates the "intelligence is what the test measures" paradigm, and researchers often do not seem to realize that in the absence of a guiding theory and corresponding operationalization, a fitting psychometric model supports only the data, not an attribute theory, which did not exist in the first place. If theories are not fully developed, one constructs items based on the available theory, complemented with intuition (what seems to be reasonable?), tradition (how were similar tests constructed?), and conformity (what do colleagues do or think?).

The abundant absence of well-developed theory may encourage a view on test and questionnaire construction as engineering and arbitrary sets of items as useful measurement instruments. One could argue that an instrument engineered this way could provide the cues that later inspire the theory of the attribute. If intuition and other knowledge sources are not completely arbitrary, and why should they be, then one cannot exclude the possibility that instruments constructed this way may provide a start for theory construction. However, this strategy is dubious and raises the question why one should not walk the route using systematic experimentation, slowly building theory, revising parts, constructing preliminary tests, returning to the theory, and so on. That is, substantive theory must guide instrument construction, and, in turn, the results of instrument construction must help revising the theory (e.g., Bouwmeester et al., 2007; Embretson & Gorin, 2001; Jansen & Van der Maas, 1997; Janssen & De Boeck, 1997; Raijmakers, Jansen, & Van der Maas, 2004).

This Monograph

In this monograph, we do two things. First, in Chapter 1, we notice that the state of psychological theorizing about its main attributes, and of the attributes in other research areas such as social science and health science, varies strongly across attributes, well-developed attribute theories probably being a minority. This state of theory development stands in the way of the construction of valid measurement instruments and leaves many of the attempts too far from perfect, perfect of course being an unattainable state. We do not and cannot have the pretention to be able on our own to lift theory development for attributes to a higher level and solve all problems we perceive. This enterprise requires a concerted effort by researchers from the social, behavioral, and health sciences and psychometricians alike, which we have not seen yet. The most we can do is to identify the problem, ask for attention, and occasionally contribute to theory development and instrument construction using methods that match the requirements the theory of the attribute posits.

Second, in Chapters 2–6, we discuss measurement models, some of which do not overstretch the possibilities the substantive theory of attributes allows. This has led us to focus primarily on models that imply the (approximate) ordering of people using the simple sum score or the classification of people based on their pattern of item scores. We also discuss more-demanding measurement models that are highly interesting but may provide serious mismatches with the data. These models nevertheless are useful in data analysis, because they clarify where the data deviate from the models' features and thus provide an abundance of interesting information one can use in the feedback loop in Figure 1.2. However, it is fair to say that also the less demanding measurement models often seem to ask too much of the data, especially when data are based on theoretical notions that provided not enough guidance for operationalization. The most we felt we could do is to critically discuss each of the measurement models and continuously ask what particular assumptions and their implications mean for the data and the attribute one intends to measure.

In each of the Chapters 2–5, we discuss one or two data examples. Each data set resulted from a theory or at least one or more hypotheses about the attribute the test or the questionnaire intended to measure. The reader will notice this gives the data much structure, which we rather easily identified using a particular measurement model. Data sets based more on exploration than intention ask for an exploratory approach. We do not discuss such approaches but emphasize that we do not oppose exploratory research leading to instrument construction if attribute theory is absent or weak. We understand well that one has to deal with the options that are available but wish to encourage the theoretical development of attribute theory. We cannot emphasize enough the importance of investing resources in theory development to have a guide for operationalization and data collection, data analysis and scale construction, and feedback on the attribute theory.

We have written this book for graduate and PhD students, researchers in the areas of the social, behavioral, and health sciences, and for fellow psychometricians. We have decided not to leave out derivations and proofs if we consider them important, but to help the reader who wishes to avoid the most arduous derivations and proofs, we have included a couple of them in boxes that are easily identifiable. The book may be of help in academic courses on measurement and as a guide for practical instrument construction. If the book raises discussion among peers about the problematic but infinitely intriguing topic of measurement in the human sciences, we have more than reached our goal.

The next chapters contain the following topics. Chapter 2 contains a discussion of CTT, not popular with advanced psychometrics anymore but in applied research, still the most frequently used method. Despite its age, the misunderstandings about CTT are as numerous as they are persistent. We discuss several of them and provide clarifications, in the hopes of creating some order in the huge literature. CTT is a theory of measurement error, but unfortunately, the approach does not specify what it is one measures with error. Rather, CTT leaves the composition of the test performance undefined and calls the error-free part of it simply the true score. CTT is mainly about determining the reliability of the error-ridden test performance. Because reliability has been and continues to be the core psychometric concept in the practice of instrument construction, we devote quite some attention to discussing different methods to estimate reliability and hope to resolve a few of the problems and misunderstandings that characterize the topic. We also discuss the distinction between the group property of reliability and the precision of the measurement of individuals.

Interestingly, if one knows a test score's reliability, one still does not know whether the item set constitutes a scale. CTT has nothing to say about this key topic, but researchers have resolved this omission *ad hoc* by using oblique multiple-group factor analysis, principal component analysis, and confirmatory factor analysis to study the item set's factor structure and select items in subsets used as scales based on their factor loading patterns. We discuss these approaches. The emptiness of the true score—or if one prefers, its unlimited abundance—inspired factor analysis specialists to reformulate CTT in terms of the factor-analysis model, redefining the reliability concept using factors replacing true scores, and systematic error introduced next to random measurement error. We also discuss the factor-analysis approach.

Chapter 3 discusses nonparametric IRT. Like CTT, nonparametric IRT uses sum score X [Equation (1.1)] to quantify people's test performance. The crucial difference with CTT is that nonparametric IRT defines the model of monotone homogeneity in which a unidimensional latent variable explains the relationships between the items and which defines a positive, monotone relation of individual items with the unidimensional latent variable. From the model's assumptions, one can derive for dichotomous-item tests that ordering people by their sum score X implies that one simultaneously orders them using the model's latent variable. Hence, the model of monotone homogeneity provides the mathematical justification for using sum score X, implying ordinal person measurement, and thus provides a decisive progression compared to CTT that used the sum score X *by fiat*. A second model called the model of double monotonicity not only implies an ordinal person scale but also assumes that, with the exception of ties, the ordering of the items is fixed for all scale values, thus implying an ordinal item scale. We discuss both nonparametric IRT models of monotone homogeneity and double monotonicity, both for dichotomous-item tests and for polytomous-item tests.

The goodness of fit of a model to the data is essential for the conclusion that one has constructed a scale for the target attribute and for the conclusion that the scale is ordinal, for persons, items, or both. We discuss the selection of items from an experimental item set into one or more subsets representing preliminary scales, each measuring a different aspect of the attribute. The item subsets are approximately consistent with the assumptions of the model of monotone homogeneity. Because they are approximately consistent with the model, it is advisable to investigate next in more detail whether the item subsets are consistent with each of the model's assumptions. We discuss several methods to investigate the goodness of fit of the two nonparametric IRT models to the data.

Chapter 4 discusses parametric IRT models. Parametric IRT models define the relation between the item and the latent-variable scale by means of a parametric function that includes item parameters, for example, defining an item's scale location and an item's potential to distinguish scale values on both sides of the item's scale location. Many parametric models fall into mainstream statistics, thus allowing the estimation of the item parameters and the person parameters using well-known estimation methods. As with any measurement model, for parametric models the goodness of fit to the data is essential for the useful interpretation of the model parameters and the responsible use of scales. A fitting model implies a scale for person measurement on which the items are also located. The latent-variable scale of parametric IRT has interesting applications in educational measurement. Because numerous books and journal articles have discussed this topic, we prefer to refrain from it and advise the interested reader to consult the vast educational-measurement literature. We discuss several parametric IRT models, their properties, and estimation methods and devote special attention to goodness-of-fit methods.

Next, we discuss a hierarchical framework for polytomous-item IRT models, showing that the parametric IRT models are mathematical special cases of three nonparametric IRT models. We also discuss hierarchical relationships for nonparametric and parametric dichotomous-item IRT models. In each of the hierarchical frameworks, because nonparametric IRT models are more general, properties of nonparametric IRT models generalize to parametric IRT models. For example, because all parametric IRT models for dichotomous-item tests are special cases of the more general nonparametric IRT model of monotone homogeneity, each of the parametric IRT models implies the ordering of people by means of the sum score on the latent variable scale. Weaker ordering properties apply to parametric models for polytomous items. These and other ordering properties justify using the simple sum score in the context of parametric IRT. We also discuss models that decompose the item parameter into linear combinations of basic parameters, explain the person parameter from observable covariates, and relate the response probability to multiple latent variables.

Chapter 5 discusses the LCM, which originated from sociology, and a special case of the LCM that may be of special interest to psychology, which is the case of cognitive diagnostic models (CDMs). LCMs are of interest because they impose even fewer assumptions on the data than nonparametric IRT models, a feature that may go well with attribute theories that are still in their infancy. The unrestricted LCM describes a data set as a collection of latent classes of persons, where the persons in a class have the same item-response probabilities for the items in the test or the questionnaire, whereas the item-response probabilities for the same item are different between classes. We discuss the unrestricted LCM and several restricted versions of the LCM. The main difference between unrestricted and restricted LCMs resides in whether one estimates model probabilities freely or under restrictions, the latter possibility comparable to using restrictions in confirmatory factor analysis. We briefly discuss estimation and in more detail goodness-of-fit methods. A variation of the LCM assumes an ordering of latent classes and is of interest here as a tool for investigating properties of nonparametric IRT models. LCMs developed to account for varying multivariate models across latent classes, including IRT models, such as used in the Bouwmeester et al. (2007) study, are beyond the scope of this book, which focuses on constructing scales.

LCMs are interesting for measurement if cognitive or other processes producing test performance distinguish people in unordered groups, thus leading to nominal scales. Cognitive diagnostic modeling offers a variety of structures explaining how a hypothesized set of attributes, different items requiring different subsets of the set of attributes for their solution, combine to produce a response to an item. For example, one CDM assumes that one needs to possess all attributes to solve the item with high probability, whereas the absence of at least one attribute implies item failure with high probability. Another CDM assumes that the absence of all attributes implies item failure with high probability, but possession of at least one attribute produces item success with high probability. Other CDMs define different constellations of attributes producing item success or failure. Cognitive diagnostic models allow the estimation of person attribute parameters and item parameters and identify the class based on patterns of attributes to which a person belongs. We discuss the basics of CDMs and explain that they have features of the LCM, nonparametric IRT, and parametric IRT.

Chapter 6 summarizes pairwise comparison, proximity, response time, and network approaches to assessing attributes. These approaches are interesting but not often used in scale construction in social science, psychology, and health science. Of course, this may change over time. The pairwise comparison model uses data based on the comparison of pairs of stimuli or items on a known dimension, for example, U.S. presidents with respect to their attitude toward human rights. Items are scaled in a first data collection and data analysis round, and the scaled items are administered to persons to determine their scale values in a second data collection and data analysis round. Proximity models quantify a response process based on the proximity of a person's scale location to that of the item. Items and persons are scaled based on one data collection and data analysis round. Response time models analyze dichotomous item scores together with the response times people take to answer items. Network models for psychometrics study networks of symptoms and identify the relationships between symptoms, assuming that symptoms affect one another. Pairwise comparison, proximity, and response time models are latent variable models much comparable in principle to the IRT models and LCMs we discuss in Chapters 3, 4, and 5. Network models only contain the observable symptoms acting as causal agents upon one another but lack one or more latent variables serving that purpose.

2

Classical Test Theory and Factor Analysis

Errare humanum est, sed in errare perseverare diabolicum.

—Seneca

Historical Introduction

Classical test theory (CTT) is a predecessor of modern measurement models used for psychological measurement. CTT is a theory of measurement error. Edgeworth (1888; also, Stigler, 1986, Chap. 9) was an early measurement pioneer who recognized that ratings of examinations are liable to error, and he identified several error sources, explicitly excluding clerical errors. Examples are the minimum degree to which raters make distinctions when assessing students' answers to questions, the *minimum sensibile*—in modern language, the just noticeable difference between stimuli—and the individual rater's tendency to push up or down the scale of marks. These and other error sources vary across raters. Edgeworth used his judgment based on personal experience to estimate the different error sources' effect on the marks and aggregated their effect, correcting for the number of questions in the examination. He noticed that errors were normally distributed, or in his words, their distribution followed the *gens d'armes' hat* (Figure 2.1), and noticed that as one presented and rated more questions, the influence of error on the ratings decreased. Edgeworth paid considerable attention to the influence that the presence of error in ratings has on classifying people in an honors category and the category just below, and estimated the numbers of false classifications. Interestingly, Edgeworth focused on the raters as sources of error and explicitly ruled out candidates being *out of sorts* on the day of examination as additional error sources. In his own words, he discussed a Theory of Error, which was part of what he called the Calculus of Probabilities, and what he did was identify error sources and apply the insight that, in modern parlance, the standard error of a mean is smaller than the standard deviation of the original variable by a factor $N^{1/2}$.

Edgeworth's ideas about measurement error were groundbreaking but still belonged to an older era. Spearman (1910) was the first to propose a formal theory of measurement error closer to what we acknowledge today

FIGURE 2.1
Edgeworth's French police officers' (gens d'armes, literally men of arms) hat, meant to illustrate the distribution of test scores at a time when this was not common knowledge. Artwork by Lars de Boer.

as measurement theory and introduced the reliability coefficient to identify the part of the measurement variance that was systematic; that is, not due to measurement error. Much of Spearman's work was practical and aimed at the measurement of intelligence (Spearman, 1904), for which he introduced the g-factor that all intelligence tests for sub-attributes of intelligence have in common and the s-factors for specific sub-attributes. Especially relevant to psychometrics, his work led to the proposal of factor analysis, and in addition to the reliability concept, he introduced many of the seminal psychometric concepts, such as the attenuation correction of correlations and the relation of reliability to test length (Spearman, 1910; also, Brown, 1910). Another example of the breadth of his contributions comes from his thoughts about the development of partial correlation (Spearman, 1907). Spearman's contributions were of invaluable significance to psychology and psychometrics alike, and it is impossible to overrate their importance, even today.

Almost simultaneously to the development of a theory of measurement error, the measurement of psychological attributes went through a developmental landslide from measurement procedures and attributes that seem almost unrecognizable from our perspective to psychological testing as we know it. We mention a few of the older measurement procedures to give the reader an idea of how they were different from today's psychological testing. Cattell (1890) discussed a series of tests measuring psychophysical properties, which he assumed were related to mental attributes. For example, dynamometer pressure recorded the pressure exercised by the squeeze of the hand and was considered the expression of bodily and mental energy that could not be separated. Particularly interesting to psychology was the relation between dynamometer pressure and volitional control or emotional excitement. The subject was trained how to hold the dynamometer to exercise maximum pressure and was allowed two trials with each hand. Another example is the rate of movement that Cattell assumed to be connected to the four classical temperaments of the sanguine, choleric, melancholic, and phlegmatic personality types. Baldwin, Cattell, and Jastrow (1898) discussed three categories of tests for mental anthropometry. They were tests for the senses, the motor capacities, and the more complex mental processes, such as memory, association, attention, and imagination. Especially the complex mental processes proved extremely difficult to measure given the habit of using physical measurements, not psychological measurements, assumed to correlate with these mental processes. Nothing in the way late-nineteenth-century psychologists measured attributes reminds one of the multi-item tests referring directly to psychological phenomena that we use today. Binet and Simon (1905) provided the breakthrough to modern measurement procedures of psychological rather than psychophysical attributes. They introduced the modern format of a psychological measurement instrument as a set of items referring to the practical problems that students encountered in school, administered under standardized conditions while one summarized a student's performance by one score, equaling the number of correct answers. Binet in particular is considered as important for psychological measurement as Spearman, the latter being the theorist and the former the practitioner.

Consistent with the older developments, huge influences on the development of CTT and the use of standardized tests came from education where one assessed achievement with respect to admission and placement decisions and the need in organizations to assess applicants for jobs, initially with respect to psychomotor and psychophysical skills more than personality traits. World War I posed the challenge for the US Army to select tens of thousands of recruits for the army units where they were best placed, and as is often the case, this circumstance greatly pushed science and technology, in our case, the theory and development of psychological testing, forward. Several psychometric publications stem from the period starting at the beginning of the twentieth century until World War II, addressing various topics, such as general psychometrics (Guilford, 1936, 1954) and factor analysis (Thomson, 1938, 1951; Thurstone, 1931, 1935). In the 1940s and 1950s, Cronbach (1949) published his monograph on psychological testing, and Gulliksen (1950) published his standard work on CTT.

In the 1940s and 1950s, the first signs of what later became known as item response theory (IRT) emerged. Lawley (1943) and Lord (1952) are important pioneers, and Birnbaum published several influential technical reports and articles in the second half of the 1950s, referred to and discussed in Birnbaum (1968). Rasch (1960) published his highly influential book on what developed into the one-parameter logistic model or, simply, the Rasch model. In 1968, Lord and Novick (1968) published their seminal book on CTT, obviously mathematically much more advanced and consistent than the earlier work on CTT and also including four chapters by Birnbaum (1968) on IRT, in particular, the two-parameter logistic model or Birnbaum model. From then on, IRT was unstoppable and among mainstream psychometrics pushed CTT backstage if not off stage. Interestingly, test constructors and researchers kept using CTT as their mainstream psychometric model, and although IRT slowly is gaining territory, the use of CTT in practical test construction and assessment continues to emphasize the gap with theoretical psychometrics embracing IRT that became visible in the previous decades. This is different in educational measurement, where IRT has taken over command due to its undeniable facilities to equate scales and tailor testing to the individual ability level and several other, practically highly relevant applications.

In this monograph, we take the stance that CTT is relevant irrespective of the existence of other formal measurement models. Because it is older, it cannot do all the things that statistically more advanced IRT and other models can (e.g., Lumsden, 1976), but its emphasis on reliability and validity in particular are valuable in their own right and capture the two main concepts that dominate psychometrics. One could even argue that modern measurement models specify and further elaborate many of the concepts that CTT introduced previously, although they undeniably have opened gates to novel topics, such as test equating (Kolen & Brennan, 2014; Von Davier, Holland, & Thayer, 2004) and computerized adaptive testing (Van der Linden & Glas, 2010; Wainer, 2000).

Given that the focus of this monograph is on models and scale construction, we skipped many important topics that CTT traditionally deals with, such as attenuation correction, correction for restriction of range, norming of test scores, test length considerations, and factors affecting reliability and measurement precision. Validity of tests is another major topic, traditionally addressed in the CTT context but obviously pivotal in all of social, behavioral, and health measurement. Probably due to its complexity, the attention given to validity remained marginal relative to modeling and estimation. The small number of books on validity reflects this marginal, albeit misguided position, but see Wainer and Braun (1988), Lissitz (2009), and Markus and Borsboom (2013), and related work on psychological measurement by Michell (1999) and Borsboom (2005). Mislevy (2018) provided a comprehensive study on understanding measurement models from a sociocognitive psychological perspective, focusing on the educational measurement context.

The Classical Test Model

Measurement Level and Norm Scores

CTT assumes that test scores have interval measurement level. Following Chapter 1, let X be a test score, usually the sum of the scores on the J items in the test or the questionnaire [Equation (1.1)], then linear transformations $Y = aX + b$, where a and b are scalars, are allowed and do not change the measurement level of a scale and the indices expressing the quality of the scale. However, unlike modern measurement models we discuss in later chapters, in CTT the assumption of interval measurement level lacks both a theoretical and an empirical basis, meaning that it is just an assumption without proper justification. The assumption of interval level is mainly a practical choice facilitating the use of parametric statistics. This pragmatic approach, lacking a theoretical and an empirical basis, however has not prevented test score X from being useful for predicting students' most appropriate future education, applicants' success in a particular job, and patients' suitability for a particular therapy. One may consult Michell (1999) and Borsboom (2005) for critical discussions about CTT and psychological measurement ambitions in a more general sense.

Before we discuss CTT, we devote some time to discussing how the practice of test administration traditionally has dealt with measurement values. To understand a person's scale position relative to another person's scale position, ideally one would use the interval between her test score and the other person's test score, but test scores lack interval-level properties and in addition a zero point, both omissions reflecting the absence of the concept of quantity. The practical use of test scores circumvents this problem by deriving test scores' meaning from their position in a distribution of test scores, called norm distributions. Useful norm distributions result from linear or nonlinear transformations of the sum score. Before we discuss the best-known norm scores, we emphasize that they are based on the sum score X, and because the measurement level of X is unknown, the measurement level of the norm scores, which are transformations of X, is also unknown. Norm-score distributions or norm distributions, for short, facilitate the interpretation of test performance, because they contain information about the tested person's test performance relative to the norm distribution, not because the norm scores have a known measurement level.

Examples of linear transformations into well-known norms are Z-scores, T-scores, and IQ-scores. Let μ_X denote the mean of sum score X, and let σ_X denote the standard deviation of X. Then Z-scores are defined using linear transformation $Z = aX + b$, in which $a = \sigma_X^{-1}$ and $b = -\mu_X \sigma_X^{-1}$, so that

$$Z = \sigma_X^{-1} X + \left(-\mu_X \sigma_X^{-1}\right) = \frac{X - \mu_X}{\sigma_X}, \quad \text{with } \mu_Z = 0 \text{ and } \sigma_Z = 1. \tag{2.1}$$

Knowing that someone has a score $Z = 1$ means knowing that she is one standard deviation above group average. Provided the distribution is approximately normal, this is quite informative about her standing relative to the group to which she belongs; that is, approximately 84% have a lower sum score. Next, we define T-scores by presenting them as linear transformations of Z-scores, using $a = 10$ and $b = 50$, so that

$$T = 10Z + 50, \quad \text{with } \mu_T = 50 \text{ and } \sigma_T = 10. \tag{2.2}$$

T-scores allow an interpretation comparable to that of *Z*-scores but with different mean and variance. However, a score equal to $T = 60$ has precisely the same meaning as $Z = 1$. We define the well-known *IQ*-scores also as linear transformations of *Z*-scores, using $a = 15$ and $b = 100$, so that

$$IQ = 15Z + 100, \quad \text{with } \mu_{IQ} = 100 \text{ and } \sigma_{IQ} = 15. \tag{2.3}$$

Someone with $IQ = 115$ scores one standard deviation above average and if the distribution were transformed to *Z*-score or *T*-scores, this person would have scores $Z = 1$ or $T = 60$. *IQ*-scores are well-known to the public and have a public status comparable to that of physical measurements, but their theoretical underpinnings are considerably weaker.

Examples of well-known non-linear norms are stanines and percentiles. Stanines, shorthand for "standard nines", run from 1 to 9. Score 1 corresponds to the area under the normal curve equal to $(-\infty; -1.75]$, covering approximately 4% of the lowest test scores, stanines 2, ..., 8 correspond to half-unit intervals under the normal curve, $(-1.75; -1.25]$, $(-1.25; -0.75]$, ..., $(1.25; 1.75]$, covering approximately 7%, 12%, 17%, 20%, 17%, 12%, and 7% of the scores, and stanine 9 corresponds to $(1.75; \infty)$, corresponding to the final 4% of the scores. Someone interchangeably scoring $Z = 1$, $T = 60$, and $IQ = 115$ (assuming *IQ* follows a normal distribution, which is not entirely correct but highly convenient for the example), falls in the interval $(0.75; 1.25]$ and thus receives a stanine score equal to 7. The percentile score runs from 1 to 100, and irrespective of the distribution of sum score *X*, a percentile score 84 means that 84% of the group had a lower test score. For a normal distribution, $Z = 1$ corresponds to percentile score 84, a result we already encountered.

Linear transformations, such as *Z*-scores, *T*-scores, and *IQ*-scores, and nonlinear transformations, such as stanines and percentiles, facilitate the interpretation of test scores and are typical end products of test construction, meant to help the test practitioner to provide meaning to test scores that she can discuss with the tested person. *Z*-scores, *T*-scores, and *IQ*-scores are used as interval measurements, and stanines and percentiles as ordinal measurements, but, like their mother test score *X*, neither use has a theoretical or an empirical basis. With the exception of *Z*-scores, one rarely uses the other transformed scores in computations that serve to assess the psychometric properties of the test. Their only purpose is to facilitate the understanding of test performance. Finally, we mention different methods of norming. Criterion-referenced norming refers to comparing the test score to an external standard, such as a cut-score, often used to make a decision, for example, about passing or failing an exam, accepting or rejecting someone for a job, or admittance or non-admittance to a therapy. Such decisions require empirical evidence of their correctness, sometimes referred to as a test's predictive validity. Standard setting seeks to define levels of proficiency on a scale corresponding to cut-scores that enable the classification of students in categories that are psychometrically defendable and consistent with education policy (e.g., Bejar, 2008).

Model Assumptions

The main accomplishment of CTT is the insight that any observable variable can be decomposed into an unobservable, true-score part, denoted *T*, not to be confused with the *T*-scores sometimes used as norm scores, and an unobservable, random measurement error part, denoted *E*, such that

$$X = T + E. \tag{2.4}$$

Variable *X* can be anything. For example, *X* can be the time an athlete accomplishes at the 100-meter dash, the weight read from the scale in one's bathroom, or a rating a psychologist assigns to an applicant whom she assesses with respect to team-building behavior in an assessment center. Variable *X* can also be the sum score a child obtains on a multi-factor intelligence test, a transformed sum score such as an *IQ*-score, a sum score on a single-factor spatial orientation test, which is part of an intelligence test battery, and the score on an individual item. According to CTT any observable variable has a true-score component. The true score has been given several meanings. For example, the Platonic interpretation suggests that test score *X* is an imperfect reflection of the score that one actually wishes to observe but that measurement error obscures from observation. In this conception, an individual possesses a true score *T*, which actually exists and which the psychologist should discover by choosing the right test, a task destined to fail because of measurement error. These and other definitions produced formal problems in CTT results, and Lord and Novick (1968, p. 30; also, see Novick, 1966) replaced the older definitions by an operational definition, consistent with modern statistics.

The modern version of the true score is defined as follows. Consider a person, indexed *i*, and imagine that one administers the same test repeatedly to person *i*, producing a distribution of test scores for person *i* known as person *i*'s propensity distribution (Lord & Novick, 1968, p. 30). We call repeated administrations independent replications, indexed *r*, and define the true score as the expectation of X_i,

$$T_i = \mathcal{E}_r(X_{ir}). \tag{2.5}$$

The true score is a hypothetical entity, without a chance of ever being observed, simply because it is a mathematical construction that helps to make CTT consistent. Its interpretation is the mean test score across an infinite number of hypothetical, independent replications that are clearly impossible to obtain for many practical reasons, such as memory, fatigue, and refusal on the part of person *i* to be tested repeatedly. Two misunderstandings one sometimes encounters need to be tackled.

First, the true score is the statistical expectation of an observable variable, but CTT does not restrict the observable variable's dimensionality in any way. Consequently, observable variable X can be a score—any score, not necessarily the number-correct score or another sum score—reflecting the performance on an analytical reasoning test, an analytical reasoning test embedded in practical, everyday problems also requiring verbal abilities, or an intelligence test battery that includes an analytical reasoning subtest. The three test scores resulting from these examples have different meanings—analytical reasoning, everyday analytical reasoning, and general intelligence. The three data sets produced by a sample of persons will have different factor structures. However, CTT's main goal is to establish the degree to which the test score—again, any test score one finds interesting and also the ones one finds uninteresting—is repeatable if testing were repeated. This is the main enterprise of CTT.

Second and related to the first issue, whereas latent variable modeling, including IRT, factor analysis, and structural equation modeling, contains one or more latent variables that simplify the relations between the observable variables in the models, true score T does not fulfil this function in CTT. That is, in latent variable models the items have one or more latent variables in common that explain the relations between the items, but an item's true score is the first moment—the expected value—of the item score's propensity distribution, and unlike a latent variable, it is item-dependent, hence unable to explain the relations between different items. The absence of a common latent variable in CTT does not prevent the items from sharing variance, hence from being related, but the model does not contain a common cause that explains the inter-item relations. We notice that it has proven difficult for people to accept that CTT is at ease with not defining one or more latent variables that the items have in common. To see that this omission is reasonable or at least a possibility and not an aberrancy to be rejected, one needs to focus on the goal of CTT, which is to estimate the influence that random measurement error has on test performance or any other measurement. Typically, CTT is an error model rather than a true-score model that splits the true score into, for example, a common component and an item-specific component. One could argue that choosing not to model the true score but define it as the first moment of the item-dependent propensity distribution is also a choice, albeit not the one that is prominent in much of modern measurement. This chapter will clarify that the consequence of not modeling the true score is that CTT revolves around estimating reliability of measurement. Focusing on reliability implies that CTT strives to estimate the proportion of variance in the test scores obtained in a group that is not measurement-error variance. If one prefers to model the true score, this has serious consequences for the estimation of reliability. We emphasize that this is a legitimate possibility and there is nothing wrong with it, but one has to understand that by doing this one leaves CTT and enters, for example, a common-factor model approach (e.g., Bollen, 1989), as we will see later in this chapter.

Given the definition of the true score, the measurement error component in Equation (2.4) is defined as

$$E = X - \mathcal{E}(X),\tag{2.6}$$

and because it is random, across a large number of independent replications, a person's mean measurement error equals 0,

$$\mathcal{E}_r(E_{ir}) = 0.\tag{2.7}$$

The existence of random measurement error was inferred from the alleged inaccuracy of human judgment (e.g., Edgeworth, 1888). We mention a couple of events inside and outside of the tested person that may occur unpredictably and that may influence her test performance. With each event we use a sign between parentheses to indicate the possible direction of the measurement error when it happens, but we emphasize that, by the assumption of CTT, the events themselves are unpredictable and so are their effects. All we wish to do is make plausible that human behavior has an unpredictable side to it, which is also active in performance when tested, and the examples may help the reader to imagine some of the causes. Then, causes of random measurement errors in maximum-performance testing, together with a possible effect and a plausible direction they may have on test performance, may be: A flu that came up the night prior to being tested (–; i.e., the effect may be a lowering of the sum score by a negative measurement error), a good night's sleep (+), a lapse of concentration (–), a sudden insight (+), a distraction due to noise (–), a happy feeling due to a forthcoming vacation (+), a sad feeling about a sick parent (–), a temporary attention deficit (–), a TV show one recently saw (+), a (test) anxiety attack (–), a feeling of self-confidence (+), an intrusive thought (–), and a clear mind (+). With typical-behavior testing one may imagine other causes exercising different effects, for example, causes having to do with self-image, perceived other-image of oneself, or the purpose of being tested, such a job selection. All of these phenomena are error sources when they intervene with the measurement process and cause a person to perform below or above expectation on the test in an unpredictable way. Person i's propensity distribution thus has mean equal to

$$T_i = \mathcal{E}_r(X_{ir})\tag{2.5}$$

and variance

$$\sigma_{E_i}^2 = \mathcal{E}_r\left\{\left[X_{ir} - \mathcal{E}_r(X_{ir})\right]^2\right\}.\tag{2.8}$$

CTT does not restrict these means and variances for different persons. For different measurement values, Figure 2.2a shows propensity distributions

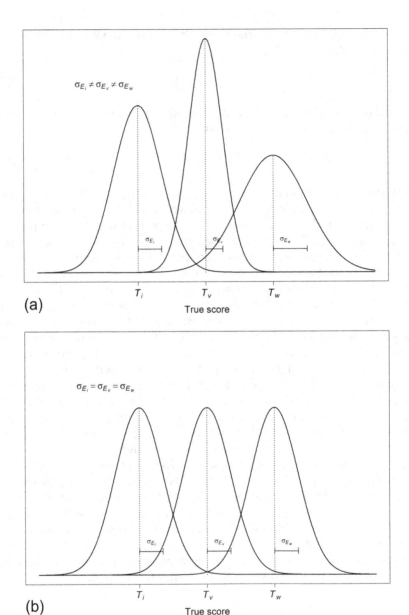

FIGURE 2.2
(a) True score and propensity distribution for three persons under an IRT model, which allows different error variances across persons, and (b) a CTT model, which requires that all error variances are equal across persons. Note: i, v, w = person indexes.

having different means and variances. Hence, if possible, given an estimate of someone's true score, an estimate of the standard deviation σ_{E_i} provides a measure of measurement precision (Mellenbergh, 1996) and can be used to estimate a confidence interval for the true score. Varying measurement precision may be adequate, because the test may not be an equally effective instrument for everybody. For example, if the items in a maximum performance test are too easy or too difficult for an individual, she will succeed or fail almost all items and one still does not know much about the attribute level, meaning great uncertainly, expressed by large σ_{E_i}. Indeed, several authors have made proposals for estimating σ_{E_i} at specific score levels (Feldt, Steffen, & Gupta, 1985), some of which we discuss later in this chapter. However, practitioners of CTT use a mean standard error of measurement, implying that each true-score estimate is assessed using one precision estimate for all. Figure 2.2b shows propensity distributions following the "one size fits all" approach, where the mean may vary but the standard deviation is the same for each measurement value. We notice that CTT does not assume or imply equal standard deviations for all measurement values, but that it is a practical decision to use the standard error of measurement. Parametric IRT models do allow a standard error that depends on the scale value and in this respect have a great advantage over CTT (Figure 2.2a was drawn based on an IRT model). This topic is taken up in Chapter 4.

Repeatability of Test Scores: Reliability

Because independent replications of test scores provided by the same people usually are unavailable, we consider a group of people rather than an individual and derive CTT results. Across group members, using Equation (2.7) for individuals, we find that

$$\mathcal{E}_i \left[\mathcal{E}_r \left(E_{ir} \right) \right] = 0. \tag{2.9}$$

We notice that $X_{ir} = T_i + E_{ir}$, so that the mean testscore in a group, μ_x, equals

$$\mu_X = \mathcal{E}_i \left[\mathcal{E}_r \left(X_{ir} \right) \right] = \mathcal{E}_i \left(T_i \right) + \mathcal{E}_i \left[\mathcal{E}_r \left(E_{ir} \right) \right] = \mu_T. \tag{2.10}$$

That is, in a group of persons, the mean measurement error equals 0 [Equation (2.9)] and the mean test score at group level equals the mean true score at group level [Equation (2.10)].

Equation (2.4) applies also to item scores,

$$X_j = T_j + E_j, \tag{2.11}$$

and adding all J item scores yields $X = T + E$ or equivalently,

$$\sum_{j=1}^{J} X_j = \sum_{j=1}^{J} T_j + \sum_{j=1}^{J} E_j. \tag{2.12}$$

Measurement error is random, hence correlates 0 with other variables Y of which it is not a component, so that the product-moment correlation

$$\rho_{EY} = 0. \tag{2.13}$$

For example, $\rho_{ET} = 0$ ($Y = T$) and, letting j and k index items, $\rho_{E_j E_k} = 0$ ($Y = E_k$). If E is part of Y, which happens if $Y = X = T + E$, then $\rho_{EX} > 0$. To see this, notice that $\rho_{EX} = \sigma_{EX}/(\sigma_E \sigma_X)$, and develop $\sigma_{EX} = \sigma_{E(T+E)} = \sigma_{ET} + \sigma_{EE}$. Because from Equation (2.13) it follows that $\sigma_{ET} = 0$, and because $\sigma_{EE} = \sigma_E^2$, we have that

$$\rho_{EX} = \frac{\sigma_E^2}{\sigma_E \sigma_X} = \frac{\sigma_E}{\sigma_X} \geq 0. \tag{2.14}$$

The interesting result is that in a group, people having higher sum scores tend to have positive measurement error and people having lower sum scores tend to have negative measurement error. This looks contradictory to the assumption of *random* measurement error but simply follows from the assumption. The point is that people having the higher sum scores X have been lucky, on average, and people having the lower sum scores have been unlucky. Upon repetition, one expects good luck and bad luck to spread randomly across the whole group. Someone having an extremely high sum score thus is *expected* to obtain a somewhat lower sum score, but not necessarily so, and so on: Random error is unpredictable at the individual level. However, at the subgroup level, one expects, for example, the low-scoring subgroup to attain a higher mean sum score. As a subgroup, they are expected to come nearer to the overall-group mean. This is the well-known regression to the mean effect, and one can consider it a pure random effect.

From the correlation behavior of measurement error [Equation (2.13)] and properties of sums of variables, also known as linear combinations, one can derive that

(1) The group variance of the test score equals

$$\sigma_X^2 = \sigma_T^2 + \sigma_E^2 + 2\sigma_{TE} = \sigma_T^2 + \sigma_E^2. \tag{2.15}$$

(2) The variance of the error component E of the test score X can be written as the sum of the J item-error variances,

$$\sigma_E^2 = \sum_{j=1}^{J}\sigma_{E_j}^2 + \sum\sum_{j\neq k}\sigma_{E_jE_k} = \sum_{j=1}^{J}\sigma_{E_j}^2. \tag{2.16}$$

Reliability is defined as the product-moment correlation between two independent replications of test scores in a group of people, and to define reliability, one needs to provide a mathematical definition of independent replications. Lord and Novick (1968, pp. 46–50) defined replications as two tests with test scores X and X' that are interchangeable in all respects except their random measurement error; that is, $E \neq E'$. Such tests are parallel (ibid., p. 48). Rather than enumerating the properties that are equal, the next two characteristics are sufficient to summarize all those properties, because they can be shown to imply all properties (no proof given here):

Definition: Two tests with test scores X and X' are parallel if

(1) For each person i, $T_i = T_i'$;

this implies that at the group level $\sigma_T^2 = \sigma_{T'}^2$, and

(2) At the group level, $\sigma_X^2 = \sigma_{X'}^2$. $\qquad\qquad(2.18)$

(2.17)

Reliability can now be defined (Lord & Novick, 1968, p. 61) as follows.

Definition: The reliability of a test score, or any observable variable for which this is deemed useful, is the product-moment correlation between two sets of parallel test scores in a group of persons and is denoted $\rho_{XX'}$.

Reliability $\rho_{XX'}$ can be written as the correlation between $X = T + E$ and $X' = T' + E'$. Using the results that (1) the covariance between two sum variables or linear combinations equals the sum of all the covariances between the separate variables, (2) by parallelism, $T = T'$, and (3) measurement errors correlate 0 with any other variable Y that does not include the measurement error [Equation (2.13)], reliability can be shown to equal

$$\rho_{XX'} = \frac{\sigma_T^2}{\sigma_X^2} = \frac{\sigma_{T'}^2}{\sigma_{X'}^2}, \tag{2.19}$$

and

$$\rho_{XX'} = 1 - \frac{\sigma_E^2}{\sigma_X^2} = 1 - \frac{\sigma_{E'}^2}{\sigma_{X'}^2}. \tag{2.20}$$

Equations (2.19) and (2.20) show that parallel tests have the same reliability, and that their correlation for both tests also expresses the proportion of test-score variance that is true-score variance [Equation (2.19)] or, equivalently, 1 minus the proportion of test-score variance that is error variance [Equation (2.20)]. In what follows, we use the first parts of the definitions that refer to test score X and score components T and E.

In the reliability definition [Equation (2.20)], numerator σ_E^2 can be replaced by the sum of the item-error variances [Equation (2.16)], resulting in

$$\rho_{XX'} = 1 - \frac{\sum_{j=1}^{J}\sigma_{E_j}^2}{\sigma_X^2}. \tag{2.21}$$

This definition of test-score reliability is key to the remainder of this chapter. However, any form in which reliability is written suffers from the problem that it contains too many unknowns. On the left-hand side of Equations (2.19), (2.20), and (2.21) one has the desired but unknown reliability $\rho_{XX'}$, whereas the right-hand sides provide the unknowns $\sigma_{E_j}^2$ [$j = 1, \ldots, J$; Equation (2.21)], σ_T^2 [and $\sigma_{T'}^2$; Equation (2.19)], and σ_E^2 [and $\sigma_{E'}^2$; Equation (2.20)].

One could avoid the problem of the unavailability of these quantities by simply estimating the correlation between two real replications of the same test or between two parallel test versions. However, it is unlikely that real replications of test scores obtained from the same group of people are independent, because if people are prepared to take the same test twice, they will remember several of the items and the answers they gave, and their memory will influence the second test administration so that results are dependent on the first administration. Likewise, if time elapses between two administrations, people learn and develop between administrations and this will influence results on the second administration. Even if the administrations of the parallel tests take place in one test session, some persons may profit from the exercise that the first test offers and improve their performance at the second test, again introducing dependence, also when items from the two tests are administered alternately. If replications are operationalized as parallel tests, they often turn out not to be parallel. Whether they are parallel can be investigated using the equality that for any variable Y_h, $h = 1, \ldots, \infty$, for two parallel tests

$$\rho_{XY_h} = \rho_{X'Y_h}, \tag{2.22}$$

where the number of variables running to infinity suggests that in principle the task is never finished (Zegers & Ten Berge, 1982). However, in practice it makes sense that a finite number, perhaps not even a large number, of well-chosen and theoretically meaningful variables Y_h will provide useful information about the possible parallelism of the tests. Thus, in addition to administering two allegedly parallel tests, the test constructor must collect scores on a number of third variables, Y_h, to find out whether the tests are parallel. Prior to this challenging task, constructing two test versions that are interchangeable except for random measurement error is difficult. For example, in constructing two parallel test versions one probably follows the strategy of constructing pairs of items that look interchangeable, for example, "$17 + 28 = \ldots$" and "$18 + 27 = \ldots$", and then assigns the items from such pairs to the two test versions. However, for several attributes, such "twin items" may be difficult to construct and when they are constructed may unexpectedly trigger different cognitive processes, thus producing differences in Equation (2.22). In addition to such problems, test constructors are often not motivated to construct two test versions when they need one and prefer using the sparse time available for testing to administer another test providing information additional to the first test.

In practice, test constructors have data available based on a single test administration or the administration of one test version. CTT has devoted much of its effort to developing methods for approximating Equation (2.21). This has produced a wealth of interesting methods, but it also created an incredible confusion among researchers and psychometricians alike. We provide the main results and weed out some of the misunderstandings.

Methods for Estimating Reliability

Psychometrics has seen a long tradition of reliability estimation. We start this section with four methods that have become traditional standards. They are the parallel-test method, the retest method, the split-half method, and the internal consistency method. Next, we discuss the best-known internal consistency method, known as coefficient α (Cronbach, 1951), and five other methods that, together with coefficient α, form a collection of six methods that Guttman (1945) proposed. Two of these methods are the first two terms of an infinite series of coefficients (Ten Berge & Zegers, 1978). We end this section with the greatest lower bound to the reliability (e.g., Woodward & Bentler, 1978). At the end of this chapter, we return to reliability in the context of the factor analysis model.

Methods Commonly Used in Test-Construction Practice

Parallel-Test Method

The parallel-test method aims at constructing two tests that are interchangeable except for the random measurement errors they elicit. The product-moment correlation between the sum scores on two parallel tests equals the reliability of the sum score on either test, hence, the notation $\rho_{XX'}$ for the reliability. We noticed already that constructing truly parallel test versions is almost impossible, and for this reason, the terminology of alternate-forms method is sometimes used. Because we already discussed problems with respect to this method and because test constructors rarely attempt this method in practice, we refrain from further discussion.

Retest Method

The retest method has at least three applications. The first application is to mimic two parallel tests by administering the same test twice to the same group of people, thus producing test scores X_1 and X_2, and taking the product-moment correlation $r_{X_1 X_2}$ as an estimate of reliability $\rho_{XX'}$. Obviously, memory and practice, to name just two possibilities, are realistic threats to an unbiased estimate, and may readily produce a positive bias suggesting reliability is higher than the true reliability one wishes to estimate. The next two applications actually do not concern reliability as defined in Equations (2.19), (2.20), and (2.21) but are often assumed to produce reliability estimates or estimates of quantities that are assumed to represent a different kind of reliability. It is for these reasons that we discuss the two applications.

The second application is to administer the test with a time interval that suits the test constructor's goal, which may be, for example, to estimate the correlation between test scores people produced directly before therapy started and test scores they produced directly after therapy ended. Assuming that the therapy's target was to influence people's attribute level, for example, reducing their anxiety level, what the test constructor wishes is that on both occasions the test recorded the person's attribute level at the time of testing. Ideally, the test constructor recorded the person's anxiety level before therapy and once more after therapy, on the second occasion assuming that only the therapy and nothing else influenced the person's anxiety level. However, it does not take much imagination to think of other influences appearing between the two test administrations, exerting their effects on the anxiety measurement after therapy, such as one's general psychological condition at the end of therapy and the attention one receives by engaging in therapy. In addition, also memory may play a role and influence the second test score, for example, because the person wants to be consistent or convey the impression that she improved. Only if such influences additional to a therapy effect were absent or if they were the same for all persons would they have no differential effect on test score X_2 and leave the correlation unaffected. Otherwise, the additional influences would contaminate the correlation between the test scores X_1 and X_2 in unpredictable ways.

Finally, one may be interested purely in the degree to which a particular attribute is volatile across time including all the influences that happen spontaneously in people's lives. We tend to interpret the resulting correlation coefficient as a validity feature of the attribute, informing one about one of the attribute's typical properties, here, its volatility.

Because it seems safe to assume that, in general, test scores obtained from two administrations of the same test to the same group can never pass for parallel test scores, Cronbach (1951) rightly called the correlation resulting from the retest method the *coefficient of stability*. This interpretation not only allows for instability caused by random error, which is what one is interested in when estimating reliability, but also includes systematic changes in people's true scores. Such systematic changes in true scores may be brought about by, for example, therapy, but also other manipulations such as training of managers aimed at improving their skill to mitigate conflicts between colleagues, and remedial teaching programs aimed at improving students' arithmetic ability.

Split-Half Method

The split-half method splits the test into two halves, computes the correlation between the test scores on the two halves as an estimate of the reliability of a test of half the length of the intended test, and then uses a prophecy formula to estimate the reliability of a test of whole length. Cronbach (1951) called this prophecy reliability the *coefficient of equivalence*. At first sight, this is an amazingly simple and attractive method that aims at approximating parallel tests by test halves that might pass for parallel half tests, but in the practice of test construction, the method did not come for free. Indeed, test constructors used the method a lot but noticed two problems. First, one could split a test into many different test halves, and each produced a unique reliability. This raised a choice problem that researchers were unable to solve. From the present-day perspective, one could argue that surely many reliabilities were not significantly different, producing clusters of values, thus limiting the number of choices considerably. However, prior to World War II, statistical testing was still in its infancy and sampling distributions were unavailable or were unknown to many researchers; hence, one considered sample results to be the first and final material to work with. Second, the Spearman–Brown prophecy formula (Brown, 1910; Spearman, 1910), which is in use nowadays to predict the reliability of a lengthened or shortened test, was then considered too restrictive, and several alternatives were proposed. This raised a second choice problem for test constructors, and again they did not know which method to choose. The contribution Cronbach (1951) made to the practice of test construction was that he had the insight that a method that had already been around for a while (Hoyt, 1941; Guttman, 1945; Kuder & Richardson, 1937), and which he called coefficient α, alleviated the two choice problems that made the split-half method a topic of constant debate. In particular, Cronbach reasoned that across the different splits, the split-half values obtained using a particular correction method showed little spread whereas the mean equaled coefficient α. Thus, Cronbach proposed that coefficient α, which so efficiently summarized the distribution of split-half values, should replace the split-half method.

Internal Consistency Method

Coefficient α succeeded the split-half method and became an unprecedented success in the history of applied psychometrics. Following Cronbach's advice, researchers called it a *coefficient of internal consistency*. The name was based on Cronbach's line of reasoning that if two test-halves had different factor compositions, this had little effect on the split-half value and hence, on coefficient α. He restricted the factor compositions to be what one typically found in real tests. Cronbach also reasoned that the test's general or dominant factor was the main influence on coefficient α, whereas group factors had a much smaller impact. Based on these insights, Cronbach argued that coefficient α was a measure of a test's internal consistency. By this, he meant (ibid., p. 320) that the test is "psychologically interpretable" although this does not mean "that all items be factorially similar". He suggested to quantify internal consistency by means of the mean inter-item correlation of all the item pairs in the test and related it to psychological interpretability; also, see Davenport, Davison, Liou, and Love (2015).

Cronbach initially proposed coefficient α as a summary of the mean split-half value, but it became famous as a coefficient of internal consistency. This emphasis on internal consistency diverted researchers' awareness away from coefficient α's historical roots, which are as an approximation of reliability, to the composition of the test, and it became quite common to identify high α values with successful attribute measurement, which may be reliable as well. Sijtsma (2009) argued that the interpretation of internal consistency is incorrect and demonstrated that a relationship between values of α and the items' factor composition is absent. Specifically, it proves simple to construct one-factor and multiple-factor item sets producing the same α, and examples concerning a particular factor structure such that α is either high or low. To summarize, the message is that $\alpha = .5$, which is considered low, can refer to highly different item-factor structures, either one-factor or multiple-factor, and so can $\alpha = .9$, which is considered high. Hence, coefficient α is an inadequate measure of effective attribute measurement or internal consistency.

Another phenomenon the authors have noticed is that different reliability methods have begun to lead a life of their own, in the sense that people believe they represent different kinds of reliability and are sensitive to different varieties of errors. While this is true for the retest method, which in practice actually functions as a measure of attribute stability, we emphasize that the methods we discussed and the methods we will discuss in the next section all serve to approximate $\rho_{XX'}$ in Equations (2.19)—(2.21). Let an approximation be denoted ρ_p, where p indexes approximations to $\rho_{XX'}$; then surely under most conditions, ρ_p is a parameter that is different from $\rho_{XX'}$, and under some other conditions, $\rho_p = \rho_{XX'}$, but the only reason ρ_p was proposed is that it is an approximation to $\rho_{XX'}$. For example, coefficient α is a mathematical lower bound to reliability $\rho_{XX'}$; that is, $\alpha \leq \rho_{XX'}$, and when items or other variables for which α is determined are essentially τ-equivalent, τ referring to the true score, meaning that for a pair of items, $T_j = T_k + a_{jk}$, a_{jk} is a scalar (Lord & Novick, 1968, p. 50), then $\alpha = \rho_{XX'}$ (Novick & Lewis, 1967). Our point is that uses of coefficient α and other reliability methods other than as approximations to the reliability are a matter of the researcher's interpretation but do not necessarily follow from CTT or another statistical model. For example, to call coefficient α a measure of internal consistency is a matter of choice, perhaps motivated by α's dependence on the inter-item covariances, but not a necessity implied by the mathematics of the method. We prefer to stay close to the mathematics and keep that separated from substantive interpretations of mathematical results.

Coefficient α has become by far the most frequently used reliability method, but the literature on α is about a lot more than α being a lower bound to the reliability. In addition to having become the standard of test reliability, coefficient α also is the subject of discussion and criticism (e.g., Bentler, 2009; Cho & Kim, 2015; Cortina, 1993; Falk & Savalei, 2011; Green, Lissitz, & Mulaik, 1977; Schmidt, 1996; Sijtsma, 2009; Zinbarg, Revelle, Yovel, & Li, 2005).

In the next subsection, we discuss methods for approximating test-score reliability based on a data matrix consisting of N rows and J columns, and filled with item scores from a test that a group of people took once. Coefficient α is one of these methods. The methods have the advantage that they avoid the practical problems associated with mimicking independent replications of the same test by constructing alternate test versions, administering the same test twice with an appropriate time interval, and splitting the test into two disjoint and exhaustive parts. Just like the parallel-test method, the retest method, and the split-half method, the methods based on one test administration we discuss are all lower bounds to the reliability; that is, the methods provide different approximations to reliability, each of which is theoretically smaller than $\rho_{XX'}$. In ways unique to the particular lower bound method, each method uses the inter-item covariances for all item pairs in the test to approximate the reliability from below with less effort than each of the standard methods does. Thus, we discuss the methods because they are convenient. We do not know unbiased methods for estimating reliability.

Reliability Methods Based on One Test Administration

The inter-item covariances for all J items in the test are collected in the so-called covariance matrix. In this subsection, we need three different covariance matrices for the J items. For the observable item scores, the covariance matrix is denoted Σ_X and is square of order $J \times J$. The item-score variances are on the main diagonal and are denoted $\sigma_{X_j}^2 = \sigma_j^2$, and the inter-item covariances are in the off-diagonal cells and are denoted $\sigma_{X_j X_k} = \sigma_{jk}$. One may notice that $\sigma_{jk} = \sigma_{kj}$. For the item true-scores, matrix Σ_T has item true-score variances on the main diagonal and covariances σ_{jk} in the off-diagonal cells. These covariances equal the covariances between the item true-scores T_j and T_k, which can be derived using the CTT model Equation (2.11) for items, written as $T_j = X_j - E_j$ and $T_k = X_k - E_k$, and the assumption that $\rho_{EY} = 0$ [Equation (2.13)], yielding

$$\sigma_{T_j T_k} = \sigma_{jk}. \tag{2.23}$$

For the item error-scores, matrix Σ_E is diagonal with main-diagonal elements $\sigma_{E_j}^2$ and off-diagonal elements $\sigma_{E_j E_k} = 0$, because of assumption $\rho_{EY} = 0$.

In CTT for items, the assumption of measurement errors from test score X correlating 0 with any random variable Y that does not include the measurement error [Equation (2.13)], is replaced with its counterpart for items, $\rho_{E_j Y} = 0$, and the test-score variance decomposition in Equation (2.15) is replaced with $\sigma_j^2 = \sigma_{T_j}^2 + \sigma_{E_j}^2$. Then, a well-known result in CTT is that $\Sigma_X = \Sigma_T + \Sigma_E$, based on the variance decomposition in Equation (2.15) for the summation of main-diagonal elements and Equations (2.13) and (2.23) for the summation of off-diagonal elements. We write this matrix equation in full to produce

$$
\begin{bmatrix} \sigma_1^2 & \cdots & \sigma_{1J} \\ \vdots & \ddots & \vdots \\ \sigma_{J1} & \cdots & \sigma_J^2 \end{bmatrix} = \begin{bmatrix} \sigma_{T_1}^2 & \cdots & \sigma_{1J} \\ \vdots & \ddots & \vdots \\ \sigma_{J1} & \cdots & \sigma_{T_J}^2 \end{bmatrix} + \begin{bmatrix} \sigma_{E_1}^2 & \cdots & 0 \\ \vdots & \ddots & \vdots \\ 0 & \cdots & \sigma_{E_J}^2 \end{bmatrix} \tag{2.24}
$$

Next, we discuss seven methods that approximate reliability in Equation (2.21) in more detail by providing upper bounds for the sum of the J item-error variances, $\sum_j \sigma_{E_j}^2$. Each of these upper bounds contains only observable quantities and thus can be estimated from the data. Let us denote the upper bounds by U_m, m indexing the method, then $U_m \geq \sum_j \sigma_{E_j}^2$. Let a reliability method be denoted λ_m, then replacing $\sum_j \sigma_{E_j}^2$ in Equation (2.21) with U_m, yields

$$
\lambda_m = 1 - \frac{U_m}{\sigma_X^2} \leq 1 - \frac{\sum_{j=1}^J \sigma_{E_j}^2}{\sigma_X^2} = \rho_{XX'}. \tag{2.25}
$$

Equation (2.25) shows that, because U_m is an upper bound to the sum of the item error-variances, $\sum_j \sigma_{E_j}^2$, method λ_m is a lower bound to the reliability. We will see that the different upper bounds U_m mathematically exploit different ways to approximate $\sum_j \sigma_{E_j}^2$ from above but that different conceptions of measurement error are not involved. Some approximations are closer on target than others and sometimes but not always is there a hierarchy of the methods reflecting which method approximates $\sum_j \sigma_{E_j}^2$ and thus reliability $\rho_{XX'}$ best. The seven methods we discuss are Guttman's methods λ_1 through λ_6 (Guttman, 1945; Jackson & Agunwamba, 1977) and the greatest lower bound (GLB) to the reliability (Bentler & Woodward, 1980; Ten Berge & Sočan, 2004; Woodward & Bentler, 1978).

Method λ_1

From Equation (2.24), we have that $\sigma_j^2 = \sigma_{T_j}^2 + \sigma_{E_j}^2$, hence, $\sigma_j^2 \geq \sigma_{E_j}^2$, and across items $\sum_j \sigma_j^2 \geq \sum_j \sigma_{E_j}^2$. Thus, the first upper bound for the error variance in

Equation (2.21) equals $U_1 = \sum_j \sigma_j^2$, and this is the first lower bound for $\rho_{XX'}$, called Guttman's λ_1:

$$\lambda_1 = 1 - \frac{\sum_{j=1}^{J} \sigma_j^2}{\sigma_X^2}. \tag{2.26}$$

Table 2.1 shows two variance-covariance matrices Σ_X and the corresponding λ_1 values together with seven different reliability values obtained by means of methods we discuss later in this chapter. The left-hand matrix represents the typical one-factor structure, and the right-hand matrix represents a typical two-factor structure. Corresponding to these matrices, we also determined reliability $\rho_{XX'}$. For this purpose, we needed the unknown item-error variances or item true-score variances. In real data, one does not have access to these unobservable quantities, but it is no problem to choose

TABLE 2.1

Two Covariance Matrices Representing a One-Factor Structure (Left) and a Two-Factor Structure (Right) and the Resulting Values of λ_1, $\lambda_3(\alpha)$, λ_2, λ_4, λ_5, λ_6, the GLB, and Method ω_t

$$\begin{pmatrix} 1 & .6 & .6 & .6 & .6 & .6 \\ .6 & 1 & .6 & .6 & .6 & .6 \\ .6 & .6 & 1 & .6 & .6 & .6 \\ .6 & .6 & .6 & 1 & .6 & .6 \\ .6 & .6 & .6 & .6 & 1 & .6 \\ .6 & .6 & .6 & .6 & .6 & 1 \end{pmatrix} \qquad \begin{pmatrix} 1 & .6 & .6 & .0 & .0 & .0 \\ .6 & 1 & .6 & .0 & .0 & .0 \\ .6 & .6 & 1 & .0 & .0 & .0 \\ .0 & .0 & .0 & 1 & .6 & .6 \\ .0 & .0 & .0 & .6 & 1 & .6 \\ .0 & .0 & .0 & .6 & .6 & 1 \end{pmatrix}$$

$$\lambda_1 = 1 - \frac{6}{24} = .75 \qquad\qquad \lambda_1 = 1 - \frac{6}{13.2} \approx .5455$$

$$\lambda_3(\alpha) = \frac{6}{5} \times \left(1 - \frac{6}{24}\right) = .90 \qquad \lambda_3(\alpha) = \frac{6}{5} \times \left(1 - \frac{6}{13.2}\right) \approx .6545$$

$$\lambda_2 = 18 + \frac{\sqrt{\frac{6}{5} \times 10.8}}{24} = .90 \qquad \lambda_2 = 7.2 + \frac{\sqrt{\frac{6}{5} \times 4.32}}{13.2} \approx .7179$$

$$\lambda_4 = .90 \qquad\qquad\qquad \lambda_4 \approx .7273$$

$$\lambda_5 \approx 1 - \frac{2.6833}{24} \approx .8618 \qquad\qquad \lambda_5 \approx .6740$$

$$\lambda_6 \approx 1 - \frac{2.8253}{24} \approx .8823 \qquad\qquad \lambda_6 = 1 - \frac{3.3}{13.2} = .75$$

$$\rho_{GLB} = .90 \qquad\qquad\qquad \rho_{GLB} \approx .8182$$

$$\omega_t = .90 \qquad\qquad\qquad \omega_t \approx .6819$$

these quantities in an artificial example. For both matrices, we chose item-error variances (left-hand matrix: $\sigma_{E_j}^2 = .35$ for $j=1, ..., 6$; right-hand matrix: $\sigma_{E_j}^2 = .1925$ for $j=1, ..., 6$) such that they produced the same reliability, equal to .9125. Method λ_1 equaled .75 (one-factor structure) and .5455 (two-factor structure). Method λ_1 underestimates reliability more when data are not one-factorial, a result that consistently returns for the other methods.

Method λ_3

Because methods λ_1, λ_3, and λ_2 are mathematically ordered such that $\lambda_1 \leq \lambda_3 \leq \lambda_2$, we discuss method λ_3 prior to method λ_2. Method λ_3 is equivalent to coefficient α (Cronbach, 1951). Because coefficient α is so well known, we use notation $\lambda_3(\alpha)$ when we mean either method λ_3 or coefficient α. Method $\lambda_3(\alpha)$ in combination with the lower bound result can be derived in different ways (Jackson & Agunwamba, 1977; Novick & Lewis, 1967; Ten Berge & Sočan, 2004). One interesting derivation we present in Box 2.1 is an adjustment of Appendix A in Ten Berge and Sočan (2004). As an aside, we notice that in each derivation one considers, it is important to realize that method $\lambda_3(\alpha)$ is a matter of definition.

BOX 2.1   Proof That Method $\lambda_3(\alpha)$ Is a Lower Bound to the Reliability

First, we write the reliability as

$$\rho_{XX'} = \frac{\sigma_T^2}{\sigma_X^2} = \frac{\sum_{j=1}^{J}\sigma_{T_j}^2 + \sum\sum_{j \neq k}\sigma_{jk}}{\sigma_X^2} = \frac{\sum_{j=1}^{J}\sigma_{T_j}^2}{\sigma_X^2} + \frac{\sum\sum_{j \neq k}\sigma_{jk}}{\sigma_X^2}. \qquad (2.27)$$

Next, we take the last fraction on the right-hand side and rewrite it as

$$\frac{(J-1)\sum\sum_{j \neq k}\sigma_{jk}}{(J-1)\sigma_X^2} = \frac{J\sum\sum_{j \neq k}\sigma_{jk}}{(J-1)\sigma_X^2} - \frac{\sum\sum_{j \neq k}\sigma_{jk}}{(J-1)\sigma_X^2}. \qquad (2.28)$$

Readers familiar with method $\lambda_3(\alpha)$ will recognize its equation in the first fraction on the right-hand side of Equation (2.28); that is,

$$\lambda_3(\alpha) = \frac{J}{J-1}\frac{\sum\sum_{j \neq k}\sigma_{jk}}{\sigma_X^2}. \qquad (2.29)$$

Now we take the two fractions on the right-hand side of Equation (2.27), replace the second fraction with the difference in Equation (2.28), and insert Equation (2.29) in the resulting equation, so that we obtain, after permuting terms,

$$\rho_{XX'} = \lambda_3(\alpha) + \frac{\sum_{j=1}^J \sigma_{T_j}^2}{\sigma_X^2} - \frac{\sum\sum_{j \neq k} \sigma_{jk}}{(J-1)\sigma_X^2}$$

$$= \lambda_3(\alpha) + \frac{(J-1)\sum_{j=1}^J \sigma_{T_j}^2 - \sum\sum_{j \neq k} \sigma_{jk}}{(J-1)\sigma_X^2}. \tag{2.30}$$

We continue with the numerator of the fraction on the right-hand side of Equation (2.30) and notice that it equals

$$(J-1)\sum_{j=1}^J \sigma_{T_j}^2 - \sum\sum_{j \neq k} \sigma_{jk} = \sum\sum_{j < k} \sigma_{(T_j - T_k)}^2. \tag{2.31}$$

This term equals 0 when all variances $\sigma_{(T_j - T_k)}^2$ on the right-hand side equal 0, and this is true when item true-scores are linear transformations of one another, such that $T_j = T_k + a_{jk}$, meaning that the items are essentially τ-equivalent (Lord & Novick, 1968, p. 50). In that case, the fraction on the right-hand side of Equation (2.30) equals 0, and $\rho_{XX'} = \lambda_3(\alpha)$. Finally, we replace the numerator on the right-hand side of Equation (2.30) with the right-hand side of Equation (2.31) and obtain

$$\rho_{XX'} = \lambda_3(\alpha) + \frac{\sum\sum_{j < k} \sigma_{(T_j - T_k)}^2}{(J-1)\sigma_X^2}. \tag{2.32}$$

From Equation (2.32) we learn two important results about method $\lambda_3(\alpha)$ or coefficient α:

1. $\lambda_3(\alpha) \leq \rho_{XX'}$; this follows from Equation (2.32), because the fraction on the right-hand side, which contains variances in both numerator and denominator, is non-negative; and

2. $\lambda_3(\alpha) = \rho_{XX'}$ if the items are essentially τ-equivalent.

The equation for method $\lambda_3(\alpha)$ thus is given by

$$\lambda_3(\alpha) = \frac{J}{J-1} \frac{\sum\sum_{j \neq k} \sigma_{jk}}{\sigma_X^2}. \tag{2.29}$$

Now that we know that $\lambda_3(\alpha)$ is a lower bound to the reliability, $\rho_{XX'}$, that is, $\lambda_3(\alpha) \leq \rho_{XX'}$, it is of interest to see how we can rewrite $\lambda_3(\alpha)$ in Equation (2.29) to resemble Equation (2.21), so that it is clear how $\lambda_3(\alpha)$ defines upper bound U_3 for $\sum_j \sigma^2_{E_j}$ in Equation (2.21). First, using the variance of linear combinations, $\sigma^2_X = \sum_j \sigma^2_j + \sum\sum_{j\neq k} \sigma_{jk}$, we rewrite Equation (2.29) in the form well-known to most researchers as

$$\lambda_3(\alpha) = \frac{J}{J-1}\left(1 - \frac{\sum_{j=1}^J \sigma^2_j}{\sigma^2_X}\right). \tag{2.33}$$

Let $\bar{\sigma}$ denote the mean covariance across all $J(J-1)$ item pairs j and k, $j \neq k$. If one equates Equation (2.33) with the desired form $1 - U_3/\sigma^2_X$, after some algebraic manipulation one obtains

$$\lambda_3(\alpha) = 1 - \frac{\sum_{j=1}^J \sigma^2_j - J\bar{\sigma}}{\sigma^2_X}, \tag{2.34}$$

with upper bound $U_3 = \sum_j \sigma^2_j - J\bar{\sigma}$ for $\sum_j \sigma^2_{E_j}$ in Equation (2.21).

Comparing method λ_1 [Equation (2.26)] with method λ_3 yields the relationship

$$\lambda_3(\alpha) = \frac{J}{J-1}\lambda_1. \tag{2.35}$$

For $J \to \infty$, one finds $\lambda_3(\alpha) = \lambda_1$, and for finite J, one finds $\lambda_1 < \lambda_3(\alpha)$. Thus, for real tests, $\lambda_3(\alpha)$ is closer to reliability $\rho_{XX'}$ than λ_1. Interestingly, method λ_1 has smaller variance than method $\lambda_3(\alpha)$. This follows from the result in Equation (2.35), which shows that, because $J/(J-1)$ is a constant, $\lambda_3(\alpha)$ is a linear transformation of λ_1, and because $[J/(J-1)] > 1$, we have that

$$\sigma^2_{\lambda_3(\alpha)} = \left(\frac{J}{J-1}\right)^2 \lambda_1 > \sigma^2_{\lambda_1}.$$

However, the fact that λ_1 is the smallest lower bound of the two methods renders it primarily of historical interest.

For the two covariance-matrices representing the one-factor and two-factor structures, Table 2.1 shows the $\lambda_3(\alpha)$ values: .90 and .6545, respectively. The $\lambda_3(\alpha)$ values can also be computed from λ_1 by means of Equation (2.35). Again, we found that for the one-factor structure lower bound values are higher than for the two-factor structure. This result may suggest that $\lambda_3(\alpha)$ is a measure for internal consistency (Cronbach, 1951), but the conclusion would be misleading based only on this example. Different examples readily produce different conclusions. A closer look at $\lambda_3(\alpha)$ shows how it relates to the item structure of the test. We do this by noticing that

$$\sum_{j\neq k}\sum \sigma_{jk} = J(J-1)\bar{\sigma},$$

and rewriting Equation (2.29) to obtain

$$\lambda_3(\alpha) = \frac{J^2\bar{\sigma}}{\sigma_X^2}. \tag{2.36}$$

Equation (2.36) shows that for a fixed number of items J and a fixed test-score variance σ_X^2, method $\lambda_3(\alpha)$ depends only on the mean inter-item covariance $\bar{\sigma}$. The mean inter-item covariance $\bar{\sigma}$ reflects the central tendency of the distribution of inter-item covariances but not the variance, let alone the structure of the covariances. Thus, the factor structure, which is based on the covariance structure, only appears in $\lambda_3(\alpha)$ through its mean $\bar{\sigma}$. Reversely, $\lambda_3(\alpha)$ cannot provide information about the covariance structure of the items other than the mean $\bar{\sigma}$.

Table 2.2 shows two covariance matrices based on equal numbers of items and having the same test-score variance and the same mean inter-item covariance, $\bar{\sigma} = .3$. Consequently, in both cases, $\lambda_3(\alpha) = .72$, but the factor structures differ greatly. Sijtsma (2009) provided several examples and discussed method $\lambda_3(\alpha)$. Davenport et al. (2015) discussed reliability, dimensionality, and internal consistency; Green and Yang (2015) and Sijtsma (2015) provided comments.

Method λ_2

Guttman's methods use different pieces of information from the covariance matrices in Equation (2.24) (Jackson & Agunwamba, 1977). We borrow Jackson and Agunwamba's shorthand notation, $t_j = \sigma_{T_j}^2$, to keep equations

TABLE 2.2

Two Covariance Matrices Representing a One-Factor Structure (Left) and a Two-Factor Structure (Right), Both Resulting in $\lambda_3(\alpha) = .72$

1	.3	.3	.3	.3	.3		1	.75	.75	.0	.0	.0
.3	1	.3	.3	.3	.3		.75	1	.75	.0	.0	.0
.3	.3	1	.3	.3	.3		.75	.75	1	.0	.0	.0
.3	.3	.3	1	.3	.3		.0	.0	.0	1	.75	.75
.3	.3	.3	.3	1	.3		.0	.0	.0	.75	1	.75
.3	.3	.3	.3	.3	1		.0	.0	.0	.75	.75	1

in the following derivations readable. Method λ_1 uses the information from the principal submatrices, σ_j^2, of Σ_X. Method $\lambda_3(\alpha)$ uses the information from the 2×2 principal submatrices of Σ_T with diagonal elements t_j and t_k and off-diagonal elements $\sigma_{jk} = \sigma_{kj}$. Method λ_2 uses the non-negative definite property of these submatrices, meaning that their determinant must be non-negative,

$$t_j t_k \geq \sigma_{jk}^2, \quad j \neq k. \tag{2.37}$$

Except for a few details, Jackson and Agunwamba (1977) presented the next derivation of method λ_2 that we discuss in Box 2.2.

BOX 2.2   Proof That Method λ_2 Is a Lower Bound to the Reliability

First, we present some preliminaries, and notice that

$$2(J-1)\left(\sum_{j=1}^{J} t_j\right)^2 = 2(J-1)\sum_{j=1}^{J} t_j^2 + 2(J-1)\sum\sum_{j \neq k} t_j t_k$$

$$= \sum\sum_{j \neq k}\left(t_j - t_k\right)^2 + 2J\sum\sum_{j \neq k} t_j t_k \tag{2.38}$$

$$\geq 2J\sum\sum_{j \neq k} t_j t_k. \tag{2.39}$$

Then, it also holds that

$$\left(\sum_{j=1}^{J} t_j\right)^2 \geq \frac{J}{J-1}\sum\sum_{j \neq k} t_j t_k,$$

and taking the square root, we have

$$\sum_{j=1}^{J} t_j \geq \left[J(J-1)^{-1}\sum\sum_{j \neq k} t_j t_k\right]^{1/2}. \tag{2.40}$$

We turn to finding an upper bound for $\sum_j \sigma_{Ej}^2$ and notice that

$$\sum_{j=1}^{J} \sigma_{Ej}^2 = \sum_{j=1}^{J} \sigma_j^2 - \sum_{j=1}^{J} t_j. \tag{2.41}$$

In Equation (2.41), we replace $\sum_j t_j$ with $\left[J(J-1)^{-1}\sum\sum_{j\neq k}t_j t_k\right]^{1/2}$ [Equation (2.40)], in which we subsequently replace $t_j t_k$ with σ_{jk}^2 [Equation (2.37)], so that

$$\sum_{j=1}^{J}\sigma_{E_j}^2 \leq \sum_{j=1}^{J}\sigma_j^2 - \left[J(J-1)^{-1}\sum\sum_{j\neq k}t_j t_k\right]^{1/2}$$

$$\leq \sum_{j=1}^{J}\sigma_j^2 - \left[J(J-1)^{-1}\sum\sum_{j\neq k}\sigma_{jk}^2\right]^{1/2}. \tag{2.42}$$

Hence, the right-hand side of Equation (2.42) is an upper bound for $\sum_j \sigma_{E_j}^2$ in the reliability definition in Equation (2.21), and replacing $\sum_j \sigma_{E_j}^2$ in Equation (2.21) with this upper bound, we obtain

$$\lambda_2 = 1 - \frac{\sum_{j=1}^{J}\sigma_j^2 - \left[J(J-1)^{-1}\sum\sum_{j\neq k}\sigma_{jk}^2\right]^{1/2}}{\sigma_X^2}. \tag{2.43}$$

Let $\overline{\sigma^2}$ denote the mean squared covariance of the $J(J-1)$ pairs of items j and k ($j\neq k$), then

$$\lambda_2 = 1 - \frac{\sum_{j=1}^{J}\sigma_j^2 - J\left(\overline{\sigma^2}\right)^{1/2}}{\sigma_X^2}, \tag{2.44}$$

with $U_2 = \sum_j \sigma_j^2 - J\left(\overline{\sigma^2}\right)^{1/2}$ as the upper bound for the sum of the item-error variances, $\sum_j \sigma_{E_j}^2$, in Equation (2.21). Hence, method λ_2 is a lower bound to reliability $\rho_{XX'}$.

The equation for method λ_2 thus equals

$$\lambda_2 = 1 - \frac{\sum_{j=1}^{J}\sigma_j^2 - \left[J(J-1)^{-1}\sum\sum_{j\neq k}\sigma_{jk}^2\right]^{1/2}}{\sigma_X^2}. \tag{2.43}$$

Now that we know that method λ_2 is a lower bound to the reliability, $\rho_{XX'}$, that is, $\lambda_2 \leq \rho_{XX'}$, we investigate the relationship between methods $\lambda_3(\alpha)$ and λ_2 in Box 2.3.

BOX 2.3   Proof That Method $\lambda_3(\alpha)$ Is a Lower Bound to Method λ_2

First, notice that $U_3 = \sum_j \sigma_j^2 - J\bar{\sigma}$ [Equation (2.34)] contains the arithmetic mean of the inter-item covariances, $\bar{\sigma}$, and that $U_2 = \sum_j \sigma_j^2 - J\left(\overline{\sigma^2}\right)^{1/2}$ [Equation (2.44) in Box 2.2] contains the arithmetic mean of the squared inter-item covariances, also known as the root mean square or the quadratic mean, $\overline{\sigma^2}$. The relation between the two means is

$$\bar{\sigma}^2 = \overline{\sigma^2} - \mathrm{Var}(\sigma), \tag{2.45}$$

where $\mathrm{Var}(\sigma)$ is the variance of the inter-item covariances. In Equation (2.34), replacing $\bar{\sigma}$ with the square root of the right-hand side of Equation (2.45) yields

$$\lambda_3(\alpha) = 1 - \frac{\sum_{j=1}^{J}\sigma_j^2 - J\left[\overline{\sigma^2} - \mathrm{Var}(\sigma)\right]^{1/2}}{\sigma_X^2}. \tag{2.46}$$

Comparing λ_2 [Equation (2.44)] and $\lambda_3(\alpha)$ [Equation (2.46)] shows that, because $\mathrm{Var}(\sigma) \geq 0$, we have $\lambda_3(\alpha) \leq \lambda_2$, with equality $\lambda_3(\alpha) = \lambda_2$ if $\mathrm{Var}(\sigma) = 0$; that is, if all inter-item covariances are equal, $\sigma_{jk} = c, j \neq k$.

Because we know now that $\lambda_3(\alpha) \leq \lambda_2$, by equating the Equations (2.38) and (2.39) (both equations are in Box 2.2; readers who have skipped Box 2.2 may simply decide to accept our line of reasoning), we find the condition for $\lambda_2 = \rho_{XX'}$ in Box 2.4.

BOX 2.4   Derivation of Condition for Which $\lambda_2 = \rho_{XX'}$

Equality $\lambda_2 = \rho_{XX'}$ implies equality of Equations (2.38) and (2.39) so that

$$\sum_{j \neq k}\sum (t_j - t_k)^2 = 0 \Rightarrow t_j = t_k = t, \quad \forall j \neq k. \tag{2.47}$$

Equality $\lambda_2 = \rho_{XX'}$ also requires equality in Equation (2.37), so that

$$t_j t_k = \sigma_{jk}^2 \Rightarrow \sigma_{jk} = t \vee \sigma_{jk} = -t, \tag{2.48}$$

but $\sigma_{jk} = -t$ cannot be true if we already found that $\sigma_{jk} = t$. Hence, covariance matrix $\Sigma_T = ((t))$, implying that matrix Σ_T has rank 1, meaning that all items or test parts are essentially τ-equivalent: $T_j = T_k + a_{jk}$, $\forall j \neq k$. To see this is true, suppose linear transformation $T_j = b_{jk}T_k + a_{jk}$, then $t_j = b_{jk}t_k$, hence $t_j \neq t_k$, and it is readily checked that not only all variances but also all covariances have different values; hence, rank 1 does not hold. We also consider the more restrictive linear transformation $T_j = bT_k + a_{jk}, \; \forall j \neq k$. For example, for $J = 3$, one finds $t_1 = b^2 t_2$ and $t_1 = b^2 t_3$, implying $t_1 = t_3$; and further $t_2 = b^2 t_3$, which together with the previous result implies $b = 1 \vee b = -1$. The former result implies essential τ-equivalence, and the second outcome implies $T_j = -T_k + a_{jk}$, which for $J = 3$ and $a_{jk} = 0$ means $T_1 = -T_2$, $T_1 = -T_3$, and thus $T_2 = T_3$, but this contradicts essential τ-equivalence.

Hence, we have shown that

$$\lambda_2 = \rho_{XX'}, \text{ if the items or test parts are essentially } \tau\text{-equivalent.} \quad (2.49)$$

We have learned the following facts about method λ_2:

1. $\lambda_2 \leq \rho_{XX'}$; this follows from Equations (2.42) and (2.44) (both equations appear in Box 2.2; readers who have skipped Box 2.2 may simply choose to accept our line of reasoning);

2. $\lambda_3(\alpha) \leq \lambda_2$, with equality $\lambda_3(\alpha) = \lambda_2$ when all inter-item covariances are equal, $\sigma_{jk} = c, j \neq k$; and

3. $\lambda_2 = \rho_{XX'}$, if the items are essentially τ-equivalent; see Equation (2.49) and Ten Berge and Zegers (1978).

We have established that,

$$\lambda_1 < \lambda_3(\alpha) \leq \lambda_2 \leq \rho_{XX'}. \quad (2.50)$$

Table 2.1 shows the λ_2 values, which are equal to .90 and .7179. Both values are necessarily at least as large as the $\lambda_3(\alpha)$ values; in this example, $\lambda_2 = \lambda_3(\alpha)$, because for the one-factor case all inter-item covariances were chosen equal.

We gave much attention to methods $\lambda_3(\alpha)$ and λ_2, because they are among the best-known methods to approximate reliability, with method $\lambda_3(\alpha)$ clearly in the lead. Both methods are part of an infinite series of μ lower bounds, $\mu_0 \leq \mu_1 \leq \ldots \leq \mu_r \leq \ldots \leq \rho_{XX'}$, in which $\mu_0 = \lambda_3(\alpha)$ and $\mu_1 = \lambda_2$ (Ten Berge & Zegers, 1978). The generic equation for these lower bounds equals

$$\mu_r = \frac{1}{\sigma_X^2} \left(p_0 + \left(p_1 + \left(p_2 + \cdots \left(p_{r-1} + (p_r)^{1/2} \right)^{1/2} \right)^{1/2} \cdots \right)^{1/2} \right), \quad r = 0, 1, 2, \ldots, \quad (2.51)$$

with

$$p_h = \sum \sum_{j \neq k} \sigma_{jk}^{(2^h)}, \quad h = 0, 1, 2, \ldots, r-1,$$

and

$$p_h = \frac{J}{J-1}\sum_{j \neq k}\sum \sigma_{jk}^{(2^h)}, \quad h = r$$

Substituting $r=0$ and $r=1$ yields $\mu_0 = \lambda_3(\alpha)$ and $\mu_1 = \lambda_2$, respectively. Ten Berge and Zegers (1978) proved that the inequalities in the array of μ are lower bounds, proved that $\mu_r = \rho_{XX'}$, for $r = 0,1,2,\ldots$, if the items or the test parts on which μ_r is based, are essentially τ-equivalent, and showed that the series does *not* asymptotically converge to $\rho_{XX'}$ as $r \to \infty$. Based on ten real-data sets, the authors concluded that it may be rewarding to compute λ_2 rather than $\lambda_3(\alpha)$, gaining a few hundredths, but that beyond μ_1 (λ_2) the gain is negligible, with the exception of μ_2 when the number of items equals three or four and the inter-item covariances show considerable spread.

The next three methods have received little attention in the psychometric literature and are rarely used in practice. This may have to do more with habit than reason, and we will briefly discuss methods λ_4, λ_5, and λ_6 here and devote some space to their properties.

Method λ_4

Method λ_4 assesses the $\lambda_3(\alpha)$ values for all possible divisions of the set of J items in two mutually exclusive and exhaustive subsets and chooses the greatest $\lambda_3(\alpha)$ value as the value of λ_4. Formally, let the subsets have test scores Y_1 and Y_2, such that $X = Y_1 + Y_2$. Let us denote by P a particular division in two test parts $(J=2)$, so that $\alpha(P)$ is the $\lambda_3(\alpha)$ value for this division,

$$\alpha(P) = \frac{4\sigma_{Y_1 Y_2}}{\sigma_X^2}. \tag{2.52}$$

Because method $\lambda_3(\alpha)$ is a lower bound, $\alpha(P)$ also is a lower bound. The partition P that yields the greatest $\alpha(P)$ value is the λ_4 value; that is,

$$\lambda_4 = \max_P [\alpha(P)]. \tag{2.53}$$

For the sake of completeness, in Box 2.5, we derive U_4.

BOX 2.5   Derivation of U_4

Following Equation (2.34) we rewrite Equation (2.52) as

$$\lambda_3(\alpha) = 1 - \frac{\sum_{j=1}^{J}\sigma_j^2 - J\overline{\sigma}}{\sigma_X^2} = 1 - \frac{\sigma_{Y_1}^2 + \sigma_{Y_2}^2 - 2\sigma_{Y_1 Y_2}}{\sigma_X^2}.$$

One may notice that the numerator equals $\sigma_{Y_1}^2 + \sigma_{Y_2}^2 - 2\sigma_{Y_1Y_2} = \sigma_{(Y_1-Y_2)}^2$, so that

$$\lambda_3(\alpha) = 1 - \frac{\sigma_{(Y_1-Y_2)}^2}{\sigma_X^2}.$$

The item-set division that produces the greatest $\alpha(P)$ also does this by producing the smallest variance of the difference $Y_1 - Y_2$. Hence, let $\sigma^2(P) = \sigma_{(Y_1-Y_2)}^2$ denote the variance of the difference for division P, then $U_4 = \min_P [\sigma^2(P)]$.

For all possible divisions, Table 2.3 shows for each covariance matrix in Table 2.1 the $\alpha(P)$ values. For the one-factor structure, the maximum value defines $\lambda_4 = .90$, and for the two-factor structure, the maximum value defines $\lambda_4 \approx .7273$. These values are as large (one-factor structure) or larger (two-factor structure) than the values of λ_1, $\lambda_3(\alpha)$, and λ_2. This is a common finding, because method λ_4 often yields values close to or equal to the GLB (Jackson & Agunwamba, 1977), but we are unaware of formal inequality relations among λ_1, $\lambda_3(\alpha)$ and λ_2, and λ_4.

TABLE 2.3

Values of $\alpha(P)$ for All Possible Splits of Six Items into Two Sets, for the One-Factor Structure and the Two-Factor Structure in Table 2.1

First set	Second set	One factor	Two factors	First set	Second set	One factor	Two factors
1	2, 3, 4, 5, 6	.5	.3636	3, 5	1, 2, 4, 6	.8	.7273
2	1, 3, 4, 5, 6	.5	.3636	3, 6	1, 2, 4, 5	.8	.7273
3	1, 2, 4, 5, 6	.5	.3636	4, 5	1, 2, 3, 6	.8	.3636
4	1, 2, 3, 5, 6	.5	.3636	4, 6	1, 2, 3, 5	.8	.3636
5	1, 2, 3, 4, 6	.5	.3636	5, 6	1, 2, 3, 4	.8	.3636
6	1, 2, 3, 4, 5	.5	.3636	1, 2, 3	4, 5, 6	.9	0
1, 2	3, 4, 5, 6	.8	.3636	1, 2, 4	3, 5, 6	.9	.7273
1, 3	2, 4, 5, 6	.8	.3636	1, 2, 5	3, 4, 6	.9	.7273
1, 4	2, 3, 5, 6	.8	.7273	1, 2, 6	3, 4, 5	.9	.7273
1, 5	2, 3, 4, 6	.8	.7273	1, 3, 4	2, 5, 6	.9	.7273
1, 6	2, 3, 4, 5	.8	.7273	1, 3, 5	2, 4, 6	.9	.7273
2, 3	1, 4, 5, 6	.8	.3636	1, 3, 6	2, 4, 5	.9	.7273
2, 4	1, 3, 5, 6	.8	.7273	1, 4, 5	2, 3, 6	.9	.7273
2, 5	1, 3, 4, 6	.8	.7273	1, 4, 6	2, 3, 5	.9	.7273
2, 6	1, 3, 4, 5	.8	.7273	1, 5, 6	2, 3, 4	.9	.7273
3, 4	1, 2, 5, 6	.8	.7273				

Method λ_5

First, we refer to Equation (2.25), in which U_m is an upper bound for the sum of the item-error variances, $\sum_j \sigma_{E_j}^2$. Second, because in the matrices Σ_X and Σ_T the covariances are the same, $\sigma_{jk} = \sigma_{T_j T_k}$ [Equation (2.23)], from Equation (2.24), one can derive that $\sum_j \sigma_{E_j}^2 = \sum_j \sigma_j^2 - \sum_j \sigma_{T_j}^2$. This result implies that finding an upper bound for the sum of the item-error variances is equal to finding a lower bound for the sum of the item true-score variances, $\sum_j \sigma_{T_j}^2$. Third, in Equation (2.34) for method $\lambda_3(\alpha)$, one may notice that the term $J\overline{\sigma}$ plays the role of lower bound for the sum of the item true-score variances, $\sum_j \sigma_{T_j}^2$, and in Equation (2.44) for method λ_2, the term $J\left(\overline{\sigma^2}\right)^{1/2}$ plays this role. Because both terms concern the mean of the (squared) inter-item covariances, an interesting question is whether one can find a selection of the covariances that produces a greater lower bound for $\sum_j \sigma_{T_j}^2$ and thus a smaller upper bound for $\sum_j \sigma_{E_j}^2$. This is what method λ_5 aims at by focusing on the greater covariances but does not uniformly accomplish. See Jackson and Agunwamba (1977) and Verhelst (1998) for conditions under which $\lambda_5 > \lambda_2$. We derive method λ_5 in Box 2.6.

BOX 2.6   Derivation of Method λ_5

Each column k of covariance matrix Σ_T contains the $J-1$ covariances of item j $(j \neq k)$ with item k, and on the main diagonal column k contains $\sigma_{T_k}^2$, previously denoted t_k. Following Jackson and Agunwamba (1977), we first write Equation (2.37) as

$$t_j t_k \geq \sigma_{jk}^2 \Leftrightarrow t_j \geq t_k^{-1} \sigma_{jk}^2,$$

from which it follows that

$$\sum_{j:j\neq k} t_j \geq t_k^{-1} \sum_{j:j\neq k} \sigma_{jk}^2. \tag{2.54}$$

Next, we introduce an inequality for arithmetic means and geometric means that proves useful for deriving U_5 and use the result that an arithmetic mean is at least as large as the geometric mean based on the same numbers. Let a and b be positive numbers; then it holds that $1/2(a+b) \geq (ab)^{1/2}$. So, if we define $a = t_k$ and $b = \left(\sum_{j:j\neq k} \sigma_{jk}^2\right)/t_k$, then,

$$\frac{1}{2}\left(t_j + \frac{\sum_{j:j\neq k}\sigma_{jk}^2}{t_j}\right) \geq \left(\frac{t_j \sum_{j:j\neq k}\sigma_{jk}^2}{t_j}\right)^{\frac{1}{2}} = \left(\sum_{j:j\neq k}\sigma_{jk}^2\right)^{1/2}. \tag{2.55}$$

Now, we take three steps. First, we write the trace of Σ_T as $\sum_j t_j = t_k + \sum_{j:j \neq k} t_j$. Second, we use the inequality in Equation (2.54) to produce $\sum_j t_j \geq t_k + t_k^{-1} \sum_{j:j \neq k} \sigma_{jk}^2$. Third, we use the inequality in Equation (2.55) in which we multiply both sides by 2, and find that the sum of the two terms on the right in the previous inequality is at least as large as two times the square root of their product [cf. Equation (2.55)]; that is,

$$\sum_{j=1}^{J} t_j = t_k + \sum_{j:j \neq k} t_j \geq t_k + t_k^{-1} \sum_{j:j \neq k} \sigma_{jk}^2$$

$$\geq 2 \left(\sum_{j:j \neq k} \sigma_{jk}^2 \right)^{1/2}. \tag{2.56}$$

Given this result, one may write

$$\sum_{j=1}^{J} \sigma_{E_j}^2 = \sum_{j=1}^{J} \sigma_j^2 - \sum_{j=1}^{J} t_j \leq \sum_{j=1}^{J} \sigma_j^2 - 2 \left(\sum_{j:j \neq k} \sigma_{jk}^2 \right)^{1/2}, \tag{2.57}$$

and we notice that the right-hand side of Equation (2.57) is an upper bound for the sum of the item-error variances. Because the last term in Equation (2.57) pertains to the inter-item covariances in column k, it makes sense to select the column yielding the greatest of these terms, so that $U_5 = \sum_j \sigma_j^2 - \max_k \left(2 \left(\sum_{j:j \neq k} \sigma_{jk}^2 \right)^{1/2} \right)$, and define lower bound,

$$\lambda_5 = 1 - \frac{\sum_{j=1}^{J} \sigma_j^2 - \max_k \left(2 \left(\sum_{j:j \neq k} \sigma_{jk}^2 \right)^{1/2} \right)}{\sigma_X^2}. \tag{2.58}$$

Thus, lower bound method λ_5 equals

$$\lambda_5 = 1 - \frac{\sum_{j=1}^{J} \sigma_j^2 - \max_k \left(2 \left(\sum_{j:j \neq k} \sigma_{jk}^2 \right)^{1/2} \right)}{\sigma_X^2}. \tag{2.58}$$

For the two covariance-matrices (Table 2.1), Table 2.4 provides the squared covariances and below each column, the sum of the squared covariances. For each covariance matrix, we used the largest sum to compute $\lambda_5 \approx .8618$ for the one-factor structure and $\lambda_5 = .6740$ for the two-factor structure. The λ_5 values

TABLE 2.4

The Squared Covariances of the Two Covariance Matrices in
Table 2.1, and Their Column Sums

$$
\begin{pmatrix}
    & .36 & .36 & .36 & .36 & .36 \\
.36 &   & .36 & .36 & .36 & .36 \\
.36 & .36 &   & .36 & .36 & .36 \\
.36 & .36 & .36 &   & .36 & .36 \\
.36 & .36 & .36 & .36 &   & .36 \\
.36 & .36 & .36 & .36 & .36 &
\end{pmatrix}
\quad
\begin{pmatrix}
    & .36 & .36 & .0 & .0 & .0 \\
.36 &   & .36 & .0 & .0 & .0 \\
.36 & .36 &   & .0 & .0 & .0 \\
.0 & .0 & .0 &   & .36 & .36 \\
.0 & .0 & .0 & .36 &   & .36 \\
.0 & .0 & .0 & .36 & .36 &
\end{pmatrix}
$$

$$
\begin{pmatrix} 1.8 & 1.8 & 1.8 & 1.8 & 1.8 & 1.8 \end{pmatrix}
\quad
\begin{pmatrix} .72 & .72 & .72 & .72 & .72 & .72 \end{pmatrix}
$$

are smaller than the λ_2 and λ_4 values, but in other examples different results may be found.

Method λ_6

Method λ_6 is based on the multiple regression of each item score X_j on the other $J-1$ item scores. The residual variance of item score X_j, $\sigma_{\varepsilon_j}^2$, is an upper bound for the item's measurement error variance, $\sigma_{E_j}^2$, so that $\sigma_{E_j}^2 \leq \sigma_{\varepsilon_j}^2$ and also $\sum_j \sigma_{E_j}^2 \leq \sum_j \sigma_{\varepsilon_j}^2$. Hence, $U_6 = \sum_j \sigma_{\varepsilon_j}^2$, and the sixth lower bound for reliability $\rho_{XX'}$ is defined as,

$$
\lambda_6 = 1 - \frac{\sum_{j=1}^{J} \sigma_{\varepsilon_j}^2}{\sigma_X^2}. \tag{2.59}
$$

For the covariance matrices representing the one-factor and two-factor structures (Table 2.1), Table 2.5 presents the residual variances for each of the six

TABLE 2.5

Residual Item Variances of the Two
Covariance Matrices in Table 2.1

Item	One factor	Two factors
1	.4706	.55
2	.4706	.55
3	.4706	.55
4	.4706	.55
5	.4706	.55
6	.4706	.55

items when regressed on the other five items. For the one-factor structure, $\lambda_6 \approx .8823$, and for the two-factor structure, $\lambda_6 = .75$.

Greatest Lower Bound

The lower bounds discussed thus far exploit only part of the information in the data given the assumptions of CTT, which are $X_j = T_j + E_j$ [Equation (2.11)] and $\rho_{E_j E_k} = 0$ [Equation (2.13), with $E = E_j$ and $Y = E_k$]. Realizing this, the challenge is to exploit all information so that one finds an upper bound U that minimizes $\sum_j \sigma_{E_j}^2$ in Equation (2.21), hence maximizes the lower bound to $\rho_{XX'}$. This results in a lower bound that exceeds all other lower bounds we discussed so far. Ten Berge and Sočan (2004) explain the GLB as follows. Using the CTT assumptions, we have that $\Sigma_X = \Sigma_T + \Sigma_E$ [Equation (2.24)], in which the inter-item error covariance matrix Σ_E is diagonal with the main diagonal entries equal to the error variances $\sigma_{E_j}^2$ and the off-diagonal entries equal to 0. The three covariance-matrices are positive semidefinite, which means that they do not have negative eigenvalues. The GLB, denoted ρ_{GLB}, is obtained by finding the nonnegative matrix Σ_E for which $\Sigma_T = \Sigma_X - \Sigma_E$ is positive semidefinite, such that

$$\rho_{GLB} = 1 - \frac{\max\left[\operatorname{tr}\left(\Sigma_E\right)\right]}{\sigma_X^2}. \tag{2.60}$$

Because solving the problem of finding $\max\left[\operatorname{tr}\left(\Sigma_E\right)\right]$ is rather technical, we prefer not to discuss the details. However, Ten Berge, Snijders, and Zegers (1981) discussed an algorithm that solves this problem. The GLB problem is discussed extensively by Woodhouse and Jackson (1977), Jackson and Agunwamba (1977), Woodward and Bentler (1978), Bentler and Woodward (1980), and Ten Berge and Sočan (2004). The solution of the maximization problem thus yields an upper bound for $\sum_j \sigma_{E_j}^2$ in Equation (2.21), equal to $U_{GLB} = \max\left(\operatorname{tr}\left(\Sigma_E\right)\right)$. Hence, Equation (2.60) is a lower bound to the reliability, $\rho_{XX'}$, based on the assumptions of CTT, and because the procedure exploits all information from the covariance matrices, Equation (2.60) is the greatest lower bound to the reliability.

The GLB points to an interesting feature about reliability and reliability approximations that is little known but important to understand reliability approximations based on one test administration yielding one data set. The GLB splits the reliability scale into two line segments, $[0; \rho_{GLB})$ and $[\rho_{GLB}; 1]$. The first line segment contains all values that are impossible given the data. This means that, assuming by some miracle one could establish the true reliability $\rho_{XX'}$ from the data, $\rho_{XX'}$ will never have a value lower than ρ_{GLB}. The consequence is that the lower bounds λ_1 through λ_6 all yield values that cannot be $\rho_{XX'}$ values given the data. Ten Berge and Sočan (2004) provide a numerical example that shows the meaning of this result. We discuss the example in Box 2.7.

BOX 2.7   Example Showing Methods λ_1 through λ_6
Do Not Yield Reliability Values Given the Data

Let $J=3$, then an example positive semidefinite matrix Σ_X is

$$\Sigma_X = \begin{bmatrix} 3 & 1 & 2 \\ 1 & 4 & 3 \\ 2 & 3 & 6 \end{bmatrix}. \qquad (2.61)$$

The GLB solution, which Ten Berge and Sočan (2004) explained in detail for this simple $J=3$ case, equals $\sigma_{T_1}^2 = 1$, $\sigma_{T_2}^2 = 2$, and $\sigma_{T_3}^2 = 5$, and $\sigma_{E_1}^2 = 2$, $\sigma_{E_2}^2 = 2$, and $\sigma_{E_3}^2 = 1$, so that

$$\rho_{GLB} = 1 - \frac{2+2+1}{25} = .8.$$

Unlike the GLB, method $\lambda_3(\alpha)$ does not separate Σ_X in Σ_T and Σ_E but only uses the elements in Σ_X. Hence, using the item variances σ_j^2 and the inter-item covariances σ_{jk}, method $\lambda_3(\alpha)$ yields

$$\lambda_3(\alpha) = \frac{3}{2}\left(1 - \frac{3+4+6}{25}\right) = .72. \qquad (2.62)$$

Next, we consider $\lambda_3(\alpha)$ in the form of Equation (2.21), resulting in Equation (2.34), and using $U_2 = \sum_j \sigma_j^2 - J\overline{\sigma} = (3+4+6) - 3 \times 2 = 7$ as a substitute for $\sum_j \sigma_{E_j}^2$, we obtain

$$1 - \frac{\sum_{j=1}^{J} \sigma_{E_j}^2}{\sigma_X^2} = 1 - \frac{7}{25} = .72. \qquad (2.63)$$

From Equation (2.63), we conclude that coefficient $\lambda_3(\alpha)$ implies $\sum_j \sigma_{E_j}^2 = 7$ (but notice that the individual item-error variances imply $\sum_j \sigma_{E_j}^2 = 5$), and because $\sum_j \sigma_j^2 = 13$, we have

$$\sum_{j=1}^{J} \sigma_{T_j}^2 = 13 - 7 = 6, \qquad (2.64)$$

(again, notice that the individual item true-score variances imply $\sum_j \sigma^2_{T_j} = 8$). Now, we know from these results and the result $\Sigma_T = \Sigma_X - \Sigma_E$ that

$$\Sigma_T = \begin{bmatrix} ? & 1 & 2 \\ 1 & ? & 3 \\ 2 & 3 & ? \end{bmatrix},$$

with trace equal to 6. Ten Berge and Sočan (2004) defined a sum score $Y = X_1 + X_2 - X_3$. Using the inter-item covariances in Equation (2.61) and the sum of the item true-score variance in Equation (2.64), we find $\sigma^2_T = 6 + 2(1 - 2 - 3) = -2$, a result that is impossible. The negative true-score variance shows that coefficient $\lambda_3(\alpha)$ implies error variances that are too large, meaning that they produce approximations of reliability that cannot be the true reliability.

Interesting as the result from Box 2.7 may be—coefficient $\lambda_3(\alpha)$ not only is a lower bound but it also provides an unrealistic value—for practical purposes it does not seem to be really important, because knowing that, say, $\lambda_3(\alpha) = .88$, tells one that $\rho_{XX'} > .88$, which is valuable information, even if .88 is not the true reliability.

The second interval $[\rho_{GLB}; 1]$ is more interesting, because it tells us that the true reliability is somewhere in the interval but not where. Thus, the data obtained from one test administration restrict the true reliability to the interval $[\rho_{GLB}; 1]$. This means that if one finds, say, $\rho_{GLB} = .88$, then $\rho_{XX'} \in [.88; 1]$, and even perfect reliability is not excluded. The fascinating result is that no GLB value or other lower bound value excludes any of the $\rho_{XX'}$ values that are greater, including the maximum value, $\rho_{XX'} = 1$. This means that a test for which $\lambda_3(\alpha) = .60$ and $\rho_{GLB} = .67$ can have a reliability $\rho_{XX'} = .67$ but also $\rho_{XX'} = .90$ or $\rho_{XX'} = 1$. Compared to other lower bounds, the GLB reduces the interval of possible reliability values, hence the interval of uncertainty. However, the results also emphasize that test constructors better assemble tests that have large lower bound values, and knowing that $\lambda_3(\alpha) = .90$ corners $\rho_{XX'}$ in a relatively small interval that is reduced to realistic values when one knows in addition that $\rho_{GLB} = .93$. This exercise teaches us that data from a single test administration restrict the reliability to the interval $[\rho_{GLB}; 1]$, and this illustrates the price we pay for not using data sets from two independent replications or parallel tests that would yield a point estimate for $\rho_{XX'}$ rather than an interval of permissible values. Also in psychometrics, we learn that there is no such thing as a free lunch. Based on the covariance matrices in Table 2.1, $\rho_{GLB} = .90$ (one-factor structure) and $\rho_{GLB} \approx .8182$ (two-factor structure).

Special Topics Concerning Methods λ_1 through λ_6 and the GLB

Mutual Relationships of Lower Bounds and Reliability

The next array of inequalities extends Equation (2.50),

$$\lambda_1 < \lambda_3(\alpha) \leq \lambda_2 \leq \rho_{GLB} \leq \rho_{XX'}. \tag{2.65}$$

If items or the test parts on which the computations are based are essentially τ-equivalent (Lord & Novick, 1968, p. 50), the next result is obtained,

$$\lambda_1 < \lambda_3(\alpha) = \lambda_2 = \rho_{GLB} = \rho_{XX'}. \tag{2.66}$$

Methods λ_4, λ_5, and λ_6 cannot be used to extend Equation (2.65), because they either have a known relationship to one of the other methods but not to others, or they have known relationships to other methods under special conditions but not in general. Lord and Novick (1968, pp. 93–94) showed that in general, $\lambda_3(\alpha) \leq \lambda_4$. Of course, it then also follows that $\lambda_1 < \lambda_4$, but because of λ_1's marginal role, this result is only of theoretical interest. Jackson and Agunwamba (1977) derived conditions under which $\lambda_4 = \rho_{GLB}$. For real data, one can check whether the conditions are satisfied and thus whether λ_4 equals ρ_{GLB}. These authors also derived conditions under which $\lambda_2 < \lambda_5$ and $\lambda_4 < \lambda_5$, but the inequalities are not true in general. Going back to Equation (2.65), because essential τ-equivalence is not satisfied in real data (i.e., when the sample is large enough, one will always be able to reject essential τ-equivalence), in practical research Equation (2.65) can be replaced by an equation consisting of strict inequalities,

$$\lambda_1 < \lambda_3(\alpha) < \lambda_2 < \rho_{GLB} < \rho_{XX'}. \tag{2.67}$$

Discrepancy of Methods λ_1 through λ_6 and the GLB

Oosterwijk, Van der Ark, and Sijtsma (2016) addressed the question of how far methods λ_1 through λ_6 are remote from ρ_{GLB}. Based on computational examples, the authors concluded that none of the methods λ_1 through λ_6 was closest to ρ_{GLB} in all situations but that in general method λ_4 was closest. Method λ_1 was much smaller than methods λ_2 through λ_6 and could be ignored. Method $\lambda_3(\alpha)$ yielded values much smaller than ρ_{GLB}. Method λ_5 yielded smaller values than methods λ_2, $\lambda_3(\alpha)$, λ_4, and λ_6, an unexpected result given that, unlike the other methods, λ_5 deals with the variation of the inter-item covariances. Moreover, variation in inter-item covariances had little effect on the five methods.

A well-known result for the GLB is that for small to moderate samples, with $N < 1,000$, and long tests, with $J > 10$, the GLB shows large positive bias (Ten Berge & Sočan, 2004). Based on their computational results, Oosterwijk et al. (2016) suggested that in these circumstances the GLB could be replaced with methods λ_2, λ_4, and λ_6. Method λ_4 often has values similar to the GLB (Jackson & Agunwamba, 1977). However, Benton (2015) suggested that method λ_4 is most useful when reliability exceeds .85, $N > 3{,}000$, and $J < 25$. Because method λ_4 belongs to the set of lower bounds that uses optimization to arrive at a result, it is likely to suffer from chance capitalization. The seriousness of this phenomenon is the topic of the next subsection.

Overestimation of Reliability in Real Data

Each of the methods λ_1 through λ_6 requires the same sample statistics for its computation and further depends on test length J. The sample statistics are all in the sample version $\mathbf{S}_X$ of covariance matrix $\mathbf{\Sigma}_X$ and are denoted S_j^2 estimating σ_j^2, S_{jk} estimating σ_{jk}, whereas $S_X^2 = \sum_j S_j^2 + \sum\sum_{j \neq k} S_{jk}$ estimates σ_X^2. We divide methods λ_1 through λ_6 into two subsets. One subset contains methods λ_1, $\lambda_3(\alpha)$, and λ_2, each of which allows the estimation of a reliability approximation in one computational step. The other subset contains methods λ_4, λ_5, and λ_6, and each of these methods requires multiple computational steps from which the researcher chooses the optimal result according to an optimization criterion, thus introducing the risk of chance capitalization. The GLB also has this feature. It uses optimization to find the desired outcome. In what follows, we explain the computational steps needed to find sample values for λ_4, λ_5, and λ_6, and the GLB, and we explain how chance capitalization may systematically increase the sample values beyond the true reliability, $\rho_{XX'}$. First, because it is so important and abundantly present in all of statistics, we explain the phenomenon of chance capitalization in some detail.

One instance of chance capitalization occurs when one repeatedly computes a particular statistic using different subsets of the data and then chooses from all the outcomes the one that best satisfies a particular criterion. This procedure tends to produce a biased result. As an example, we first estimate a population mean without bias and then use an estimation procedure that is biased. First, we use a random sample of N people and mean sample income, $\bar{X}$, to estimate mean population income, μ_X. Statistic $\bar{X}$ is an unbiased estimate of μ_X, that is, $\mathcal{E}(\bar{X}) = \mu_X$. When sample size N is larger, the sampling distribution of $\bar{X}$ has a smaller standard error and estimation is more precise. Second, suppose that by some accepted albeit not actually practiced line of reasoning we draw several small random subsamples of size N and accept the greatest subsample mean to estimate μ_X. Obviously, the procedure is positively biased, because half of the subsample means are expected to be at least as large as their mean, which is μ_X. Taking the greatest subsample mean surely produces a mean income estimate that is too high. The bias will grow smaller as subsample size N is larger and will almost disappear if N becomes large.

Another example concerns the well-known linear regression model. One estimates the intercept and regression weights of the independent variables to minimize the mean squared deviation of the estimated dependent variable $\hat{Y}$ and the sample observations Y; that is, one minimizes the mean sum of squares,

$$N^{-1}\sum_{i=1}^{N}\left(\hat{Y}_i - Y_i\right)^2.$$

The proportion of variance that the M independent variables explain in the dependent variable Y, denoted R^2, is an index of the success of the model. Now, suppose one has access to a large number of candidate independent variables; hence, M is large. Obviously, one cannot include all variables in a regression model—the model would contain many variables contributing little if anything unique to the explanation of Y and become meaningless—and wishes to select a smaller number, denoted K, so that $K \ll M$. In an exploratory context, a researcher may follow a strategy starting with selecting the variable from the set of M candidate independent variables that explains the most variance of the dependent variable Y, thus finding the maximum R_1^2 (subscript 1 refers to the first variable selected) given the data, and include this variable in the regression model. In the next step, the researcher selects the variable from the remaining $M-1$ candidate independent variables that produces the maximum R_2^2 (subscript 2 refers to the two variables selected thus far), and includes this variable in the model. The researcher continues selecting variables until she has a K-variable multiple regression model explaining the maximum variance based on this stepwise strategy, expressed by R_M^2.

Chance capitalization likely happens with this procedure. For example, suppose that in a sample, the first variable selected only explains the maximum variance due to sampling error. This can happen because the number M of variables to choose from is large so that coincidence has several opportunities to exert its varying influence. This coincidental result influences the selection of the second variable and may be decisive in where the procedure is going. Now suppose that in another sample a different variable is chosen as the first variable in the model and that this selection result implies that the variable selected second in the first sample would not be selected second in the new sample. Then, one can see how the mechanism of sampling error can lead a selection procedure astray and realize that in each next selection step a new selection error is possible. Drawing a new sample after each selection step and checking the result, called cross-validation, would probably correct the process, but when one uses the same sample repeatedly, this correction does not occur other than by sheer coincidence. Multivariate methods based on optimization are vulnerable to capitalization on chance. We wish to know how chance capitalization operates for methods λ_4, λ_5, λ_6, and GLB.

To find λ_4, one estimates $\alpha(P)$ in a sample of size N, which involves imprecision due to sampling error, and does this for each of a huge number of divisions of the *J*-item set in two subsets, choosing $\lambda_4 = \max_P[\alpha(P)]$ [Equation (2.53)]. This procedure, which selects the highest value, resembles the income example. The problem is here with the number of splits, denoted #*s*, which increases rapidly with test length *J*; that is, for *J* even,

$$\#s = \sum_{q=1}^{\frac{1}{2}J} \binom{J}{q},$$

and for *J* uneven,

$$\#s = \sum_{q=1}^{\frac{1}{2}(J-1)} \binom{J}{q}.$$

For *J*=6, λ_4 is the greatest of 41 $\alpha(P)$ values; for *J*=7, λ_4 is the greatest of 63 $\alpha(P)$ values; and for *J*=10, λ_4 is the greatest of 637 $\alpha(P)$ values. All $\alpha(P)$ values are estimated in the same sample with size *N*. The greatest $\alpha(P)$ value, which is λ_4, is tailored to the sample but not to other, equally likely *N*-sized random samples, and it thus overestimates the true λ_4 value. Knowing this, one needs to know whether it also overestimates reliability, $\rho_{XX'}$.

To find λ_5, for each of the *J* columns of sample covariance matrix $\mathbf{S}_X$ one has to compute the quantity $2\left(\sum_{j:j\neq k} S_{jk}^2\right)^{1/2}$ and insert the greatest of the *J* outcomes in the sample version of Equation (2.58). Because which quantity is the greatest also depends on sampling error, which resembles the first step in the regression example, method λ_5 overestimates the true λ_5 value, and again one needs to know whether it also overestimates reliability, $\rho_{XX'}$.

To find λ_6, for each item one has to estimate its residual variance, $S_{\varepsilon_j}^2$, based on a multiple regression of the item on the other *J*−1 items. The sum of the *J* residual variances is inserted in the sample version of Equation (2.59). Here, the chance capitalization is with the *J* multiple regressions and happens as we explained with the linear regression problem. We expect method λ_6 to overestimate λ_6 and thus like to know whether it also overestimates reliability, $\rho_{XX'}$.

To estimate the GLB, one uses an iterative algorithm that finds the maximum trace of measurement-error covariance matrix Σ_E, and like all optimization algorithms, this one finds a solution that takes full advantage of the sample data including all peculiarities due to random sampling. We leave technical details out of consideration, but see Bentler and Woodward (1980) and Ten Berge et al. (1981).

Oosterwijk et al. (2017) studied the proportion of random samples of fixed size drawn from a bivariate normal distribution of latent variables, in which each of the methods λ_4, λ_5, and λ_6, and the GLB overestimated true reliability

$\rho_{XX'}$. For method λ_4, its estimate was denoted $\hat{\lambda}_4$ and the mean across R random samples was denoted $\overline{\hat{\lambda}}_4$, and so on for other methods. Discrepancy was defined as $\lambda_4 - \rho_{XX'}$, bias as $\overline{\hat{\lambda}}_4 - \lambda_4$, standard deviation (SD) as $R^{-1} \sum_{r=1}^{R} \left(\hat{\lambda}_4 - \overline{\hat{\lambda}}_4\right)^2$, and the proportion of samples showing overestimation of reliability was defined as $\text{Prop}(\hat{\lambda}_4 > \rho_{XX'})$, with $R = 5{,}000$. Overestimation depends on discrepancy, bias, and SD, and one expects excessive overestimation when discrepancy is small, bias is positive, and SD is large (Figure 2.3). Precise proportions depend on sample size, test length, and dimensionality of the data.

For samples smaller than 1,000 simulated cases and test length larger than ten items, Oosterwijk et al. (2017) found that method λ_4 and ρ_{GLB} excessively overestimated reliability $\rho_{XX'}$. As an illustration, using the covariance matrix on the left-hand panel of Table 2.1 as population covariance matrix Σ, for $R = 100{,}000$ replications, we sampled $N = 100$ persons from a multivariate normal distribution $N(0, \Sigma)$. Then, we computed $\hat{\lambda}_4$ (100,000 times) and estimated its density (Figure 2.3); we retrieved $\lambda_4 = .9$ from Table 2.1, and assuming that $\sigma_{E_j} = .35$ for all six items, $\rho_{XX'} = .9125$. The discrepancy was $\lambda_4 - \rho_{XX'} = -.0125$. We found that $\overline{\hat{\lambda}}_4 \approx .9183$, hence the bias was $\overline{\hat{\lambda}}_4 - \lambda_4 \approx .0183$, and the estimated probability that $\hat{\lambda}_4$ exceeds $\rho_{XX'}$ equaled $P\left(\hat{\lambda}_4 > \rho_{XX'}\right) = .6877$. $P\left(\hat{\lambda}_4 > \rho_{XX'}\right)$ increased for smaller samples and decreased for larger samples (Table 2.6,

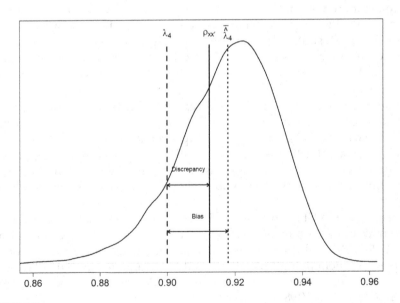

FIGURE 2.3
Estimated density of $\hat{\lambda}_4$ under the simulation conditions (see text), when $\lambda_4 = .9$, $\overline{\hat{\lambda}}_4 \approx .9183$, and $\rho_{XX'} = .9125$, resulting in a relatively small discrepancy (−.0125) and a positive bias (.0183), and therefore a large probability (.6877) that $\hat{\lambda}_4$ exceeds the true reliability.

TABLE 2.6

$P\left(\hat{\lambda}_2 > \rho_{XX'}\right)$ and $P\left(\hat{\lambda}_4 > \rho_{XX'}\right)$, for Different Sample Sizes (for Details See Text)

	Method	
Sample size	λ_2	λ_4
50	.2769	.7579
100	.1995	.6877
500	.0262	.2988
1000	.0029	.0099

second column). Furthermore, $P\left(\hat{\lambda}_4 > \rho_{XX'}\right)$ increased when bias increased, and discrepancy tended to zero. In other words, small to negligible discrepancy, positive bias, and relatively small standard deviation push an estimation method over the edge and render chance capitalization visible, causing almost consistent overestimation, even when $N > 500$. Methods λ_5 and λ_6 had greater negative discrepancies in this condition and also in general, meaning that for these methods the proportions of samples in which they overestimated the true reliability were small, but the downside was that the methods often grossly underestimated $\rho_{XX'}$.

Method λ_2 was not vulnerable to chance capitalization. It had negligible bias and large negative discrepancy. Table 2.6 (first column) shows that for all sample sizes, the probability that $\hat{\lambda}_2$ exceeds the reliability, $P\left(\hat{\lambda}_2 > \rho_{XX'}\right)$, was substantially smaller than the probability that $\hat{\lambda}_4$ exceeds the reliability, which is $P\left(\hat{\lambda}_4 > \rho_{XX'}\right)$.

The discussion on overestimation raises an interesting issue, which is whether one should be prepared to report reliability that overestimates true reliability, thus promising more than one can deliver. In general, consumers do not appreciate being sold a car that uses more energy than they are promised and patients expect a drug to be effective when the pharmacist has promised them so, and there is no reason to expect applicants will find it acceptable when they are tested at a lower reliability level than the test manual promises. This raises the question of whether one should want to use reliability approximation methods that capitalize on chance when the sample used to estimate reliability was not large or whether one should simply resort to lower bounds such as λ_2 and $\lambda_3(\alpha)$.

Confidence Intervals

A remarkable observation in applied psychometrics is that people report reliability as if it were a parameter rather than a statistic, vulnerable to sampling error and thus in need of an indication of its degree of precision (Fan &

Thompson, 2001; Oosterwijk et al., 2019). This is all the more remarkable given that several methods for estimating standard errors for the predominantly used coefficient $\lambda_3(\alpha)$ are available (Feldt, 1965; Feldt, Woodruff, & Salih, 1987; Hayashi & Kamata, 2005; Kistner & Muller, 2004; Kuijpers, Van der Ark, & Croon, 2013; Maydeu-Olivares, Coffman, & Hartmann, 2007; Padilla, Divers, & Newton, 2012; Van Zyl, Neudecker, & Nel, 2000). Sedere and Feldt (1977) studied the sampling distributions of other methods, such as method λ_2. Oosterwijk et al. (2019) investigated 1,024 reliability estimates, predominantly method $\lambda_3(\alpha)$, from a database containing assessments of 520 tests and questionnaires regularly used for diagnosis of individuals in the Netherlands and the Dutch-speaking part of Belgium and found that none of the instruments' manuals reported confidence intervals. Test manuals predominantly reported coefficient $\lambda_3(\alpha)$ and far less frequently the split-half method but rarely reported other methods. Based on the sample values, the database used cut scores to divide the tests and questionnaires into quality categories for reliability. The authors redid the exercise using confidence intervals and decided that if a reliability value did not exceed the cut score significantly, it could not stay in the category above that cut score. For example, suppose estimate $\hat{\lambda}_3(\alpha) = .82$ was not significantly greater than cut score $c = .80$ (null hypothesis: $\lambda_3(\alpha)=c$, against the alternative: $\lambda_3(\alpha)>c$), then, it was decided that statistical evidence was too weak to leave the test in the category (.8; .9], and it was downgraded. Following this procedure, the authors had to move approximately 20% of the instruments to lower categories and occasionally two categories if $\hat{\lambda}_3(\alpha)$ was not significantly greater than the second smaller cut score.

We do not discuss the available theory of estimating sampling distributions, standard errors, and confidence intervals for coefficient $\lambda_3(\alpha)$ but provide a simple example only to remind the reader of the simple truism that, like anything computed from a sample, estimate $\hat{\lambda}_3(\alpha)$ suffers from imprecision and cannot simply be considered a population value. For this purpose, we assume that the covariance matrices in Table 2.1 represent sample results based on $N=100$ cases and use a procedure Feldt (1965) proposed to compute a 95% confidence interval for coefficient $\lambda_3(\alpha)$. Nominal Type I error rate equals .05. Critical values under the F distribution are F_a with $N-1$ degrees of freedom and F_b with $(N-1)(J-1)$ degrees of freedom and are chosen such that $P(F<F_a)=.025$ and $P(F<F_b)=.975$. Based on these choices, we found $F_a \approx .7244$ and $F_b \approx 1.3374$. For the sample covariance matrix representing the one-factor case, we estimated a 95% confidence interval as

$$\left(1-[1-.9]\times F_b ; 1-[1-.9]\times F_a\right) = (.8663; .9276).$$

For the two-factor case, we estimated a 95% confidence interval as

$$\left(1-[1-.6545]\times F_b ; 1-[1-.6545]\times F_a\right) = (.5380; .7494)$$

Suppose, one approximates reliability using coefficient $\lambda_3(\alpha)$, requires that $\rho_{XX'} = c \geq .8$, c for cut score, and finds a sample value $\hat{\lambda}_3(\alpha) = .81$ for $N = 200$. Because one does not know to what degree coefficient $\lambda_3(\alpha)$ underestimates reliability $\rho_{XX'}$, a reasonable decision is to equate both quantities and seek evidence that $\hat{\lambda}_3(\alpha)$ is significantly greater than $c = .8$. What most researchers seem to do (Oosterwijk et al., 2019), is conclude that, because sample value $\hat{\lambda}_3(\alpha) > .8$, the test's reliability satisfies the requirement of minimum reliability, but the confidence interval (.7661; .8481) suggests that, say, both .77 and .8 are likely two random draws from the same sampling distribution. Hence, they are not significantly different, and the conclusion that the test has satisfactory reliability is premature. A sample value of, say, $\hat{\lambda}_3(\alpha) = .87$ is outside the confidence interval and thus would probably not have originated from the same sampling distribution as .8. The conclusion would be that the population reliability exceeds the cut score. What if one finds $\hat{\lambda}_3(\alpha) = .77$, a value smaller than the desirable .8 but not significantly smaller? Now, the evidence that the test's reliability is too small is premature, but one would make a mistake to conclude that reliability is high enough because it is not lower than .8. When $\hat{\lambda}_3(\alpha)$ is not significantly different from a target reliability value, the sample size is too small to provide conclusive evidence about its real value, which could also equal the value under the null hypothesis, and a larger sample is recommendable. Target values c for $\rho_{XX'}$ outside the confidence interval for $\lambda_3(\alpha)$ provide more certainty, but we recommend that especially when the sample is large and a small difference $\hat{\lambda}_3(\alpha) - c$ may already be significant, this is only evidence of a borderline case and one may consider lengthening the test to boost reliability. Very little is known about sampling theory for reliability approximation methods other than coefficient $\lambda_3(\alpha)$.

Reliability versus Measurement Precision

We started this chapter defining the hypothetical experiment in which one administers the same test repeatedly to the same group of people to illustrate the concept of reliability. Informally, one considers a measurement procedure reliable if it produces (approximately) the same measurement values upon repetition. This conception is not much different from expecting a physician to come up with the same diagnosis upon repetition (an experiment that suffers from precisely the lack of practicability as repeated testing does) or a car to take one from one place to the next without problems. In everyday parlance, a reliable person is someone you can rely on to show (nearly) the same behavior in similar situations. In the examples, the expectation refers to positive events, but this need not be the case. Expectation is the key to reliability, not *what* one expects. We noticed that a problem in CTT is that reliability is a group property, expressing the proportion of test-score variance in the group that is true-score variance, and thus says little if anything about the precision with which a particular person is measured. When assessing an individual's test score, we need to know how precisely her test score estimates her true score. This is known as measurement precision, and unlike reliability, measurement precision assesses individual test performance. In this subsection, we discuss measurement precision in the context of CTT, notice that its determination does not go without problems, and discuss provisional solutions.

Traditional Methods

Lord and Novick (1968, Chapter 2) discussed CTT as a double stochastic model, where first one draws a sample of individuals from a population, and second a sample of measurement values from individuals' propensity distributions. A well-known variance decomposition describes the consequence of this sampling process as

$$\sigma_X^2 = \mathcal{E}_i\left[\sigma_r^2(X_{ir} \mid T_i)\right] + \sigma_i^2\left[\mathcal{E}(X_{ir} \mid T_i)\right] = \mathcal{E}\left(\sigma_{E_i}^2\right) + \sigma_T^2. \tag{2.68}$$

Replacing σ_E^2 in Equation (2.20) with the first term on the right-hand side of Equation (2.68) yields

$$\rho_{XX'} = 1 - \frac{\mathcal{E}\left(\sigma_{E_i}^2\right)}{\sigma_X^2}, \tag{2.69}$$

which alternatively shows that by averaging individuals' error variances, reliability is a group property. Mellenbergh (1996) pointed out that CTT assumes that, contrary to the practical use of discrete scores, item scores and test scores are continuous. Assuming the propensity distribution is normal, that is, $X_i \sim \mathcal{N}\left(T_i, \sigma_{X_i}^2\right)$, the log likelihood of the test score equals

$$\ln L = -\ln\left(\sigma_{X_i}\sqrt{2\pi}\right) - \frac{\left(X_i - T_i\right)^2}{2\sigma_{X_i}^2}. \tag{2.70}$$

For person i, standard deviation σ_{X_i} is a constant, so that setting the derivative of $\ln L$ with respect to T_i equal to 0, that is, $\dfrac{d\ln L}{dT_i} = 0$, and solving for T_i yields

$$\hat{T}_i = X_i. \tag{2.71}$$

Equation (2.71) shows that, given continuity and normality, the simple test score serves as the maximum likelihood estimator of the true score. The standard deviation of the propensity distribution, $\sigma_{X_i}^2 = \sigma_{E_i}^2$, varies across persons and, given one test administration, cannot be estimated without additional assumptions. We discuss such additional assumptions and methods for estimating $\sigma_{E_i}^2$ after we have discussed the common, unconditional approach to measurement precision.

The unconditional approach to measurement precision is to replace standard deviation $\sigma_{E_i}^2$ with the standard error of measurement of the test, which is derived from test-score reliability [Equation (2.20)] as

$$\sigma_E = \sigma_X \sqrt{1 - \rho_{XX'}}. \tag{2.72}$$

For reliability $\rho_{XX'}$, in practice one inserts one of the reliability lower bounds we discussed. A systematic lower bound produces an overestimate of the true value of σ_E, thus introducing a conservative estimate. Systematic overestimates, such as method λ_4 and the greatest lower bound ρ_{GLB}, especially in smaller samples may produce underestimates, thus suggesting greater measurement precision than is justified. This is a distinct limitation of using method λ_4 and the greatest lower bound ρ_{GLB}.

Another way to estimate the true score is by means of linear regression (Lord & Novick, 1968, pp. 64–66),

$$\mathcal{E}\left(\hat{T} \mid X\right) = \beta_0 + \beta_1 X, \text{ with } \beta_0 = \mu_T - \beta_1 \mu_X \text{ and } \beta_1 = \rho_{TX} \frac{\sigma_T}{\sigma_X},$$

so that, using $\mu_T = \mu_X$, $\rho_{TX} = \sqrt{\rho_{XX'}}$, and $\frac{\sigma_T}{\sigma_X} = \sqrt{\rho_{XX'}}$, we obtain

$$\mathcal{E}\left(\hat{T} \mid X\right) = \rho_{XX'} X + \left(1 - \rho_{XX'}\right)\mu_X. \tag{2.73}$$

The standard error of estimate is obtained from $\epsilon = \hat{T} - \mathcal{E}\left(\hat{T} \mid X\right)$, as

$$\sigma_\epsilon = \sigma_T \sqrt{1 - \rho_{XX'}}. \tag{2.74}$$

Obviously, the regression estimate of true score T [Equation (2.73)] uses more information than the maximum likelihood estimate in Equation (2.71), and this is reflected in the standard error of estimate, which is a factor σ_T / σ_X smaller than the standard error of measurement,

$$\sigma_\epsilon = \frac{\sigma_T}{\sigma_X} \sigma_E. \tag{2.75}$$

Like the reliability, the standard error of measurement and the standard error of estimate are group characteristics, but one can use these quantities to assess the precision of estimated true scores provided one is willing to accept ignoring the obvious variation of the test's precision across individuals. Hence, one estimates confidence intervals for T that have equal width for all true scores. When making inferences about individual true scores, one has to assume that the measurement errors E and the residuals ϵ are normally distributed, but especially for shorter tests and questionnaires, this assumption is a rough approximation at best.

For sample results, denoting sample variance by S^2 and sample reliability by $r_{XX'}$, the standard error of measurement can be estimated as

$$S_E = S_X \sqrt{1 - r_{XX'}}. \tag{2.76}$$

To construct a confidence interval, for significance level α, let $z_{\alpha/2}$ be the standard normal deviate corresponding to the corresponding area under the normal curve that describes the individual test-score distribution. One estimates a $(1-\alpha)100\%$ confidence interval as

$$\left[\hat{T}_i - z_{\alpha/2} S_E ; \hat{T}_i + z_{\alpha/2} S_E \right]. \tag{2.77}$$

To estimate the confidence interval for the true score based on Equation (2.73), first we write $\sigma_T = \sigma_X \sqrt{\rho_{XX'}}$, and insert this result in Equation (2.74) to obtain

$$\sigma_\epsilon = \sigma_X \sqrt{\rho_{XX'}} \sqrt{1 - \rho_{XX'}}. \tag{2.78}$$

The sample version of Equation (2.78) equals

$$S_\epsilon = S_X \sqrt{r_{XX'}} \sqrt{1 - r_{XX'}}. \tag{2.79}$$

To obtain a confidence interval, one replaces S_E with S_ϵ in Equation (2.77). Because $\sigma_\epsilon \leq \sigma_E$, with equality if $\rho_{XX'} = 1$ (which one might check is a trivial case), using S_ϵ rather than S_E produces a narrower confidence interval. One may ask how σ_E and σ_ϵ compare to the theoretical lengths of the scales of the two true-score estimates, thus taking into account that a smaller confidence interval has little use when the scale is relatively short. Box 2.8 shows that based on the comparison with the theoretical scale length, use of σ_ϵ may be preferred over use of σ_E. However, one may notice that statistical criteria other than precision may be used for deciding which true-score estimation method to prefer, and an obvious candidate is estimation bias. Different statistical quality measures may counteract one another, one method being less biased and the other more precise, rendering a decision on which method to use difficult. As far as we know, the bias of the two methods represented by

Equations (2.71) and (2.73) is unknown, thus leaving the question of which of the two methods to use for person measurement open for the moment.

BOX 2.8   Based on Scale Length, Comparing σ_ϵ and σ_E

First, we consider the maximum likelihood estimate of the true score [Equation (2.71)], which we denote by $\hat{T}_{ML}$. Obviously, for sample results

$$S_{\hat{T}_{ML}} = S_X \tag{2.80}$$

and, moreover, the scale for $\hat{T}_{ML}$ coincides with that for test score X. For example, if the J items are polytomously scored, $x = 0,\ldots,M$, then similar to the scale of X, for $\hat{T}_{ML}$ we have a scale

$$0 \leq \hat{T}_{ML} \leq MJ. \tag{2.81}$$

The theoretical scale length, also known as the range and denoted TSL_{ML}, for $\hat{T}_{ML}$ equals $TSL_{ML} = MJ$ units. For the regression-based true-score estimate [Equation (2.73)], which we denote by $\hat{T}_{REGR}$, the situation is different. Using the linear regression model to estimate true score T, the model necessarily produces a standard deviation for the estimated true score that is smaller than the true-score standard deviation; that is,

$$S_{\hat{T}_{REGR}} \leq S_T. \tag{2.82}$$

Because $S_T \leq S_X$, and $S_X = S_{\hat{T}_{ML}}$, we conclude that

$$S_{\hat{T}_{REGR}} \leq S_{\hat{T}_{ML}}. \tag{2.83}$$

This result could be used to argue in favor of using the regression model for estimating T and using S_ϵ to estimate confidence intervals, were it not that the scale length for $\hat{T}_{REGR}$ also is smaller than that for $\hat{T}_{ML}$, and we have to incorporate this result in our argument. Using Equation (2.73), and inserting $X=0$ and $X=MJ$, produces scale length

$$(1-\rho_{XX'})\mu_X \leq \hat{T}_{REGR} \leq \rho_{XX'}MJ + (1-\rho_{XX'})\mu_X. \tag{2.84}$$

Subtracting the minimum value from the maximum value shows that the theoretical scale length equals $TSL_{REGR} = \rho_{XX'}MJ$ units.

The question of whether there is merit in using σ_ϵ rather than σ_E can be answered by asking whether the ratio of σ_ϵ to the theoretical scale length TSL_{REGR} is smaller than the ratio of σ_E to the theoretical scale length TSL_{ML}. If so, their difference is negative. Hence, we write

$$\frac{\sigma_\epsilon}{TSL_{REGR}} - \frac{\sigma_E}{TSL_{ML}} = \left(\frac{\sqrt{1-\rho}}{\rho} \times \frac{\sigma_T}{MJ}\right) - \left(\sqrt{1-\rho} \times \frac{\sigma_X}{MJ}\right). \qquad (2.85)$$

and notice that

$$\frac{\sqrt{1-\rho}}{\rho} \leq \sqrt{1-\rho} \text{ and } \frac{\sigma_T}{MJ} \leq \frac{\sigma_X}{MJ},$$

so that the difference in Equation (2.85) indeed is negative. Hence, based on the comparison with the theoretical scale length, use of σ_ϵ may be preferred over use of σ_E. One may notice that if one uses the scale length that a group actually employs—the empirically established scale length—rather than the theoretical scale length, the conclusion about use of σ_E and σ_ϵ would not change. We notice once more that other statistical criteria, such as bias, might produce different conclusions.

An interesting question is whether one can measure with enough precision using a scale consisting of few items but having a high reliability. This question is relevant, because in clinical practice, job selection, and marketing research there is a tendency to use short scales to reduce the burden that patients, applicants, and consumers experience when filling out long questionnaires (e.g., Kruyen, Emons, & Sijtsma, 2012). For tests consisting of 5, 10, 15, and 20 polytomously scored items with $x = 0,\ldots,4$, Figure 2.4 shows 95% confidence intervals for John's true score T_{John} in the middle of the scale, so that $T_{John} = 2J$ (open circles). We want to know whether John's test score X_{John} is significantly smaller than cut score X_c, which equals $2.2 \times J$ (massive circles). For the sets of items we used to simulate data, Figure 2.4 shows some well-known, interesting facts (technical details in the figure caption). First, as the test grows longer, test scores grow further apart, so that σ_X increases. Second, reliability increases, and lower bound estimates such as coefficient $\lambda_3(\alpha)$ also increase. Third, the standard error of measurement [and the standard error of estimate, Equation (2.74)] also increases, thus producing wider confidence intervals for true score T_{John}. Wider confidence intervals run counter to the intuition that increasing reliability should correspond with increasing measurement precision. Fourth, with increasing test length, scale length JM (based on $x = 0,\ldots,JM$) grows faster than the width of the confidence interval and not only that: The confidence interval grows more slowly

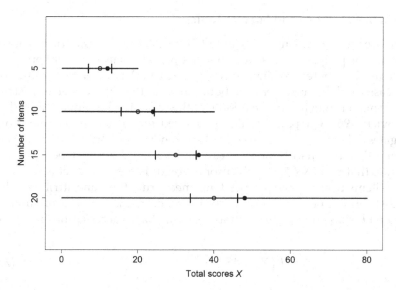

FIGURE 2.4
Confidence intervals (between vertical line segments) for John's true score (open circle), for tests consisting of 5, 10, 15, and 20 parallel items, scored $x = 0,\ldots,4$. Massive circles represent cut-scores. Note: For all items, $\sigma_{jk} = 0.5$, $\sigma_j^2 = 2$, and $\sigma_{T_j}^2 = 1.5$. Corresponding (rounded) reliabilities equal .875 ($J=5$), .923 ($J=10$), .944 ($J=15$), and .957 ($J=20$), and rounded standard errors of measurement equal 1.15 ($J=5$), 2.24 ($J=10$), 2.74 ($J=15$), and 3.16 ($J=20$).

than John's distance to the cut point, and if test is long enough, the cut point moves out of the confidence interval. This happens for 15 and 20 items.

We draw two conclusions. First, a longer test consisting of items with acceptable psychometric properties allows greater measurement precision. Measurement precision must be understood in relation to scale length and thus is a relative concept. In an absolute sense, expressed in scale units, the standard error of measurement increases, and thus "measurement precision" decreases. Second, the five-item scale has coefficient $\lambda_3(\alpha)$ equal to .875, which is acceptable to many researchers, but assuming a normal distribution, the 95% confidence interval has a width of approximately 4.6 units, thus covering $(4.6/20) = 23\%$ of the scale length. For individual measurement values, one needs the standard error of measurement or an alternative quantity. Emons, Sijtsma, and Meijer (2007) related test length to the correct and incorrect decisions based on the test score and demonstrated that tests containing 6–12 items, even if the items have the best quality encountered in practice, based on people's propensity distributions, repeatedly misclassify a large part of the group. The classification results improved when test length increased. See Harvill (1991) for more discussion about measurement precision.

Alternative Methods and Special Topics

If one uses the Equations (2.72) and (2.74) to estimate symmetric confidence intervals for people's true scores, one assumes the measurement errors are normally distributed. Wainer and Thissen (2001) noticed the assumption may easily fail, because near the boundaries of the scale the error distribution is smaller than it is in the middle of the scale. The binomial model (Lord & Novick, 1968, Chap. 23), which is defined for dichotomous item scores, mitigates this problem, because unlike the normal distribution, it does not assume that the distribution's mean and variance are independent. One assumes that $\tau = \mathcal{E}(X / J \mid \tau)$ is the proportion of 1-scores on a set of randomly drawn items from a large pool of items measuring the same attribute. Then, the distribution of sum score X can be conceived as a count of successes based on J trials—item administrations—that have success probability τ; or

$$P\left(X = x \mid \tau\right) = \binom{J}{x} \tau^x \left(1 - \tau\right)^{J-x}. \tag{2.86}$$

The distribution of sum score X is widest in the middle of the scale and narrows down toward the scale boundaries, where it is skewed to the right (low sum score) or skewed to the left (high sum score). The estimation of confidence intervals is rather complicated and uses the result that the sum score follows a negative hypergeometric distribution (Johnson & Kotz, 1969; Wainer & Thissen, 2001). Although interesting and important, we refrain from further pursuing this path.

Feldt et al. (1985) discussed five methods to estimate measurement precision conditional on sum score X. We discuss two possibilities.

The Thorndike (1951) method is based on the assumption that the test can be split in two parallel half tests with sum scores X_1 and X_2, so that $X_1 = T_1 + E_1$ and $X_2 = T_2 + E_2$, with $T_1 = T_2$ and $\rho_{E_1 E_2} = 0$. It follows that $\sigma_E = \left(\sigma_{E_1}^2 + \sigma_{E_2}^2\right)^{1/2}$. The difference between the half-test scores equals $X_1 - X_2 = E_1 - E_2$, so that, using $\sigma_{E_1 E_2} = 0$, one obtains

$$\sigma_{X_1 - X_2} = \sigma_{E_1 - E_2} = \left(\sigma_{E_1}^2 + \sigma_{E_2}^2\right)^{1/2} = \sigma_E. \tag{2.87}$$

The traditional equation for the standard error of measurement [Equation (2.72)] contains standard deviation, σ_X, so that conditioning on sum score $X = x$ yields $\sigma_X = 0$ and $\rho_{XX'}$ undefined, and thus σ_E undefined. Hence, conditioning on $X = x$ cannot produce a conditional standard error of measurement, but using Equation (2.87) can accomplish this result as follows: First, select the group with $X = x$; then, for each individual in this subgroup, determine their scores on the two half tests, X_1 and X_2; and, finally, compute

$$S_{X_1 - X_2} = \hat{\sigma}_{E \mid X = x}. \tag{2.88}$$

First, because not all people with the same sum score, $X = x$, will have the same difference score, $X_1 - X_2$, one will likely find $\hat{\sigma}_{E|X=x} > 0$. Second, the conditional standard error of measurement in Equation (2.88) can be expected to vary across different values of X, thus rendering the standard-error-of-measurement value conditional on sum score X. Third, low-frequency sum-score groups yield imprecise estimates, $\hat{\sigma}_{E|X=x}$. Joining adjacent sum-score groups until a preset group size is attained may mitigate imprecision. Finally, failure to satisfy the strong assumption of parallel half tests may distort estimates of $\sigma_{E|X=x}$. However, because the traditional standard-error-of-measurement methods rely on underestimates of reliability $\rho_{XX'}$, the traditional standard error of measurement produces overestimates of σ_E. As far as we know, little is known about the Thorndike method, and more research may be in order.

Joining adjacent sum-score groups to increase statistical precision produces fewer conditional standard error of measurement estimates, thus losing finer-grained measurement precision. The polynomial method (Mollenkopf, 1949) avoids the joining of adjacent sum-score groups as follows. In a linear regression model, we use the squared difference between the sum scores on two parallel test halves, $Y = (X_1 - X_2)^2$, as the dependent variable, and estimate

$$\mathcal{E}(Y^2 \mid X, X^2, X^3, \ldots) = \mathcal{E}\left[(X_1 - X_2)^2 \mid X, X^2, X^3, \ldots\right]$$

$$= \beta_0 + \beta_1 X + \beta_2 X^2 + \beta_3 X^3 + \cdots \tag{2.89}$$

For parallel test halves,

$$\mathcal{E}\left[(X_1 - X_2)^2 \mid X, X^2, X^3, \ldots\right] = \mathcal{E}\left[(E_1 - E_2)^2 \mid X, X^2, X^3, \ldots\right]. \tag{2.90}$$

The mean error difference equals 0, that is, $\mu_{(E_1 - E_2)} = 0$; hence, the right-hand side of Equation (2.90) equals the conditional variance of the error difference, and from the Thorndike method we know that its square root, which is the conditional standard deviation of the error difference, equals the conditional standard error of measurement. Thus, for a fixed value $X = x$ we have that

$$\left\{\mathcal{E}\left[(E_1 - E_2)^2 \mid X, X^2, X^3, \ldots\right]\right\}^{1/2} = \sigma_{E|X=x}. \tag{2.91}$$

The polynomial method has the effect of smoothing the estimates obtained using the Thorndike method, producing a more stable result. This is what Feldt et al. (1985) found in a study in which they compared the two methods with three other methods from CTT and one method from IRT (Chapter 4). Further, the authors found that for the large-sample data they used, in the middle of the sum-score distribution $\hat{\sigma}_{E|X}$ was approximately twice the magnitude it attained in the tails of the distribution, where bottom and ceiling effects occurred.

We have discussed the measurement precision of one test score, but one may also be interested in at least three different questions. First, one may ask whether a test score differs significantly from a cut score X_c used for admitting a person to a course or a therapy. Second, one may ask whether the scores of different persons, X_v and X_w, are different. Third, one may ask whether person v's test score has changed as a result of an intervention, such as education or therapy, in which case the person is tested before and after the intervention, resulting in test scores $X_{v(pre)}$ and $X_{v(post)}$.

The first question can be answered by checking whether cut score X_c, which does not contain measurement error, is in the confidence interval for T_v. If no, then one concludes that T_v is different from X_c; if yes, then one concludes that there is insufficient statistical evidence that T_v is different from X_c. By appropriately adapting the significance level, one can test either one- or two-sidedly.

The second question requires considering the difference score, $D_{vw} = X_v - X_w$, and testing the null hypothesis $H_0 : \Delta_{vw} = T_v - T_w = 0$, either one- or two-sidedly. Given the null hypothesis of no change, one may assume that the standard error of measurement, σ_E, is the same on both measurement occasions, whereas measurement errors correlate 0 between occasions. Then, one can derive (e.g., Sijtsma & Emons, 2011) that the standard error of measurement for difference score D equals

$$\sigma_{E(D)} = \sqrt{2}\sigma_E. \tag{2.92}$$

The confidence interval for Δ_{vw}, using $\hat{\Delta}_{vw} = D_{vw}$ and $S_{E(D)} = \sqrt{2}S_E$ [S_E from Equation (2.76)], equals

$$\left[\hat{\Delta}_{vw} - z_{\alpha/2}S_{E(D)}; \hat{\Delta}_{vw} + z_{\alpha/2}S_{E(D)} \right].$$

One may check whether value $\Delta_{vw} = 0$ lies in the interval; if no, one assumes $T_v \neq T_w$; if yes, following the logic of hypothesis testing, one cannot reject the null hypothesis, which in practice is taken as evidence that T_v and T_w do not differ. Using the appropriate significance level, one may test one- or two-sidedly.

The third question requires evaluating the null hypothesis $H_0 : T_{v(post)} - T_{v(pre)} = 0$, to test whether person v has changed as a result of taking a course, undergoing a therapy, and so on. Several significance tests exist, but the reliable change (RC) index (Jacobson & Truax, 1991) is claimed to have the best statistical properties (Atkins, Bedics, McClinchey, & Beauchaine, 2005; Bauer, Lambert, & Nielsen, 2004) and is the most frequently used method in the practice of change assessment based on two measurement occasions. For person v, the RC index is defined as

$$RC_v = \frac{X_{v(\text{post})} - X_{v(\text{pre})}}{S_{E(D)}}. \tag{2.93}$$

The standard error of measurement for difference score D, $S_{E(D)}$, can be obtained by inserting the sample standard error of measurement, S_E [Equation (2.76)], for parameter σ_E in Equation (2.92). The RC index is assumed to be a standard normal deviate. Statistical testing can be done one- or two-sidedly. Gu, Emons, and Sijtsma (2018) provided a critical review of persistent negative beliefs about the usefulness of change scores for assessing treatment effectiveness with individuals, and using a multilevel framework for analyzing the beliefs they concluded that these beliefs can be renounced convincingly.

Constructing Scales in the Classical Test Theory Context

CTT is a theory of random measurement error that applies to any measurement value. The measurement value may stem from a procedure that is well-defined, but from the perspective of CTT, it does not matter whether the procedure is sound or whether it results from an educated guess or worse. The true score from CTT is the expected value of the observable measurement value collected in independent replications of the measurement procedure. Consequently, the composition of the true score remains undefined and can be anything, having a dimensionality ranging from one dimension to any composition of multiple dimensions. CTT is blind to the composition of the true score, and it is for this reason that its central concern is to determine the reliability of measurement, not its dimensionality.

For test constructors, determining only reliability is not sufficient, because they need to know whether the initial set of items from which they intend to construct a measurement instrument is adequate for attribute measurement or whether they need to adapt the item set. Examples of adaptations are to delete some items, replace some items by other items, divide the item set into subsets each representing different aspects of the attribute, or a combination of these options. Because CTT was not concerned with these issues, it did not offer the tools for this, and researchers went their own way and used research strategies most people know nowadays. We discuss three strategies for item assessment and test construction regularly encountered in publications reporting the assessment of the dimensionality of an item set. The methods are the inspection of corrected item-total correlations and the oblique multiple group method, principal component analysis (PCA), and factor analysis.

Both PCA and factor analysis are based on the inter-item correlation (or covariance) matrix. Nunnally (1978, pp. 141–146) explained how differences between the distributions of discrete item scores affect the product-moment correlation. The greater the difference between two discrete item-score distributions, the more the difference reduces the maximum value that the product-moment correlation between the items can attain, and the effect can be rather dramatic. Because this effect of discrepancy of distributions strongly affects the structure of the inter-item correlation matrix, it also strongly affects the result of PCA and factor analysis. An extreme case is difficulty factors that group items because they have similar means (Gorsuch, 1983, pp. 292–295) rather than because they share a common attribute. Several authors have considered the problem of factor analysis of discrete items (e.g., Dolan, 1994; Forero, Maydeu-Olivares, & Gallardo-Pujol, 2009; Muthén & Kaplan, 1985; Rhemtulla, Brosseau-Liard, & Savalei, 2012). A summary of the literature is the following. For approximately symmetrical distributions based on at least five ordered item scores, one can use factor analysis as if the item scores were continuous and normal. When the number of item scores is smaller than five or when the item-score distributions are markedly skew, one assumes that underlying the discrete variables are continuous variables that were discretized to obtain the discrete, ordered score categories. Based on this assumption, for binary item scores one uses tetrachoric correlations and for polytomous item scores one uses polychoric correlations, both estimated to express the linear association between the underlying continuous normal variables.

Corrected Item-Total Correlations and Oblique Multiple Group Method

The precursor to the oblique multiple group method is to compute for each item in the initial item set its product-moment correlation with the total score on the other $J-1$ items. This correlation is known as the item-rest correlation or the corrected item-total correlation. Let j index the item of interest and let the rest score be defined as $R_{(j)} = \sum_{k:k \neq j} X_k$; then, one computes $r_{X_j R_{(j)}} = r_{j R_{(j)}}$, $j = 1, \ldots, J$. The idea is that items belonging together and presumably measuring the same attribute correlate highly positively, but items assessing attributes other than the attribute that the majority of items assess correlate lowly with this majority, resulting in a low corrected item-total correlation. Such items are candidates for removal from the initial item set, whereas items showing relatively high correlations are maintained.

Because

$$\rho_{j R_{(j)}} = \frac{\sigma_{j R_{(j)}}}{\sigma_j \sigma_{R_{(j)}}} = \frac{\sum_{k:k \neq j} \sigma_{kj}}{\sigma_j \sigma_{R_{(j)}}}, \tag{2.94}$$

clearly items having high corrected item-total correlations contribute predominantly high inter-item covariances (right-hand side) to reliability lower-bound coefficient $\lambda_3(\alpha)$ [Equation (2.29)], and items having low corrected item-total correlations contribute little and their presence may even lower coefficient $\lambda_3(\alpha)$. Some software programs (e.g., SPSS; IBM Corp., 2016) offer options to the user to select a subset of items to optimize coefficient $\lambda_3(\alpha)$ by some criterion, such as "alpha if item deleted" (also, see Raykov, 2008).

The oblique multiple group method (Gorsuch, 1983; Holzinger, 1944; Nunnally, 1978; Stuive, 2007) provides a specialization of the corrected item-total method and consists of the next steps. First, the items are divided into Q distinct and exhaustive subsets, indexed $q = 1,\ldots,Q$, each subset representing items that the researcher hypothesizes to measure an aspect of the attribute or a separate attribute, different from the aspects of attributes that the other items assess. Second, in the sample, for each item subset one computes the subset-total scores. Third, for each item, one computes the correlation with each of the subset-total scores and collects the correlations in a correlation matrix having J rows for items and Q columns for item subsets. Items should have the highest correlation with the subset to which they were assigned, and items may be re-assigned to another subset with which they correlate highest. Correlations between subset total-scores may suggest combining highly correlating subsets or deleting one to obtain a shorter test.

Correlations between item scores and subset-total scores are inflated, because item j also is part of the total score for the subset q to which it is assigned. Inflation is easily corrected by removing item j from this item subset and correlating the item with the rest score $R_{q(j)}$, denoted $r_{jR_{q(j)}}$. Correlation $r_{jR_{q(j)}}$ suffers from another problem, which is that the item-subset length affects the correlation, and longer item subsets tend to produce higher correlations. Stuive (2007, pp. 6–7, 134–135) suggested correcting the item-subset correlation for item-subset length. Because nuances like this, however useful, are beyond the scope of this monograph, we do not discuss the correction method.

Principal Component Analysis

Researchers use PCA frequently to find out whether the items in the initial item set belong together in the same test or whether several subtests need to be distinguished or individual items deleted or replaced. PCA summarizes the information from the J item scores in weighted linear combinations of the J item scores as efficiently as possible. We denote the qth weighted linear combination by C_q, with $q = 1,\ldots,Q$ and $Q \leq J$, to be explained shortly, and we denote weights by w_{jq} so that

$$C_q = \sum_{j=1}^{J} w_{jq} X_j. \tag{2.95}$$

The weights of the first linear combination, C_1, are chosen under restrictions to guarantee a unique solution, such that the variance of C_1 is maximal. This is the first principal component. The weights of the second linear combination, C_2, are chosen such that C_2 explains the maximum variance of the residuals of the item scores from their regression on C_1. This guarantees that C_2 is orthogonal to C_1 and that it only explains variance from the items that C_1 does not explain. Next, C_3 is the weighted linear combination of the item-score residuals that remain after items are regressed on C_1 and C_2, and that explains the maximum variance, and so on for the next principal components. Because they are based on residuals, all principal components are orthogonal. Each next principal component explains less variance than the previous one, and $Q = J$ principal components explain all the variance from the J items; hence, the number of principal components is J. The explained variances of the principal components correspond to the eigenvalues of the inter-item correlation matrix, $\mathbf{R}_X$.

The meaning of PCA is that it piles up the maximum amount of statistical information, identical to explained variance, from the J items in the first weighted sum variable, the maximum amount of the remaining information in the second sum variable, and so on. Metaphorically, it is as if one packs as many travel requisites in the first, biggest available suitcase, as many of the remaining travel requisites in the second biggest suitcase, and so on. Suppose, one can take as many suitcases (technically, with a maximum equal to the number of travel requisites) on the plane as one wishes, but the extra price one pays for each next suitcase is larger than the previous, even though it contains fewer travel requisites. Assuming one does not want to pay the maximum price by bringing all suitcases, how does one choose the number of suitcases to bring on the plane? This is approximately the problem researchers face when they have to decide how many principal components to retain as summaries of the J items. One might first ask why one would want to retain principal components at all, but the answer to this question comes later.

Two rules of thumb are popular for retaining principal components. The first method is graphical and is known as the scree test (Cattell, 1966). The graph shows the eigenvalues as a function of the principal components' rank numbers where straight lines connect points [coordinates are (rank number, eigenvalue)], and by definition the function is monotone decreasing and convex by approximation. Usually, the plot descends steeply and after a few points more or less levels off. One retains the principal components before leveling off starts for further analysis, because they seem to explain enough variance to be of interest. The second method retains all principal components that have an eigenvalue greater than 1 (Kaiser, 1960), because these are the variables that explain more variance than one item, when all items are standardized having variance equal to 1. Whereas the scree plot requires judgment, the eigenvalue-greater-than-1 rule is objective but easily produces too many principal components when their eigenvalues are only a little greater than 1, overestimating the number of useful principal components. Using data simulation, parallel analysis (Horn, 1965; also, Garrido, Abad, &

Ponsoda, 2013; Timmerman & Lorenzo-Seva, 2011) provides a statistical perspective on the deterministic eigenvalue-greater-than-1 rule by assessing the influence of sampling error on the eigenvalues.

The correlation of an item with a principal component, known as the item's loading, informs one which items predominantly define a principal component. However, because the principal components were constructed not with the goal to obtain an intelligible pattern of loadings but to explain the maximum amount of variance, in real data the resulting matrix of J items by Q principal components containing the loadings does not have an intelligible interpretation. Hence, the F ($F \leq Q$) principal components one retains are factor analyzed to obtain a new loadings matrix that enables a better opportunity for interpretation. Factor analyzing the principal components to an intelligible factor structure is called rotation. Geometrically, one can imagine that the items are located in an F-dimensional space that is described by an F-dimensional coordinate system, and rotation literally means that one turns the system around its fixed origin to another position. The positions of the items in the F-dimensional space are fixed; hence, they have different coordinates when the coordinate system rotates to a new position. Algebraically, the system's axes are the factors and the items' coordinates are the loadings. Given the freedom one has in rotating the coordinate system, one can search for the constellation that provides the loadings matrix having the most intelligible interpretation in terms of which items load high and which items load low on a factor. This gives a factor its meaning. Many rotation methods have been proposed (Gorsuch, 1983, pp. 203–204), and axes can be orthogonal or oblique, algebraically corresponding to null-correlating factors or correlating factors, respectively. The best-known method is Varimax rotation, which seeks the orthogonal solution that approximates the ideal of simple structure the best; that is, each item has a loading of 1 on one factor and 0s on all other factors, and this ideal is assumed to separate different factors maximally.

Rotation leaves the total amount of variance that the F principal components explain in the J item scores intact. With orthogonal factors, the sum of the variances that each factor explains equals the total amount of explained variance. With oblique factors, one cannot simply add the amounts of explained variance, but together they explain as much variance as an orthogonal factor solution.

PCA followed by rotation is typically used when the researcher does not have a clear-cut expectation of the structure of her item set. One can easily vary the number of principal components to retain for rotation and try different rotation methods and accept the item-subset structure that is best interpretable. This flexibility may produce useful results but also is the Achilles heel of the method, because it invites endless trial and error where one would like to encourage the researcher to focus on the theory of the attribute. The method has gained enormous popularity in ability and trait research. Examples are the application of PCA to study the structure of intelligence test batteries (e.g., Kroonenberg & Ten Berge, 1987) and to study the structure of the Big Five personality traits (e.g., De Raad & Perugini, 2002). Factors are usually taken as scales, and the sum score on the items defining a factor is the person's measurement value on the scale.

Factor Analysis

The classical test model for two items equals

$$X_j = T_j + E_j \text{ and } X_k = T_k + E_k, \tag{2.96}$$

but the two true scores may have nothing in common; CTT does not model their relationship, so we cannot know what the relationship is. This indeterminacy is often badly understood, and some people even posit that CTT is a one-factor model or equate the true score with a factor or a latent variable. This is incorrect; the key to CTT is that it is an error model, which has nothing to say about the observable measure other than that its observation is clouded by random measurement error. CTT's quest is to estimate the proportion of the observable variance that is not error variance, that is, reliability.

CTT as a formal measurement model is blind to test constructors' intentions that items in the test should measure the same attribute or different aspects of the same attribute. Factor analysis (e.g., Bollen, 1989) models the item scores as weighted linear combinations of Q latent variables or factors, denoted ξ_q ($q = 1, \ldots, Q$; $Q \leq J$), with intercept b_j, loadings of items on latent variables denoted a_{jq} and residual δ_j, so that for two items the model equations are

$$X_j = b_j + \sum_{q=1}^{Q} a_{jq}\xi_q + \delta_j \text{ and } X_k = b_k + \sum_{q=1}^{Q} a_{kq}\xi_q + \delta_k, \tag{2.97}$$

and it is clear that together they assess at most Q latent variables. The factor model explains the correlation between the items by means of the latent variables they have in common and leaves room for different items assessing partly different sets of latent variables. Residual δ_j is the part of item score X_j that the Q latent variables ξ_q cannot linearly explain and consists of two additive parts, an item-specific part or unique (to the item) error, denoted U_j, and random measurement error, E_j, as in CTT, so that $\delta_j = U_j + E_j$. The item-specific part U_j may represent an attribute that is unique to item j. For example, one item in an arithmetic test that embeds the arithmetic problems in a substantive problem may require knowledge of geography whereas the other items do not. Knowledge of geography is unique to the item in the sense that people having the same arithmetic ability level but differing in their knowledge levels have different probabilities of solving the arithmetic problem correctly, but because it is unique to one item, the item-specific attribute does not contribute to correlations of the item with the other items. In factor analysis, one considers item-specific latent variables unique errors that are usually unknown.

It is assumed that unique error U_j covaries 0 with the latent variables ξ_q and with random measurement error E_j. Taking the expectation of item score X_j in Equation (2.97) yields the true score,

$$T_j = b_j + \sum_{q=1}^{Q} a_{jq}\xi_{jq} + U_j. \tag{2.98}$$

Clearly, modeling the true score by means of a factor structure has serious consequences for defining reliability, as we will see in the next section. As a prerequisite, we notice that reliability methods based on factor models usually assume that each item has a loading on at least one latent variable and that a latent variable has at least two items loading on it. Hence, these reliability methods assume that unique components are superfluous, and they thus disappear from the model. Bollen (1989, pp. 220–221) posited that adopting multiple latent variables to some extent compensates for the absence of unique variables (Bollen, 1989, pp. 220–221). Another simplification of the model is assuming intercept $b_j=0$, for all items. Then, test score X can be written as

$$X = \sum_{j=1}^{J} X_j = \sum_{j=1}^{J} \sum_{q=1}^{Q} a_{jq}\xi_{jq} + \sum_{j=1}^{J} E_j. \tag{2.99}$$

We return to this model in the next section discussing reliability based on the factor model, but first we consider the exploratory and confirmatory factor analysis strategies for studying the test composition and the definition of one or more scales.

Like PCA, exploratory factor analysis (EFA) typically aims at reducing the number of J items by replacing them with a much smaller number of factors. Whereas PCA defines the linear combination of J standardized item scores that has the greatest variance possible [Equation (2.95)] and similarly for consecutive residual item scores, EFA finds a small number of latent variables that enable one to conceive of item scores as weighted sums of these latent variables plus a residual [Equation (2.97)]. Hence, the two approaches are different, PCA being a typical data-reduction technique whereas (exploratory) factor analysis is a measurement model assuming that one or more latent variables represent attributes driving responses to items. Technically, the differences are smaller yet important. PCA determines the eigenvalues (i.e., percentages of explained variance) and the eigenvectors (i.e., loadings) of the inter-item correlation matrix $\mathbf{R}_X$, whereas EFA replaces the 1s appearing on the main diagonal of $\mathbf{R}_X$ with the item communalities, yielding matrix $\mathbf{R}_X^c$. The communality of item j is the proportion of variance the Q latent variables explain of item score, X_j. Item communalities approximate but are not equal to item true-score variances. Thus, whereas PCA analyzes $\mathbf{R}_X$, not distinguishing item true-score variance and item error-score variance, EFA analyzes $\mathbf{R}_X^c$ using the item communalities as approximations of item true-score variances. As a starting value for the item communality, one may use the squared multiple correlation of X_j on the other $J-1$ item scores.

EFA shares with PCA that it seeks to reduce the number of J item-score variables and may determine the eigenvalue and eigenvectors, albeit of $\mathbf{R}_X^c$ rather than $\mathbf{R}_X$. Eigenvalues greater than 1 determine the number of factors retained, and loadings are used to interpret the factors. Given its purpose, which is usually to determine the number of factors to retain, and the way the model is estimated, some limitations become apparent (e.g., Bollen, 1989, p. 232). For example, the researcher can determine the number of factors but has to accept the loadings as the data determine them and thus is unable to use her knowledge to *a priori* fix loadings to 0 or to equate particular groups of loadings. In addition, the method does not allow for correlated errors, such as those resulting from repeated measurement in a longitudinal design. Finally, all factors are either correlated or uncorrelated.

With confirmatory factor analysis (CFA), the limitations of EFA do not exist. Instead, based on her hypothesis of which items assess which latent variable, the researcher chooses the number of latent variables Q, and if $Q=1$, she chooses a one-factor model. By specifying which item loads on which factor, the researcher posits a factor model as a hypothesis representing a theory of the attribute that the item set measures. The parameters of the model can be estimated using maximum likelihood methodology or other estimation methods (e.g., Bollen, 1989, Chap. 4), and the basic strategy of goodness-of-fit analysis is to reconstruct the inter-item correlations based on the estimated factor model. The discrepancy between observed and estimated correlations provides an impression of the adequacy of the model for the data. Hu and Bentler (1999) discussed several goodness-of-fit measures and proposed rules of thumb for the interpretation of the values these measures can attain. In the data example at the end of this chapter, out of the large number of fit indices that quantify the degree of model fit we used the root mean squared error of approximation (RMSEA) and the Tucker-Lewis Index (TLI). The RMSEA is an absolute fit index expressing how well the factor model reproduces the data. Rules of thumb for interpretation are that *RMSEA* < .05 means the factor model fits well, and *RMSEA* < .08 means that model fit is acceptable. The TLI is an incremental fit index expressing the proportionate improvement of model fit when comparing the factor model of interest with a baseline model. TLI values in excess of .90 suggest acceptable fit. Bollen (1989, pp. 296–305) discussed procedures that help the researcher to adapt her model by suggesting which parameters to restrict or set free and thus estimate an alternative model. In addition to CFA, many different factor models exist (Gorsuch, 1983; Harmann, 1976; Mulaik, 1972), but CFA has become particularly popular for test construction.

Obviously, CFA provides the researcher with a more powerful tool for testing hypotheses about the latent variables underlying item performance than EFA and PCA. In particular, the possibility to *a priori* fix or equate particular item loadings and to allow for correlated errors and factors renders the method more appropriate for hypothesis testing. However, we wish to emphasize that whether research is confirmatory, testing a model as a null hypothesis formulated prior to setting up a design and collecting the data, or exploratory, using a data set to try several possible hypotheses *a posteriori* is a matter of strategy rather than method. Our point is that, even when one intends the method for confirmatory use there is nothing that prevents researchers from trying as many adaptations of their confirmatory factor model as they wish and then choosing the one they like best. This strategy of using CFA would qualify as exploratory. Likewise, one could test the expectation that one latent variable underlies the data using EFA, and a large first eigenvalue would provide the support one was expecting. In our view, even if EFA does not formally test a null hypothesis, when used to assess whether an expectation is correct it qualifies as a confirmatory strategy. Hence, it is the researcher's research plan that determines whether the research is confirmatory or exploratory, not whether the method used uses formal statistical testing of a model. PCA, EFA, and CFA are methods that some researchers tend to use to exploit their data in search of an interpretable solution. The literature on questionable research practices discusses what can go wrong when blindly pursuing hypothesis testing with a particular goal in mind, such as finding a fitting model or exploiting one's data until one finds the desired outcome (Ioannidis, 2005; MacCallum, Roznowski, & Necowitz, 1992; Masicampo & Lalande, 2012; Wagenmakers, Wetzels, Borsboom, Van der Maas, & Kievit, 2012).

Factor-Analysis Approach to Reliability

In the previous section, we discussed factor analysis as one of the methods researchers regularly use to identify the items that together constitute a scale for an attribute or a sub-attribute. Once the items defining a scale have been identified, one estimates the test score's reliability, and the methods used for this purpose are from CTT, method $\lambda_3(\alpha)$ (i.e., coefficient α) being the most popular by far (e.g., Oosterwijk et al., 2019; Sijtsma, 2009). However, in factor analysis, methods for estimating reliability exist that are tailored to the measurement model, which relates the item score to one or more latent variables that replace CTT's true score. The resulting reliability methods are technically equal to the classical reliability, expressing the proportion of the test-score variance that is of interest to the researcher, but the underlying models differ between CTT-based and factor-analysis based methods, that is, true score versus factor structure. In this section, we consider factor models for the reliability of test score X.

As we saw, CTT defined a true score, T, that absorbs latent variables and systematic error sources without revealing the structure explaining how these variables and error sources collaborate to cause people to vary systematically on the test score, X. This has led to criticism, for example, that coefficient $\lambda_3(\alpha)$ is not a lower bound to the reliability but can also be positively biased (e.g., Bentler, 2009; Kelley & Cheng, 2012; Raykov, 2001). However, this criticism is only correct when one adopts a model for the true score, such as a factor model, so that results with respect to coefficient $\lambda_3(\alpha)$ and other methods are based on a factor model different from CTT, which only posits true score T without further hypothesizing its structure. Based on the assumptions of CTT, coefficient $\lambda_3(\alpha)$ is a lower bound to reliability $\rho_{XX'}$, but if one replaces these assumptions with the assumptions of factor analysis, the lower bound result is no longer correct.

One-Factor Model

In what follows, we use the terms factor and latent variable interchangeably. Bollen (1989, pp. 218–221) discussed the next model, consistent with Equation (2.97) but with $Q = 1$, so that, dropping index q, we have

$$X_j = b_j + a_j\xi + \delta_j. \tag{2.100}$$

We recall that $\delta_j = U_j + E_j$ in which the unique error U_j usually has unknown variance, $\sigma_{U_j}^2$. Bollen (1989, pp. 220–221) suggested the possibilities to either assume $\sigma_{U_j}^2 = 0$ or ignore $\sigma_{U_j}^2$, two options that are practically indistinguishable and produce underestimates of test-score reliability. Another possibility that circumvents the problem of unknown unique variance is to define test-score reliability as the squared multiple correlation, R^2, resulting from the linear model that includes all the latent variables and perhaps also observable variables that influence the item score, and that further includes error variance but not unique variance. Henceforth, we assume $\sigma_{U_j}^2 = 0$ and in this sense join the practice of factor analysis.

In Equation (2.100), we consider δ_j simply to be equal to the random error component, so that $\delta_j = E_j$, and the one-factor model (OFM) is

$$X_j = b_j + a_j\xi + E_j. \tag{2.101}$$

Relating CTT to Equation (2.101) yields

$$T_j = b_j + a_j\xi. \tag{2.102}$$

In Equation (2.102), the true score depends on item parameters b_j and a_j, and latent variable ξ is a quantity that the different items in the test have in common. This is a principal difference with CTT, where the true score depends on the item, but different items may measure unrelated attributes or unrelated sets of attributes [Equation (2.96)].

Based on Equation (2.102), noticing that $T = \Sigma_j T_j$ and using the classical definition [Equation (2.19)], we define the reliability of test score X as

$$\rho_{XX'}^{OFM} = \frac{\sigma_T^2}{\sigma_X^2} = \frac{\sigma^2 \left[\Sigma_{j=1}^J \left(b_j + a_j \xi \right) \right]}{\sigma^2 \left[\Sigma_{j=1}^J \left(b_j + a_j \xi \right) \right] + \sigma_E^2} = \frac{\sigma_\xi^2 \Sigma_{j=1}^J \Sigma_{k=1}^J a_j a_k}{\sigma_\xi^2 \Sigma_{j=1}^J \Sigma_{k=1}^J a_j a_k + \sigma_E^2} \tag{2.103}$$

(Bollen, 1989, p. 220). In Equation (2.103), we have simply equated the true score with the one-factor definition. In real data, this equality will not hold and the right-hand side of Equation (2.103) will not equal classical reliability, hence the notation $\rho_{XX'}^{OFM}$. Mellenbergh (1994, 1998) studied an approach leading to the definition of the reliability of the estimated factor score rather than the true score estimated by means of the test score [Equation (2.71)]. We briefly outline Mellenbergh's approach.

The OFM in Equation (2.101) is the point of departure. Equation (2.101) is also known as the model for congeneric measures (Jöreskog, 1971), allowing items to vary with respect to item parameters b_j and a_j, and population error variance, thus allowing $\sigma_{E_j}^2 \neq \sigma_{E_k}^2$ (Bollen, 1989, p. 208; in Bollen's definition, b_j is absent). Because variable ξ is unobservable, its origin and unit need to be defined to estimate the item parameters b_j and a_j. An arbitrary but common choice is to define ξ in the population as a standardized variable with mean 0 and variance 1, that is, $\mathcal{E}(\xi) = 0$ and $\sigma_\xi^2 = 1$.

For person i, the OFM assumes that across hypothetical, independent replications of the same measurement procedure, item score X_j has a normal distribution with variance assumed to be equal for all persons; that is, item-score variance, which equals error variance, is equal for all persons: $\sigma_{X_j}^2 = \sigma_{E_j}^2$. Taking the expectation of the item score across persons and for each person across replications, one can derive the following results for the OFM. The mean item score equals

$$\mathcal{E}(X_j) = b_j; \tag{2.104}$$

that is, given a standardized latent variable, ξ, the mean item score equals item parameter b_j. The variance of the item score equals

$$\sigma_{X_j}^2 = \sigma_{T_j}^2 + \sigma_{E_j}^2 = a_j^2 + \sigma_{E_j}^2. \tag{2.105}$$

For two items j and k, for which according to Equation (2.102), $T_j = b_j + a_j \xi$ and $T_k = b_k + a_k \xi$, the covariance equals

$$\sigma(X_j, X_k) = a_j a_k; \tag{2.106}$$

hence, due to the common latent variable, different items from the test are related by the product of their loadings. Higher loadings produce a higher covariance.

Comparable with the test-score reliability from CTT, a reliability coefficient can be defined in which latent-variable variance, σ_ξ^2, replaces true-score variance, σ_T^2, and the variance of the estimated latent-variable score, $\hat{\xi}$, conditional upon latent-variable score ξ, $\sigma_{\hat{\xi}|\xi}^2$, also known as the residual variance, replaces error variance, σ_E^2. Mellenbergh (1994) showed that the conditional variance of the estimated latent-variable score equals

$$\sigma_{\hat{\xi}|\xi}^2 = \left(\sum_{j=1}^{J} \frac{a_j^2}{\sigma_{E_j}^2} \right)^{-1}. \tag{2.107}$$

The reliability of the estimated latent-variable score is defined as

$$\rho_\xi^{OFM} = \frac{\sigma_\xi^2}{\sigma_\xi^2 + \sigma_{\hat{\xi}|\xi}^2}, \tag{2.108}$$

and because we defined $\sigma_\xi^2 = 1$, reliability reduces to

$$\rho_\xi^{OFM} = \left[1 + \sigma_{\hat{\xi}|\xi}^2 \right]^{-1}. \tag{2.109}$$

Mellenbergh (1994, 1998) discussed a simple method for estimating latent-variable scores ξ and variance $\sigma_{\hat{\xi}|\xi}^2$, thus allowing the estimation of reliability ρ_ξ^{OFM}. Equation (2.109) provides values for the reliability that are different from the values that Equation (2.19) provides, because latent variable ξ is a variable shared by the items in the test, whereas the composition of the item true-scores is unspecified. This difference is crucial but sometimes ignored. Readers interested in an adaptation of coefficient $\lambda_3(\alpha)$ to congeneric measures may consult Raykov (1997, 2001).

Multi-Factor Model

We introduced the multi-factor model (MFM) in Equation (2.97) in the context of CFA, and we repeat the first part of it for convenience,

$$X_j = b_j + \sum_{q=1}^{Q} a_{jq} \xi_q + \delta_j. \tag{2.97}$$

If available, a well-founded attribute theory posits the number of latent variables, Q, and specifies which items load on a particular latent variable. Based on the theory, loadings can be fixed, for example, to be equal, $a_{jq}=a_{kq}$, and then estimated freely under this constraint, or some loadings can be fixed at a particular value, for example, $a_{jq}=0$, expressing that a particular attribute represented by latent variable ξ_q is not relevant for item j, and other loadings are estimated freely.

To keep the MFM simple, we assume that each item loads on at least one latent variable and that at least two items load on the same latent variable. These restrictions rule out unique components. Adopting multiple latent variables to some extent compensates for the absence of unique variables (Bollen, 1989, pp. 220–221). Relating CTT to multiple-factor analysis yields

$$T_j = b_j + \sum_{q=1}^{Q} a_{jq}\xi_q. \tag{2.110}$$

The next simplification is assuming intercept $b_j=0$ for all items. We write test score X as

$$X = \sum_{j=1}^{J}X_j = \sum_{j=1}^{J}\sum_{q=1}^{Q}a_{jq}\xi_q + \sum_{j=1}^{J}E_j, \tag{2.111}$$

with

$$T = \sum_{j=1}^{J}\sum_{q=1}^{Q}a_{jq}\xi_q. \tag{2.112}$$

and

$$E = \sum_{j=1}^{J}E_j, \tag{2.113}$$

We use Equation (2.19) to define reliability and write CTT reliability of test score X as

$$\rho_{XX'}^{MFM} = \frac{\sigma_T^2}{\sigma_X^2} = \frac{\sigma^2\left(\sum_{j=1}^{J}\sum_{q=1}^{Q}a_{jq}\xi_q\right)}{\sigma^2\left(\sum_{j=1}^{J}\sum_{q=1}^{Q}a_{jq}\xi_q\right)+\sigma_E^2} \tag{2.114}$$

(Raykov & Shrout, 2002), which is also known as coefficient ω (McDonald, 1999; Revelle & Zinbarg, 2009). As with reliability based on the OFM, Equation (2.114) equates classical reliability with multiple-factor reliability, but this is only correct if the factor model explains the true score completely. In real data, this will not be the case, and the equality fails. Concretely, either ignoring latent variables or including irrelevant latent variables, misspecifying the loading structure, or a combination of these errors, produces a misfit of the model to the data and a biased estimate of reliability $\rho_{XX'}$.

Revelle and Zinbarg (2009) rephrased Equation (2.114) using an item's communality, denoted h_j^2, which is the proportion of variance of item score, X_j, explained by the common latent variable and the group variables, which are factors on which subsets of items load but not all items. Let the total number of relevant latent variables be Q, and assume they are orthogonal, then,

$$h_j^2 = \sum_{q=1}^{Q} a_{jq}^2. \tag{2.115}$$

Complement $d_j^2 = 1 - h_j^2$ is the proportion of unexplained variance, which is used as an approximation to the item's proportion of error variance. For standardized item scores, one uses these proportions to define coefficient ω_t,

$$\omega_t = 1 - \frac{\sum_{j=1}^{J}\left(1 - h_j^2\right)}{\sigma_X^2} = 1 - \frac{\sum_{j=1}^{J} d_j^2}{\sigma_X^2}, \tag{2.116}$$

(McDonald, 1999; Revelle & Zinbarg, 2009).

Except for their different numerators, Equation (2.116) appears like Equation (2.21), but because the models underlying the two equations are different, relating Equation (2.116) to the classical reliability definition in Equation (2.21) might be misleading. A different factor-model based method, denoted coefficient ω_h, is based on only the common latent variable (Revelle & Zinbarg, 2009) and simplifies Equation (2.116) to Equation (2.103) by letting $Q = 1$. Raykov and Shrout (2002) estimated coefficient ω by means of a structural equation modeling approach. In Table 2.1, for the covariance matrix representing a one-factor structure, $\omega_t = .9$, and for the matrix representing a two-factor structure, $\omega_t \approx .6819$.

Real-Data Example: The Type D Scale14 (DS14)

The Type D Scale-14 (DS14; Denollet, 2005) is a diagnostic instrument for the assessment of the Type-D personality. Type D (D for distressed) is defined as the joint tendency towards negative affectivity (e.g., worry, irritability, gloom) and social inhibition (e.g., reticence and a lack of self-assurance). Negative affectivity consists of three sub-traits: Dysphoria, anxious apprehension, and irritability. Social inhibition also consists of three sub-traits: Discomfort in social situations, reticence, and lack of social poise (Figure 2.5). DS14 contains 14 items, each consisting of a statement (Table 2.7) and five ordered response categories (0 = completely disagree, 1 = disagree, 2 = agree

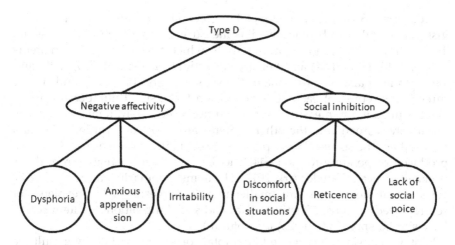

FIGURE 2.5
Theoretical structure of Type D personality.

TABLE 2.7

The Items of the DS14, Sub-Traits and the Mean and Standard Deviation (SD) of the Item Scores

Item	Statement	Sub-trait	Mean	SD
Si1*	I make contact easily when I meet people	Lack of social poise	1.277	1.176
Si3*	I often talk to strangers	Lack of social poise	1.806	1.262
Si6	I often feel inhibited in social interactions	Discomfort in soc. sit.	1.213	1.176
Si8	I find it hard to start a conversation	Discomfort in soc. sit.	1.270	1.227
Si10	I am a closed kind of person	Reticence	1.458	1.331
Si11	I would rather keep other people at a distance	Reticence	1.564	1.141
Si14	When socializing, I don't find the right things to talk about	Discomfort in soc. sit.	1.177	1.133
Na2	I often make a fuss about unimportant things	Anxious apprehension	1.865	1.306
Na4	I often feel unhappy	Dysphoria	0.896	1.107
Na5	I am often irritated	Irritability	1.671	1.236
Na7	I take a gloomy view of things	Dysphoria	0.963	1.184
Na9	I am often in a bad mood	Irritability	0.939	1.058
Na12	I often find myself worrying about something	Anxious apprehension	1.824	1.342
Na13	I am often down in the dumps	Dysphoria	0.871	1.125

Note: *=negatively worded item. Discomfort in soc. sit.=discomfort in social situations. Mean and SD were computed after recoding.

nor disagree, 3 = agree, 4 = completely agree). Depending on the psychologist's goal, DS14 can be conceived as measuring either the unidimensional Type D trait or two unidimensional traits, which are social inhibition (items 1, 3, 6, 8, 10, 11, and 14) and negative affectivity (items 2, 4, 5, 7, 9, 12, and 13). Each item also assesses one of the six sub-traits (Table 2.7), and two or three items represent each sub-trait. Given these small numbers of items, DS14 is unsuited to measure the six sub-traits separately. Items 1 and 3 are negatively worded, and the other 12 items are positively worded. The data we used are the scores $N = 541$ persons (87.4% males, average age 58.7 years) produced responding to the DS14. The data are freely available from the R package mokken (Van der Ark, 2012). The items' popularity varies (Table 2.7), yet all mean item scores are smaller than 2, suggesting low prevalence of self-reported negative affectivity (Na) and social inhibition (Si). Item scores have similar spread (Table 2.7, last column).

First, we considered corrected item-total correlation and oblique multiple group method results. The corrected item-total correlations ranged between .325 and .669 (Table 2.8, first column). Most researchers would take these values as support for all items constituting a single scale. The oblique multiple group method suggests the following. Both the Si and Na items correlate highly with the sum score on the other items in their item group (Table 2.8, second and third column, bold face values), whereas the Si and Na items

TABLE 2.8

Results from Item-Rest Correlation and Oblique Multiple Group Method, and Principal Component Analysis

Item	Item-rest correlation, and Oblique multiple group method			Principal component analysis	
	IRC	Group 1	Group 2	RPC 1	RPC 2
Si1*	.505	**.707**	.169	**.816**	.019
Si3*	.325	**.531**	.040	**.710**	−.121
Si6	.669	**.615**	.471	**.639**	.421
Si8	.624	**.732**	.321	**.789**	.218
Si10	.566	**.689**	.274	**.766**	.159
Si11	.495	**.594**	.239	**.685**	.134
Si14	.566	**.644**	.303	**.720**	.215
Na2	.399	.137	**.551**	−.022	**.673**
Na4	.597	.324	**.675**	.200	**.754**
Na5	.460	.189	**.597**	.031	**.711**
Na7	.646	.360	**.720**	.228	**.786**
Na9	.526	.263	**.621**	.124	**.717**
Na12	.548	.263	**.671**	.120	**.751**
Na13	.620	.306	**.744**	.166	**.809**

Note: IRC = item-rest correlation, RPC = rotated principal component, * = negatively worded item.

correlate lower with the sum scores in the other item groups (Table 2.8, second and third column, normal font). These results suggest a subdivision of the 14 items into two groups.

Second, PCA produced the following results. Based on the scree-plot (Figure 2.6), we selected two principal components and rotated the two-component solution by means of Varimax rotation. The two components explained 58.1% of the variance that the 14 items represented. The rotated loadings (Table 2.8, fourth and fifth column) show a two-dimensional structure. The high loadings of the Si items on the first factor (i.e., a principal component by definition explains the maximum amount of variance, but after rotation this is no longer the case, and it becomes a factor) justify the interpretation of the item setting up a social-inhibition factor. The high loadings of the Na items on the second factor justify the interpretation of a negative-affectivity factor. Item Si6 also had a high cross loading on the "other" factor.

Third, confirmatory factor analyses using the R package lavaan (Rosseel, 2012), and based on the model in Figure 2.5, were used to estimate a one-factor model (Type D personality), a two-factor model (negative affectivity and social inhibition), and a six-factor model (sub-traits). One should interpret the solutions with caution, because we treated the discrete item scores as continuous, and for several items, the histograms of the (highly discrete) item scores did not approximate a normal distribution. The factor loadings for the first item were fixed to 1 to identify the model.

The one-factor model did not fit well. The RMSEA, which is one of the most reported absolute fit indices, equaled .183, whereas as a rule of thumb, values

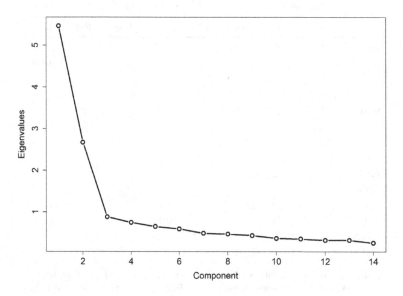

FIGURE 2.6
The eigenvalues of the 14 items of the DS14.

smaller than .10 are associated with a good fit. The TLI, which is another frequently reported comparative fit index that compares the model under consideration with the null model of uncorrelated items, equaled .535, whereas values greater than .90 or .95 are associated with good fit. The fit of the two-factor model with correlated factors was better (RMSEA = .094, TLI = .878). Table 2.9 (first and second column) shows the factor loadings. The estimated correlation between the factors of social inhibition and negative affiliation was $r = .434$. The six-factor (sub-trait) model fitted best (RMSEA = .062; TLI = .946). The sub-traits that fall under negative affectivity are highly correlated, and so are the sub-traits that fall under social inhibition. The correlations between sub-traits that fall under different traits are lower (Table 2.10). Both the fit of the six-factor model and the correlation structure of the sub-traits support the validity of the theory for Type D personality as the DS14 operationalizes it.

All analyses suggested that it is reasonable to use the negative affectivity and social inhibition scales of DS14. For both scales, Table 2.11 shows the reliability estimates discussed in this chapter. For the sake of completeness, Table 2.11 also shows the reliability estimates for the entire DS14 scale. Because the sample was rather small, method λ_4 and the GLB are likely to overestimate the reliability, so that method λ_2 is a safe alternative here.

TABLE 2.9

Factor Loadings (note, not correlations) of the Two-Factor Model and the Six-Factor Model

	Two-factor model		Six-factor model					
	Si	Na	Dys	AA	Irr	DSS	Ret	LSP
Si1*	1*							1*
Si3*	0.826							0.834
Si6	0.978					1*		
Si8	1.169					1.160		
Si10	1.159						1*	
Si11	0.851						0.734	
Si14	0.962					0.963		
Na2		1*		1*				
Na4		1.241	1*					
Na5		1.033			1*			
Na7		1.382	1.096					
Na9		0.978			0.916			
Na12		1.353		1.272				
Na13		1.360	1.096					

Note: Si = Social inhibition, Na = Negative affectivity, Dys = Dysphoria, AA = Anxious apprehension, Irr = Irritability, DSS = Discomfort in social situations, Ret = Reticence, LSP = Lack of social poise. Empty cells were fixed to 0. 1* = value fixed to 1.

TABLE 2.10

Correlation Matrix of the Six Sub-Traits

	Dys	AA	Irr	DSS	Ret	LSP
Dys	1					
AA	.813	1				
Irr	.777	.775	1			
DSS	.517	.398	.437	1		
Ret	.371	.332	.344	.894	1	
LSP	.225	.141	.109	.733	.816	1

Note: Dys = Dysphonia, AA = Anxious apprehension, Irr = Irritability, DSS = Discomfort in social situations, Ret = Reticence, LSP = Lack of social poise.

TABLE 2.11

Reliability Estimates for the Negative Affectivity Scale, the Social Inhibition Scale, and the Entire DS14

Scale				Method				
	λ_1	λ_2	$\lambda_3(\alpha)$	λ_4	λ_5	λ_6	GLB	ω_t
Na	.748	.876	.873	.894	.848	.872	.911	.914
Si	.745	.871	.869	.901	.845	.869	.904	.911
DS14	.812	.883	.874	.933	.853	.905	.937	.914

Note: Na = Negative affectivity, Si = Social inhibition.

Discussion

As promised in Chapter 1, we have focused on the conceptual aspects of CTT, its main characteristics, possibilities and limitations, and the construction of scales. We noticed that CTT is an error model. Hence, central to CTT are the concepts of reliability and measurement precision. By exclusively concentrating on error, CTT does not have any interest in the composition of the measurement. This has caused quite some confusion among psychometricians and researchers using CTT, probably because it is difficult to grasp that CTT is at peace with the idea that the measurement value can represent anything. CTT's central question is, to what degree do we find the same measurement values, irrespective of what they signify, if the administration procedure is repeated? Because of CTT's focus, we have discussed methods to estimate reliability and measurement precision. The selection of items for constructing a scale formally is not part of CTT, because CTT does not say anything about the composition of the true score, but researchers have found their way using mainly PCA and factor analysis for this purpose. This is the reason why we also gave some attention to these methods.

The literature on CTT and reliability is immense. It was immense some 20 years ago (e.g., Li, Rosenthal, & Rubin, 1996), and it continues to grow at an enormous pace. We believe the magnitude of the literature reflects several things. One is the importance that researchers constructing tests and questionnaires assign to the reliability concept for their measurement instruments. Another is that apparently, the knowledge about reliability is incomplete and psychometricians see cause to continue investigating properties of existing methods and proposing new methods. However, there is also reason for concern, given that the expanding literature shows persistent misunderstanding of the reliability concept. We mention three such problems that we have tried to tackle in this chapter.

First, many researchers estimate and report reliability without being fully aware of what it represents. Only relatively few articles make a sharp distinction between reliability as a group property and measurement precision as a property of an individual measurement value. A test having a reliability of .90 can imprecisely measure John's test performance, and a test with a low reliability can still measure some people with precision. Perhaps most importantly, knowing that one's test has sufficient reliability is not enough for the responsible use of the test for measuring individuals. Few people seem to understand or even know the distinction between reliability and measurement precision, and clearly, this nuance must be taught well and with great persistence.

Second, Cronbach (1951) very successfully propagated the use of coefficient $\lambda_3(\alpha)$, which at the time was a highly fortunate course of action given the state in which reliability estimation found itself. However, this made test constructors and researchers blind to the alternatives that already were available (e.g., Guttman, 1945), albeit not in a mathematically easily accessible form and without the availability of user-friendly software. However, in the next decades, little changed with respect to the preference for particular reliability methods and the neglect of others, and this is still true today. Coefficient $\lambda_3(\alpha)$ remains the most popular method for reliability estimation in the face of better methods for which user-friendly software is nowadays available. Market leaders are always difficult to beat, but unfamiliarity with the abundance of available methods leaves researchers with what they know, which is coefficient $\lambda_3(\alpha)$, and prevents the optimal use of reliability methods.

Third, given the CTT model [Equation (2.4)] and its assumptions, coefficient $\lambda_3(\alpha)$ is a lower bound to the reliability [Equation (2.32)]. However, if one adopts a factor-analysis perspective on reliability estimation, the model changes [e.g., Equation (2.103)], and coefficient $\lambda_3(\alpha)$ is not a lower bound to the reliability. This is true, because one has adopted another model for the test score, but it should come as no surprise that theorems cease to be true if the assumptions on which they were based are replaced with other assumptions. We hope that our discussion of both perspectives on modeling test scores, the classical perspective, which prefers to consider measurement as error-ridden but not to model the true score, and the factor-analysis perspective, which adopts the error approach and models the true score, made clear that both approaches have their merits but produce different results (e.g., Revelle & Condon, 2019). This difference means that the approaches use different models, not that one or the other is wrong.

In the next chapter on nonparametric IRT, we discuss a reliability approach tailored to nonparametric IRT (Mokken, 1971, pp. 142–147). In the chapter on parametric IRT, we discuss measurement precision using the latent variable. The CTT and CTT-related literature discusses many more methods, but these methods have not risen beyond their theoretical treatise and have largely remained unnoticed by test constructors. This does not rule out that they might be useful, and therefore we mention a few. Jackson and Agunwamba (1977) proposed the methods λ_7, λ_8, and λ_9 as possible improvements of methods λ_1 through λ_6. Verhelst (1998) studied these methods theoretically. Schulman and Haden (1975) proposed reliability methods for ordinal scores. Van der Ark, Van der Palm, and Sijtsma (2011; also, see Van der Palm, Van der Ark, & Sijtsma, 2014) proposed using the latent class model to approximate parallel item versions as closely as possible. Kristof (1974) and Bartholomew and Schuessler (1991) proposed additional approaches to estimating reliability. A more influential approach is generalizability theory proposed by Cronbach, Gleser, Nanda, and Rajaratnam (1972) and discussed by, for example, Shavelson and Webb (1991) and Brennan (2001). Generalizability theory estimates reliability after having corrected for error sources that affect relative or absolute interpretations of persons' test scores. Sijtsma and Van der Ark (2015) compared CTT, factor analysis, and generalizability theory approaches to reliability.

3

Nonparametric Item Response Theory and Mokken Scale Analysis

In many domains of psychological measurement, the existence of a rank ordering among examinees on [the latent variable; *the authors*] is plausible even when no clear unit of measurement is established.

—**Roger E. Millsap**

Introduction

In this chapter, we introduce and discuss nonparametric IRT models, especially the models Mokken (1971, 1997) proposed. Mokken's models can be considered foundational, and they are on par with other approaches to nonparametric IRT (e.g., Douglas, 1997, 2001; Junker, 1991, 1993; Ramsay, 1991; Stout, 1990, 2002). These approaches include research investigating the assumptions that are the least restrictive but still allow measurement at an ordinal level (e.g., Ellis & Junker, 1997; Holland & Rosenbaum, 1986; Junker & Ellis, 1997; Rosenbaum, 1984, 1987a, b). In this monograph, the discussion of nonparametric IRT precedes the discussion of the wealth of parametric IRT models in a general hierarchical framework (Chapter 4), because we think that first studying nonparametric IRT facilitates a better understanding of parametric IRT models. Our point of view may not really diverge from what several other authors think, but one will not find the notion of first discussing nonparametric IRT in other monographs. Edited volumes pay attention to nonparametric IRT as one of the approaches within the broader family of IRT models (e.g., Boomsma, Van Duijn, & Snijders, 2001; Rao & Sinharay, 2007; Van der Linden, 2016a).

In statistics, the terminology of *nonparametric* is different from its use in IRT and often refers to exact distributions that directly describe features of the data, as in the Mann–Whitney U-test for the null hypothesis that two independent samples originate from the same population. One sometimes calls nonparametric tests distribution-free, which means that one does not make assumptions about the form of the distribution of the variables of interest.

Another name is ranking tests, which remind one that the values variables attain are ranks rather than numerical scores that have meaningful intervals (Siegel, 1956, p. vii). The terminology of *parametric* refers to particular distributions that are described by means of a limited number of parameters, as in Student's *t*-test for the null hypothesis that a mean is equal to a particular value or that the means of two groups are equal. The *t*-test assumes that the variable of interest has a normal distribution with parameters μ for the mean and σ² for the variance.

The delineation between nonparametric IRT and parametric IRT does not run along the distinction with respect to the distribution of variables but focuses on the relation between the probability of obtaining a particular score on a particular item and one or more latent variables that represent the attribute the item assesses. The item response function (IRF) describes this relation. Parametric IRT models use parametric functions to describe the IRF. Much used parametric IRFs are logistic functions and normal ogive functions. Only a few parameters define these functions, sometimes only one, as in the one-parameter logistic model or Rasch model (Chapter 4), where one item parameter locates a particular IRF on the scale of measurement and IRFs of different items are differently located but run parallel. Figure 3.1 (dashed curves) shows two logistic IRFs from the Rasch model, only differing with respect to their location parameters, denoted δ_j $(j = 1, ..., J)$ and having values $\delta_j = -0.5$ (left-hand IRF) and $\delta_k = 1$. Researchers estimate the item parameters but rarely assess directly whether the empirical shape of the IRF is consistent with, say, a logistic curve; however, see Molenaar (1983)

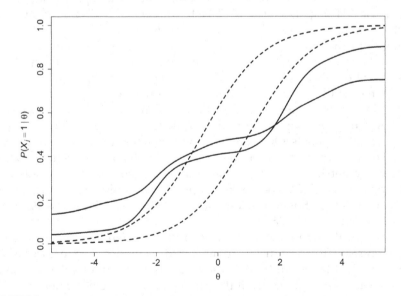

FIGURE 3.1

IRFs for two dichotomous Rasch items (dashed curves) and two dichotomous items that satisfy the monotone homogeneity model (solid curves).

and Glas and Verhelst (1995a), who proposed goodness-of-fit tests that assess the shape of the observed IRFs. Nonparametric IRT models refrain from using a parametric function but instead impose order restrictions on the IRF. For example, one may assume that the IRF is a monotone non-decreasing function of the latent variable, but given this restriction, it can have any shape (Figure 3.1, solid curves), or the mean of a set of J IRFs is monotone whereas individual IRFs at the lower aggregation level can have any shape and need not be monotone. To investigate whether the IRFs are monotone, one estimates several of their function values from the data and checks whether their progression is monotone. Here, nonparametric does not mean that one does not estimate parameters; on the contrary, one estimates function values of IRFs and item parameters differently from those estimated from logistic or other functions.

Because Mokken's approach is central to this chapter, we discuss some of his ideas prior to discussing the nonparametric IRT model assumptions. Mokken (1971, Chap. 3) discussed parametric IRT models, which he termed latent structure models, before introducing his nonparametric IRT models and noticed (ibid., p. 115) that the use of parametric IRFs requires profound knowledge of the items and the examinees. To obtain such profound knowledge, he posited that one has to set up a continuous research program that enables the accumulation of knowledge relevant to the measurement of the attribute of interest. To do this well, one has to have access to data from large numbers of items and examinees. Mokken (1971, p. 116), a political scientist, was somewhat pessimistic about the possibilities for measurement in the social sciences of which political science is part. There, measurement often concerned attitudes directed at a specific object, such as religion, abortion, capital punishment, or elections, implying research was small-scale. The specificity of the attitudes entailed use of smaller numbers of items and smaller samples than was common in behavioral and educational measurement, which were directed at broad abilities and achievement. In Mokken's view, the latter attributes justified extensive research, better facilitating knowledge accumulation and thus the use of parametric models. In addition, Mokken argued that respondent fatigue and limited questionnaire length dictated by the circumstances of the research were prohibitive for the numbers of items used with the same persons.

According to Mokken, all of this put social-science measurement research and the possibilities for practical measurement at a disadvantage compared to behavioral-science measurement research and called for different measurement models that reflected the less advanced state of social-science attribute research. Thus, in his view, the use of IRT models based on order-restricted IRFs was consistent with the lower knowledge level in many social-science areas and with the practical restrictions on measurement conditions. He believed parametric IRT models were consistent with higher knowledge levels, and in fact, the parametric IRT models were mathematical special cases of the more general and less restrictive nonparametric IRT models, thus reflecting a developmental progression of the knowledge about the attributes.

Being active in the social sciences, Mokken undoubtedly had a sharp eye for the scale differences between social science and behavioral science measurement, the former using smaller-scale, narrower-targeted attribute measurement using few items, and the latter using larger-scale, broader-targeted measurement using many items. However, his interpretation of the consequences of the scale differences of the research areas for the use of particular measurement models may have been too optimistic with respect to the behavioral sciences. The fact that psychological and, in particular, educational measurement has access to more data—longer tests and larger samples—may be true but merely reflects the relative ease with which it is possible to construct large item banks for abilities, skills, and knowledge domains, and test large student groups due to the societal need for large-scale educational assessment and selection. The result of this intensive testing, as far as we know, has not been that the development of attribute theories with respect to abilities, skills, and knowledge domains has taken off to a level that justifies the use of highly restrictive measurement models. However, educational testing has three typical characteristics that favor the use of parametric IRT and which, we emphasize, have little to do with development of attribute theory. That is, educational measurement is large scale, involving large item banks and many examinees; it is repeated periodically, for example, regularly monitoring the same students' arithmetic and language performance and annually testing different students for high school diplomas and university admission; and it often is high stakes, having important consequences for the individual and the education organization. The common measurement feature of these three characteristics is that for the same testing problem one has to use different sets of items to avoid students recognizing the same items the next time they are tested or to avoid regularly used items leaking to students who have yet to be tested. Because the scales based on the different sets of items assessing the same attribute differ, one has to link the parameters for the different sets of items so that measurement values on different scales for the same knowledge domain or attribute become comparable (Van der Linden & Barrett, 2016). Parametric IRT models are suited for linking, and this is where they prove particularly useful. Tailored testing or computerized adaptive testing, in which each examinee receives a set of items adapted to her latent-variable level, is a fine example of specialized linking of items from a large item bank (Van der Linden & Glas, 2010).

The other limitations of social science research on the scale of the research Mokken mentioned were respondent fatigue and questionnaire length, but we believe these limitations to be present also beyond the social sciences where they exercise the same effect on research possibilities. Examples are clinical psychology and work and organizational psychology, where tests are often short, with sometimes only a few items, so as not to strain patients and applicants by presenting them with long tests (Kruyen et al., 2013).

Marketing research uses short questionnaires, with as few as one item, for ad and brand assessment (Bergqvist & Rossiter, 2007), and medical and health research uses short inventories for anxiety and depression measurement, often enforced by patients' physical condition and restrictions imposed by medical-ethical committees guarding the medical and ethical justifiability of scientific research on human beings. We mention these circumstances to argue that the social and behavioral (and other human) sciences may be more similar than Mokken suspected and that the state of the attribute theories is the main determinant of the state of measurement, in all areas.

Outside of the context of educational measurement, whenever a test or questionnaire is administered only once to the same group of people and there is no need for multiple test versions or tailored testing, parametric IRT models are used regularly in all research areas in which nonparametric IRT models are also used. In this standard situation—one test and one test administration—unless a well-founded attribute theory suggests the use of a parametric model, the parametric model fits the data well, and the researcher prefers a logit scale or a similar scale, using a nonparametric IRT model is a viable alternative. The reason is that most nonparametric IRT models are less restrictive than parametric IRT models and thus tend to admit more items, each assessing interesting aspects of the attribute and together producing more reliable sum scores. A more demanding model may not accept all of these items, not because they do not add to attribute representation and reliable measurement, but because they are inconsistent with the measurement model. While each measurement model, nonparametric and parametric, serves as a gatekeeper for admitting items based on the model's requirements for admission, we think one should choose the model that admits the larger number of items to set up the scale. In Chapter 4, we will see that parametric IRT allows possibilities not facilitated by nonparametric IRT, thus allowing applications unavailable when one uses nonparametric models.

Model of Monotone Homogeneity

Prerequisites

Definitions and Notation

A test or questionnaire may contain items using a different scoring format, so that item scores are $X_j = 0, ..., M_j$. Although one regularly encounters this item score variability in practice, in the context of nonparametric IRT, we assume the same item-score format in one test or questionnaire. Hence, throughout this chapter we assume $X_j = 0, ..., M$, for $j = 1, ..., J$. The reason for this recommendation is that for items with the same score format nonparametric IRT implies an ordinal scale for the sum score X on the J items in the test. In particular, Grayson (1988) showed that sum score X based on J dichotomously scored items implies an ordinal scale. For polytomous item scores, sum score X only orders people correctly by approximation and a different but less practical ordering property is valid. Our choice for sum score X based on items having the same score format aligns well with the practice of psychological and other measurement, where researchers and practitioners alike usually take the sum score based on items having the same score format as the preferred measurement value. This preference sits well with the modest theoretical status of many attributes that asks for simplicity rather than complexity that requires a theoretical basis for differentially scoring items that usually is absent

Assumptions

The model for monotone homogeneity allows the ordering of persons based on the sum score X (Mokken, 1971; Mokken & Lewis, 1982; Sijtsma & Molenaar, 2002, 2016; Van Schuur, 2003, 2011). Hence, it is a model for ordinal person measurement. The model's three assumptions are the following.

    1—*Unidimensionality*. The items in the instrument measure one attribute. This assumption is quantified by means of a latent variable, denoted θ. One may notice that, mathematically, θ is the variable the J items have in common, but from a substantive perspective, θ may represent either a substantively simple attribute or a complex attribute, or a possibility in between these extremes. For example, in one test, θ may represent arithmetic ability defined as the ability to add integers, but in another test, θ may represent a mixture of arithmetic ability and language skills, as in constructed-response exercises, as long as this mixture presents itself mathematically such that one latent variable suffices to explain the structure of the data. Obviously, in practice the mixture will vary easily across items depending on the precise characteristics of the items and the way people react to such stimulus characteristics. The result is a non-fitting model, because one latent variable θ cannot explain the relation between the items. The reader may notice the similarity with factor analysis models (Chapter 2).

    2—*Local Independence*. We assume that items measure only one attribute represented by latent variable θ—unidimensionality—and that all other influences on item performance are unique or random. When people vary with respect to θ, this variation causes items to covary positively. If one conditions on θ, the only source of variation that causes items to covary vanishes, and the covariance is 0; that is,

$$\sigma\left(X_j, X_k \mid \theta\right) = 0, \quad \text{all } j, k; j \neq k. \tag{3.1}$$

Equation (3.1) is known as weak local independence, because it only refers to the second moment of the joint distribution of the J item scores in the test (Stout, 1990). If θ is the only cause of covariance among J items, conditioning on θ renders the J items statistically independent, which is a stronger form of independence than conditional covariances equal to 0. In psychometrics, statistical independence conditional on the latent variable is called local independence and defined as

$$P\left(X_1 = x_1,\ldots,X_J = x_J \mid \theta\right) = \prod_{j=1}^{J} P\left(X_j = x_j \mid \theta\right), \qquad (3.2)$$

so that the joint distribution of the J item scores equals the product of the J marginal distributions. If at least two items have other attributes in common causing variation on the latent variables representing these attributes, Equation (3.2) fails to hold. Another way of saying this is as follows.

Letting vector $\mathbf{X}_{(j)}$ contain $J-1$ item-score variables except X_j and let vector $\mathbf{x}_{(j)}$ contain realizations of the J item scores, then absence of latent variables additional to the common latent variable θ can be expressed as (Sijtsma & Junker, 2006),

$$P\left(X_j = x \mid \theta; \mathbf{X}_{(j)} = \mathbf{x}_{(j)}\right) = P(X_j = x \mid \theta), \quad \text{for } j = 1,\ldots,J. \qquad (3.3)$$

Equation (3.3) shows that once one knows latent variable θ, the item scores on the other $J-1$ items do not contain information for predicting the score on item j. If during the test session, the person would learn from trying to solve the previous problems before she arrives at problem j, this would mean that she would develop novel skills or improve existing ones, and the previous items would contain information relevant for her score on item j. However, local independence treats the attempt to solve or answer an item as an experiment, independent of attempts to solve or answer previous and later items in the test. Clearly, the presence of explicit cues in items for the solution of other items is inadmissible. Another way to look at local independence in Equation (3.3) is that latent variable θ is a sufficient summary of the item scores.

When test performance and the resulting data are multidimensional, one latent variable is insufficient and one may add latent variables until one has a vector $\boldsymbol{\theta}$ for which

$$P\left(X_1 = x_1,\ldots,X_J = x_J \mid \boldsymbol{\theta}\right) = \prod_{j=1}^{J} P\left(X_j = x_j \mid \boldsymbol{\theta}\right). \qquad (3.4)$$

Adding latent variables is easier said than done, and more involved practical procedures to mimic this idea are, for example, proposed by Mokken (1971, Chap. 5) and Stout et al. (1996). Mokken's search procedure discussed in this chapter selects items in clusters, the items selected in the same cluster purportedly measuring the same attribute, while different clusters represent different attributes or different aspects of a more complex attribute. Items are selected in a cluster one by one, based on the strength of their relation with the sum score of the items already selected. Clusters are also selected one by one, with the aim to select as many items as possible in a cluster, and fewer items in the next cluster. Some items may be left unselected, because their relation with the latent variables is too weak or because they measure unique latent variables. Stout's search procedure also discussed in this chapter finds the clustering of the items that minimizes a function of the conditional covariances in Equation (3.1) to approximate weak local independence as well as possible.

It is tempting to assume that if response probabilities on J-item sets (or individual items) depend on subgroup membership, then local independence fails and we have a case of differential item functioning (Holland & Wainer, 1993) on our hands. Ellis and Van den Wollenberg (1993) showed that this is incorrect. Let a subgroup variable be denoted G, with values $g = 1, \ldots, G$ indicating membership to a particular subgroup from the population of interest, such as defined by gender, ethnic identity, or education level; then, local independence does *not* imply that

$$P\left(X_j = x \mid \theta; G = g\right) = P\left(X_j = x \mid \theta\right), \text{ for } j = 1, \ldots, J \text{ and } g = 1, \ldots, G. \quad (3.5)$$

For Equation (3.5) to hold, in addition to local independence, one has to assume local homogeneity, meaning that people with the same θ value have the same joint response probability on the J items in the test, also known as the stochastic subject probability (Holland, 1990). Let person i belong to a set of people denoted S, all of whom have latent-variable value θ_s; then local homogeneity equals

$$P\left(X_1 = x_1, \ldots, X_J = x_J \mid i\right) = P\left(X_1 = x_1, \ldots, X_J = x_J \mid \theta_S\right). \quad (3.6)$$

Lord and Novick (1968, pp. 30, 47) also used this concept when they discussed the propensity distribution and random measurement error (ibid., pp. 37–38; also, see Chap. 2). Ellis and Van den Wollenberg (1993) argued one must assume local homogeneity to render a measurement model, such as the monotone homogeneity model and parametric IRT models alike, suited for the measurement of individuals. Without local homogeneity, measurement models usually implicitly assume the random sampling concept of response probability, meaning that a conditional response probability represents the proportion of the subgroup of people having a particular θ value who obtained a particular item score. Empirically testing models for local homogeneity is practically awkward. This shows a difficult if not unsolvable problem in social, behavioral, and health measurement; also, see Chapter 2, where we addressed a population of people each tested once and assumed group results were useful for individuals without proof. This is common practice in the measurement of individuals. Because of the problems involved in the investigation whether test data are consistent with local homogeneity, we adopt the practice of assuming results implied by measurement models hold for individuals, even if this practice is formally incorrect.

Finally, in Chapter 5, we discuss latent class analysis (e.g., Hagenaars & McCutcheon, 2002), which assumes that the latent variable is discrete with values $w = 1, \ldots, W$, so that $\theta = w$. Then, local independence implies

$$P(X_1 = x_1, \ldots, X_J = x_J \mid \theta = w) = \prod_{j=1}^{J} P(X_j = x_j \mid \theta = w). \qquad (3.7)$$

Given a solution for W that produces Equation (3.7), one obtains class probabilities, $P(\theta = w)$, and response probabilities, $P(X_j = x \mid \theta = w)$, that are interpreted in an effort to characterize the latent classes. Given the correct latent class structure, Equation (3.7) holds, but without this structure, local independence fails.

The psychometric literature reports competing views on the relation between unidimensionality and local independence. On the one hand, several authors (e.g., Gustafsson, 1980; Henning, 1989; Ip, 2010) argued that unidimensionality implies local independence. Hence, violations of local independence cannot happen if unidimensionality holds, and thus, when they happen, violations indicate that unidimensionality does not hold. On the other hand, unidimensionality and local independence are sometimes considered unrelated, so that unidimensionality may hold when local independence does not (e.g., Goldstein, 1980; McDonald, 1981). The difference between the two perspectives hinges on one's readiness to explain the association between items that remains unexplained by θ by introducing additional latent variables. The first perspective considers the additional latent variables genuine, so that the data are multidimensional. The second perspective considers the additional latent variables just a trick to rewrite local dependencies, implying that these variables are not genuine latent variables serving the goal of simplifying the relations between the items in a substantively intelligible way; hence, unidimensionality does not imply local independence. Because unidimensionality and local independence usually are discussed as separate assumptions of IRT models, and because the assessment of either assumption in real data proves to be rather informative of different aspects of the fit of a model to the data, we let the theoretical discussion rest. In real-data analysis, we assess the degree to which each assumption is consistent with the data.

3—*Monotonicity.* A person located higher on the θ scale than another person is more likely to obtain a higher score on any of the items. This assumption is quantified by assuming response functions that are monotone nondecreasing. Mathematically, monotonicity implies that, as θ increases, the probability of scoring at least x on item j cannot decrease, only increases or remains constant. Substantively, monotonicity reflects that a person possessing more of the attribute has a greater probability of obtaining scores typical of the higher attribute level. Mathematically, M item-step response functions (ISRFs) describe the relationship between the probability of obtaining at least score x and latent variable θ, denoted $P(X_j \geq x \mid \theta)$, for $x = 1, \ldots, M$. Monotonicity means

$$P(X_j \geq x \mid \theta) \text{ monotone non-decreasing in } \theta, \text{ for}$$

$$j = 1, \ldots, J, \text{ and } x = 1, \ldots, M. \tag{3.8}$$

For $x = 0$, the response probability equals $P(X_j \geq 0 \mid \theta) = 1$, because it says simply that one has given a response, not which one, and obtained a score. Hence, this probability is uninformative about the relationship between the item score and the latent variable. For $M = 1$, the item is characterized by one ISRF, $P(X_j \geq 1 \mid \theta)$, which is then called the IRF, already alluded to in the introduction and here defined as $P(X_j \geq 1 \mid \theta) = P(X_j = 1 \mid \theta)$. Response probability $P(X_j = 0 \mid \theta) = 1 - P(X_j = 1 \mid \theta)$ and is superfluous once $P(X_j = 1 \mid \theta)$ is known. Figure 3.2 shows ISRFs for two items j and k, both having five ordered response categories, so that $M = 4$, meaning four ISRFs each (solid and dashed curves). It may be noted that

$$P(X_j = x \mid \theta) = P(X_j \geq x \mid \theta) - P(X_j \geq x+1 \mid \theta). \tag{3.9}$$

Nonparametric IRT models restrict $P(X_j \geq x \mid \theta)$ monotonically and, depending on the two ISRFs on the right-hand side of Equation (3.9), response probability $P(X_j = x \mid \theta)$ can have different shapes, often approximately bell-shaped but not necessarily so.

Mellenbergh (1995) and Hemker, Van der Ark, and Sijtsma (2001) defined three different response probabilities, also known in parametric IRT. In Chapter 4, we discuss these classes extensively for parametric IRT models. Response probability, $P(X_j \geq x \mid \theta)$, is one of the three different response probabilities, and together with the assumptions of unidimensionality and local independence, this response probability defines the nonparametric graded response model (Hemker, Sijtsma, Molenaar, & Junker, 1996). Like $P(X_j \geq x \mid \theta)$, the other two response probabilities are assumed to be monotone non-decreasing functions of θ. One response probability is

$$P(X_j = x \mid \theta; X_j = x-1 \lor X_j = x), \quad x = 1, \ldots, M. \tag{3.10}$$

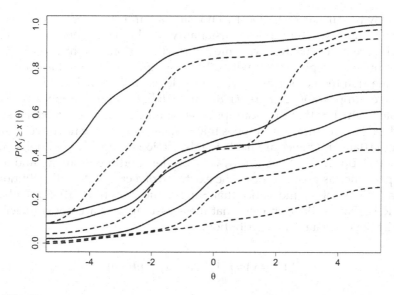

FIGURE 3.2
ISRFs for two items j (solid curves) and k (dashed curves). Each item has five ordered response categories, hence four ISRFs.

From this probability, one can derive response probability, $P(X_j = x \mid \theta)$. Together with the assumptions of unidimensionality and local independence, Equation (3.10) defines the nonparametric partial credit model (Hemker et al., 1996). The other response probability is

$$P(X_j \geq x \mid \theta; X_j = x - 1), \quad x = 1, \ldots, M. \qquad (3.11)$$

Together with the assumptions of unidimensionality and local independence, Equation (3.11) defines the nonparametric sequential model (Hemker, 1996, Chap. 6; also, see Van der Ark, Hemker, & Sijtsma, 2002). Only the nonparametric graded response model, henceforth called the model of monotone homogeneity for polytomous items, is of interest in this chapter, because it is the only one of the three nonparametric models for polytomous items developed well enough to be discussed here. All three models play an important role in Chapter 4, when we discuss a hierarchical taxonomy of polytomous IRT models for polytomous items.

Strictly and Essentially Unidimensional IRT

Our discussion of the three assumptions of unidimensionality, local independence, and monotonicity revealed that the assumptions are special cases of multidimensional model formulations. The three assumptions together define a strictly unidimensional IRT model (Junker, 1993; Sijtsma & Junker, 1996) that may be further restricted by choosing a parametric ISRF. An interesting question is whether unidimensionality, local independence, and monotonicity are necessary to have a model or whether one might weaken these assumptions or replace them with other assumptions and still have a model. The question then arises as to what a model is, and we will follow the perspective Suppes and Zanotti (1981) and Holland and Rosenbaum (1986) adopted, which is that the assumptions that together define a model restrict the data. This requirement implies that models may or may not be consistent with the data and hence are liable to falsification (Popper, 1935; English edition, 2002, pp. 17–20). Specifically, we consider unidimensionality and local independence as given, assume the existence of a cumulative distribution G for latent variable θ, and notice that Suppes and Zanotti (1981) and Holland and Rosenbaum (1986) showed that these assumptions are not sufficient to restrict the J-variate data distribution,

$$P(\mathbf{X} = \mathbf{x}) = \int \prod_{j=1}^{J} P(X_j = x_j \mid \theta) dG(\theta). \tag{3.12}$$

One may notice that inside the integral we use local independence. We take the product of response probabilities that depend on one latent variable and then integrate across $G(\theta)$ to obtain the distribution of the data. It has been shown that one can always write the product inside the integral such that the distribution of the data is perfectly reconstructed. One needs restrictions on the response probabilities, such as monotonicity (Chapter 3) or parametric functions (Chapter 4), to make it impossible to accomplish this reconstruction irrespective of the data. Another possibility is to restrict the distribution of the latent variable. The latent class model (Chapter 5) seeks the minimum number of discrete values for θ and the corresponding probabilities, hence the smallest discrete distribution that allows the reconstruction of the distribution of the data. This produces a subgrouping of the distribution of θ and a unique pattern of response probabilities in each subgroup. Because within subgroups local independence holds, one could argue that we are back with Equation (3.12), but the difference between subgroups with respect to response patterns may be informative of what the test measures, as we will see in Chapter 5.

For dichotomous items, Stout (1987, 1990) studied essentially unidimensional IRT models, and Junker (1991) generalized Stout's results to polytomous items. As the name suggests, essentially unidimensional IRT models weaken the assumptions of the monotone homogeneity model. First, essential unidimensionality replaces strict unidimensionality. Typical of essential unidimensionality is the assumption of a dominant latent variable θ, that the items have in common, and nuisance latent variables, $\theta_1, \ldots, \theta_Q$, that several but not all items may assess, and all latent variable are collected in $\boldsymbol{\theta} = (\theta, \theta_1, \ldots, \theta_Q)$. Consequently, when conditioning on only the dominant latent variable θ, local independence [Equation (3.2)] does not hold for all J items simultaneously. The idea of essential unidimensionality is that in the long run, when $J \rightarrow \infty$, the influence that the dominant latent variable exercises on the response process overshadows the influence that the nuisance latent variables have so that they lose their influence. Essential unidimensionality is reasonable, because real-data analysis shows that strict unidimensionality is unrealistic in psychological and other social, behavioral, and health science measurement, where language skills and other unwanted influences affect the measurement procedure. Essential independence replaces local independence and reflects the dominance of θ over the nuisance latent variables, and Stout (1990) defined essential independence as

$$\binom{J}{2}^{-1} \sum_{j<k} \sum \left| \sigma\left(X_j, X_k \mid \theta\right) \right| \rightarrow 0 \quad \text{if } J \rightarrow \infty. \tag{3.13}$$

Further, Stout (1987) relaxed monotonicity of individual IRFs [Equation (3.8)] to weak monotonicity. Weak monotonicity assumes that the mean of the J IRFs, also known as the test characteristic function (Lord, 1980, p. 49), is an increasing function of θ. Writing shorthand $P_j(\theta) = P\left(X_j = 1 \mid \theta\right)$, for two values of vector θ, say, θ_a and θ_b, such that $\theta_a \leq \theta_b$ in each coordinate of θ, weak monotonicity means

$$J^{-1} \sum_{j=1}^{J} P_j\left(\theta_a\right) \leq J^{-1} \sum_{j=1}^{J} P_j\left(\theta_b\right), \text{ for all } \theta_a \leq \theta_b, \text{ coordinatewise.} \tag{3.14}$$

Weak monotonicity is reasonable, because real-data analysis estimating IRFs repeatedly shows that monotonicity at the item level often is violated at least locally and often without the researcher being able to give a solid explanation. Together, Equations (3.13) and (3.14) define the essential unidimensional IRT model. This model is useful because Stout (1990) showed that, given Equations (3.13) and (3.14), sum score X is a consistent estimator of a transformation of the dominant, unidimensional latent variable, as $J \rightarrow \infty$, and Junker (1991) generalized this result to polytomous items.

It is not our intention to lead the reader further into the crypts of latent variable measurement models but to discuss the basic measurement models and their use in constructing scales. Excellent foundational work is provided by the authors already mentioned in the introduction to this chapter and further by, for example, Ellis and Van den Wollenberg (1993), Scheiblechner (1995), Junker (1998), Zhang and Stout (1999a, b), Ünlü (2007), Ellis (2014), and Zwitser and Maris (2016). The purpose we have here is to clarify the unity of latent variable measurement models. The models may best be considered variations on one basic theme, which is their potential to reveal the sensitivity with which the measurement instrument enables recording the attributes' dimensionality and is able to identify disturbing influences and differentiate group structure. The choice of the model depends on the attribute structure and the measurement properties one expects, for example, a subgroup classification, an ordinal person-scale, or item parameter estimates used for linking different scales.

An Ordinal Scale for Person Measurement

From the perspective of parametric IRT, it seems odd to define IRFs and ISRFs by means of the monotonicity restriction, because not having a parametric definition prevents the estimation of both person and item parameters from the likelihood of the data given the model. We define data matrix $\mathbf{X} = \left(\left(x_{ij} \right) \right)$ containing the scores of N persons on J items. We assume that persons are sampled i.i.d. (i.e., independently and identically distributed) from $G(\theta)$ and that the three assumptions of the monotone homogeneity models hold. Then, the likelihood of the data given the model is

$$L\left(\mathbf{X} = \mathbf{x} \mid \theta \right) = \prod_{i=1}^{N} \prod_{j=1}^{J} P\left(X_{ij} = x_{ij} \mid \theta \right). \tag{3.15}$$

Even without much knowledge of statistical estimation, one can conjecture that a product of parametric functions, $P\left(X_{ij} = x_{ij} \mid \theta \right)$, such as logistic functions, of latent variable θ and one or more item parameters can be transformed into a set of equations from which the most likely values of the latent variable and the item parameters producing the data can be solved using standard algebra. For example, inserting the Rasch model [Equation (4.11)] in Equation (3.15) yields

$$L\left(\mathbf{X} = \mathbf{x} \mid \theta \right) = \prod_{i=1}^{N} \prod_{j=1}^{J} \frac{\exp\left(\theta_i - \delta_j \right)^{x_{ij}}}{1 + \exp\left(\theta_i - \delta_j \right)}, \tag{3.16}$$

and after some algebraic manipulation, standard maximum likelihood estimates for the person and item parameters can be obtained (Chapter 4). If the particular IRT model fits the data, these θ-parameter estimates can be used to construct a scale for person measurement. Nonparametric IRT typically does not use parametrically defined IRFs and ISRFs but rather imposes order restrictions on the IRFs and the ISRFs, which do not pinpoint the exact response functions and hence do not enable the solution of latent variable θ from Equation (3.15). Thus, the question arises as to how nonparametric IRT models, in particular, the model of monotone homogeneity, produce a scale for person measurement. The answer involves the sum score, X.

First, we consider J dichotomous items. From results that Grayson (1988) proved, it follows that the monotone homogeneity model implies that sum score X orders persons stochastically on latent variable θ (also, see Huynh, 1994).

We identify two values of sum score X, say, x_a and x_b, and arbitrarily assume $0 \leq x_a < x_b \leq J$. Then, for any value t of latent variable θ, the inequality

$$P(\theta > t \mid X = x_a) \leq P(\theta > t \mid X = x_b) \tag{3.17}$$

shows the stochastic ordering of the latent trait (i.e., more generally, the latent variable) by means of the sum score, abbreviated SOL (Hemker et al., 1997). Figure 3.3 visualizes SOL for $J = 8$, hence $x = 0, \ldots, 8$, based on Equation (3.17). SOL shows that ordering persons by the observable sum scores, X, also orders them by their latent variable values, θ, but without the need to estimate θ, which would be impossible anyway. Thus, the model of monotone homogeneity implies an ordinal person-scale. This is a huge difference with CTT, where one never knows whether the items form a scale, let alone the scale level. Nonparametric IRT provides sum score X meaning as a summary of test performance.

An implication of SOL [Equation (3.17)] is

$$\mathcal{E}\left(\theta \mid X = x_a\right) \leq \mathcal{E}\left(\theta \mid X = x_b\right), \tag{3.18}$$

also see Figure 3.3 (dots at the bottom of the graph represent expected values). Like Equation (3.17), Equation (3.18) shows that test sum X orders subgroups by increasing mean θs. It is important to notice that Equation (3.18) refers to subgroups, not individuals. However, random measurement error

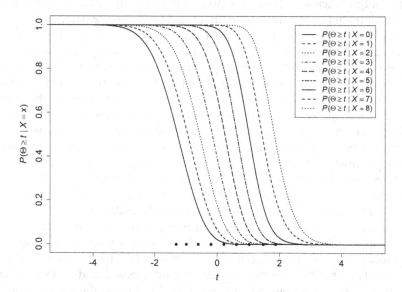

FIGURE 3.3
Visualization of SOL, showing $P(\theta > t \mid X = x)$ for $x = 0, \ldots, 8$, from left to right. The nine dots at the bottom of the graph represent $\mathcal{E}(\theta \mid X = x)$ for $x = 0, \ldots, 8$, which are ordered according to Equation (3.18) on the horizontal axis.

affects sum scores for individuals, so that for two individuals v and w one cannot simply conclude that

$$x_v < x_w \Rightarrow \theta_v \le \theta_w. \tag{3.19}$$

Because of random measurement error, the ordering based on sum score X frequently shows the wrong person order given the ordering by means of latent variable θ, but this is no different from the measurement of individuals in other IRT models. We already met Ellis and Van den Wollenberg's assumption of local homogeneity (ibid., 1993), which must be added to a measurement model to render the model suited for the measurement of individuals. Testing whether ones data are consistent with local homogeneity is practically difficult or even impossible to achieve, and we decided to adopt the practice of assuming that subgroup results also hold for individuals in this monograph.

For polytomous-item tests, unfortunately, the monotone homogeneity model—without local homogeneity, as in all of this monograph—does not imply SOL (Hemker et al., 1997), and we are faced with the problem that the model neither enables the estimation of latent variable θ nor the ordering of persons on θ by means of sum score X. This result suggests that for polytomous items, the monotone homogeneity model is not a useful measurement model, because it does not imply a scale for person measurement, and we are back in the situation in which CTT finds itself, using sum score X without justification. There are two ways out of this problem, one showing that if the monotone homogeneity model holds for the data, the ordering of people by X almost always orders them correctly by θ, the other deriving ordering implications weaker than SOL but still lending measurement by means of the monotone homogeneity credence. The first approach addresses the robustness of the model for violations of SOL, and the second approach secures analytically which possibilities for measurement using the model remain when ordering people by sum score X is not an option.

First, Van der Ark (2005) simulated latent variable distributions and ISRFs for six parametric polytomous IRT models and six more-flexible models based on nonparametric definitions. The design he used varied test length, item parameters, and the θ distribution. With few exceptions, Van der Ark found that sum score X orders people on θ, even when theoretically there is no reason to expect such a result. For the cases where the ordering of people according to sum score X reversed their true ordering with respect to latent variable θ, this happened mostly for adjacent X values. When X values were more than one unit apart, their order predominantly reflected the true θ ordering. As the number of items and the number of item scores decreased, and ISRFs were similar, the proportions of person pairs for which X showed the incorrect ordering decreased. However, small numbers of items that had, say, five ordered scores and ISRFs that varied greatly produced more ordering violations than large numbers of items having, say, three ordered scores and similar ISRFs.

The number of item scores was most influential in distorting the correct person ordering based on sum score X. In general, based on extensive simulations, sum score X is a robust ordinal estimator of latent variable θ, provided that small differences between X values or between an X value and a cut score are unimportant for decision-making based on sum scores. We think that this position is defendable in most social, behavioral, and health science research and applications, because decisions often are binary, for example, admitting people to a course or rejecting them, and measurement precision is much smaller than in the natural sciences, not permitting fine-grained distinctions anyway. These arguments also stand when we assign people to three or more categories. Researchers often boost measurement precision by not making decisions based on only one sum score but on the combination of different data sources and this seems like a good practice.

Robustness results represent a practical approach and in the case of the monotone homogeneity model mitigate a theoretical problem, but one might still argue that the failure of SOL is unacceptable, only worsening measurement precision by introducing systematic distortions in addition to measurement error. This position calls for different models implying SOL or for finding out whether the monotone homogeneity model implies weaker forms of person ordering than SOL that are still useful. Van der Ark and Bergsma (2010) proved that the monotone homogeneity model for polytomous items implies an ordering property they called weak SOL, because SOL implies weak SOL but not reversely. To discuss weak SOL for polytomous items, we assume a fixed value of sum score X, denoted x_c, such that $1 \leq x_c \leq JM$; then, the monotone homogeneity model implies

$$P(\theta > t \mid X < x_c) \leq P(\theta > t \mid X \geq x_c). \tag{3.20}$$

Equation (3.20) represents weak SOL. For $x_c = 0$, the first probability in Equation (3.20) is undefined. Van der Ark and Bergsma (2010) provided a computational example showing weak SOL is satisfied but SOL fails, demonstrating that weak SOL is a weaker ordering property than SOL. A look at Equation (3.20) reveals that splitting the group into exhaustive and exclusive subgroups, one subgroup having relatively low sum scores so that $X < x_c$ and the other subgroup having relatively high sum scores so that $X \geq x_c$, then the cumulative probability $P(\theta > t)$ is at least as high in the latter group. This result is valid for JM different divisions of the sum-score distribution in exhaustive and exclusive subgroups, that is, for all values of x_c in Equation (3.20), so that $1 \leq x_c \leq JM$.

Just like SOL [Equation (3.17)] implies an ordering property concerning the mean latent variable conditional on sum scores, weak SOL [Equation (3.20)] implies such an ordering conditional on pairs of intervals based on sum score X,

$$\mathcal{E}(\theta \mid X < x_c) \leq \mathcal{E}(\theta \mid X \geq x_c). \tag{3.21}$$

Weak SOL is a useful ordering property if the test is used to select the people having, say, the highest 20% of the sum scores. By selecting the 80% of the lowest scoring people on sum score X and the 20% of the highest scoring people on X, weak SOL guarantees that the latter group on average scores higher on the latent-variable scale than the former group [Equation (3.21)].

For the ISRFs in Figure 3.2 (i.e., $J=2$, $M=4$, $x=0,\ldots,4$, hence $X=0,\ldots,8$), and a bimodal latent variable distribution (Figure 3.4a), Figure 3.4b shows that SOL [Equation (3.17)] is violated, yet Figure 3.5 (a–h), graphically reflecting Equation (3.20), illustrates that weak SOL holds. For the computational

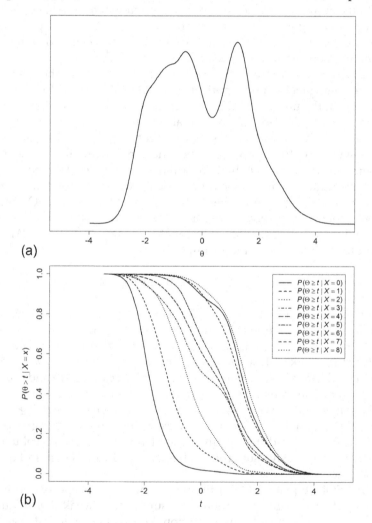

(a)

(b)

FIGURE 3.4

(a) Density of the latent variable, and (b) $P(\theta > t|X=x)$ for $x \in \{0,1,\ldots,8\}$. SOL is violated: For $1.5322 < t < 3.1947$, $P(\theta > t|X=7) < P(\theta > t|X=6)$; for $1.2175 < t$, $P(\theta > t|X=7) < P(\theta > t|X=6)$; for $1.7105 < t < 3.4211$, $P(\theta > t|X=5) < P(\theta > t|X=4)$, and for $t < -3.9670$, $P(\theta > t|X=4) < P(\theta > t|X=3)$.

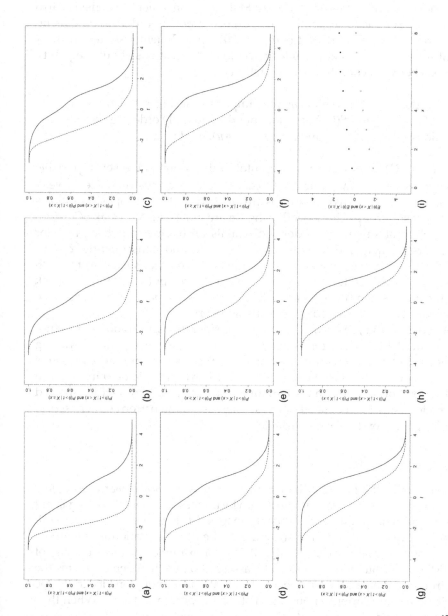

FIGURE 3.5
$P(\theta > t \mid X < x)$ (dotted curves) and $P(\theta > t \mid X \geq x)$ (solid curves) for $x = 1$ (a), $x = 2$ (b), $x = 3$ (c), $x = 4$ (d), $x = 5$ (e), $x = 6$ (f), $x = 7$ (g), $x = 8$ (h), and $\mathcal{E}(\theta \mid X < x)$ (open dots) and $\mathcal{E}(\theta \mid X \geq x)$ (solid dots) for $x \in \{0, 1, \dots, 8\}$ (i).

data used in the first eight panels (a–h) in Figure 3.5, panel (i) shows mean θs according to Equation (3.21) for the eight dichotomies based on sum score X. Next, we consider the implications of Equation (3.21).

To increase understanding of weak SOL, it is convenient to consider also what weak SOL does *not* imply with respect to person ordering on θ using sum score X. Van der Ark and Bergsma (2010, p. 277) demonstrated the next result for weak SOL that excludes particular implications about inequality relations between conditional expectations:

> Proposition: Weak SOL neither implies ordering restrictions on $\mathcal{E}(\theta \mid X < x_c)$ and $\mathcal{E}(\theta \mid X < x_c + u)$, nor does it imply ordering restrictions on $\mathcal{E}(\theta \mid X \geq x_c)$ and $\mathcal{E}(\theta \mid X \geq x_c + u)$, $u = Jm - x_c$.

Equation (3.21) shows that when a cutoff value x_c splits the set of possible sum scores $\{0, 1, \ldots, JM\}$ into two exhaustive and exclusive subsets, weak SOL implies that the two subgroups are ordered on θ. The proposition shows that when subgroups show overlap, hence are not exclusive, weak SOL does not imply that subgroups order individuals on θ. One can check that for three persons, i_1, i_2, and i_3, with $x_{i_1} < x_{i_2}$, $x_{i_2} < x_{i_3}$, and, consequently, $x_{i_1} < x_{i_3}$, for each person pair one can always find cut scores x_c, such that for each person pair weak SOL implies a pairwise ordering, but one can also check that the proposition prevents an ordering of all three persons because three subgroups based on two cut scores always overlap.

We conclude that, based on theoretical considerations, the monotone homogeneity model for polytomous items only allows pairwise person ordering but not complete person ordering. Van der Ark's (2005) computational results give us enough confidence to use X scales to order people on θ in practical applications of tests and questionnaires. Zwitser and Maris (2016) studied the model requirements needed when one wishes the pattern of item scores to order people on the latent variable.

Goodness-of-Fit Methods

Nonparametric IRT has a tradition of meticulously investigating the goodness of fit of the model of interest to the data. The impossibility of estimating latent person and item parameters probably inspired this tradition, thus necessitating an explicit focus on the structure of the data in relation to the model assumptions (e.g., Junker & Sijtsma, 2001a). For example, the monotonicity of the IRF for dichotomous items can be assessed only when one estimates the IRF from the data and studies possible violations of monotonicity to decide whether they are serious enough to disturb the person ordering. When they are, the item may be removed from the test or questionnaire. In parametric IRT, the degree to which goodness of fit is investigated varies across models and seems to depend much on the model's complexity. For example, the simple one-parameter logistic model or Rasch model has a well-developed toolkit of goodness-of-fit methods (e.g., Glas & Verhelst, 1995a, b), but other models have a more complex structure rendering goodness-of-fit research more difficult. Assessing any model's goodness of fit is paramount for interpreting the structure of the model in relation to the data and the estimated parameter values. A model of which the misfit remains undetected misguides the researcher into believing that the data are consistent with the model and that the measurement instrument for the attribute has the properties of the model.

We make a distinction between absolute and relative model fit. Nonparametric IRT models are assessed using absolute goodness-of-fit methods. We discuss relative goodness-of-fit methods in Chapter 5. In the present chapter, we discuss methods for assessing unidimensionality, local independence, and monotonicity. Together, they enable assessing the fit of the model of monotone homogeneity to the data.

Unidimensionality: Scalability and Item Selection

Scalability Coefficients and Scale Definition

We start this section with conditional association, which is an observable consequence of the model of monotone homogeneity that proves useful in what follows. Let $\mathbf{X}$ be the vector containing the J item-score variables X_j, $X_j = 0,\ldots,M$, and $j = 1,\ldots,J$. We split $\mathbf{X}$ into two exclusive but not necessarily exhaustive vectors $\mathbf{Y}$ and $\mathbf{Z}$; that is, these vectors contain different item-score variables but not necessarily all J of them. Let f_1 and f_2 be nondecreasing functions, and let h be any function. Let $\sigma(.,.)$ denote the population covariance. Holland and Rosenbaum (1986) derived that the monotone homogeneity model—unidimensionality (actually, the authors assumed multidimensional $\boldsymbol{\theta}$), local independence, and monotonicity (i.e., item response surfaces depending on $\boldsymbol{\theta}$)—implies conditional association,

$$\sigma[f_1(\mathbf{Y}), f_2(\mathbf{Y}) \mid h(\mathbf{Z}) = \mathbf{z}] \geq 0. \tag{3.22}$$

A special case of Equation (3.22) proves useful in the present section on unidimensionality. We also use Equation (3.22) and two additional special cases in the section on local independence. The special case we use here defines $f_1(\mathbf{Y}) = X_j$, $f_2(\mathbf{Y}) = X_k$, and further $\mathbf{Z} = \varnothing$, meaning vector $\mathbf{Z}$ has dimension 0 and $h(\mathbf{Z})$ is undefined, so that we can ignore $h(\mathbf{Z})$ (Rosenbaum, 1984). Using these definitions, Equation (3.22) reduces to non-negative inter-item covariances,

$$\sigma(X_j, X_k) \geq 0, \quad \text{all } j,k; j \neq k; X_j = 0,\ldots,M, j = 1,\ldots,J; \tag{3.23}$$

also, see Mokken (1971, pp. 119–120), Ellis and Van de Wollenberg (1993), and Sijtsma and Molenaar (2002, pp. 155–156). In this section, we use non-negativity of inter-item covariances to assess the scalability of IRT models by means of three scalability coefficients (Mokken, 1971, pp. 148–153; Sijtsma & Molenaar, 2002, Chap. 4). First, we define the three scalability coefficients and their properties in relation to the model of monotone homogeneity and to one another, and then we define a scale. In the next subsection, we discuss item selection based on the scalability coefficients.

For pairs of items, let $\sigma_{jk} = \sigma(X_j, X_k)$, and let $\sigma_{jk}^{\max}$ denote the maximum value of the covariance given the marginal distributions of the two item scores. An example of how marginal distributions affect the inter-item covariance for binary variables, such as dichotomous item scores, is the following. Let π_j denote the univariate proportion of people from the population that has a score 1 on item j, and let π_{jk} denote the bivariate or joint proportion of people that has a score 1 on both items j and k, then,

$$\sigma_{jk} = \pi_{jk} - \pi_j \pi_k. \tag{3.24}$$

Let π_j and π_k be equal, for example, $\pi_j = \pi_k = .6$. Then, given these values, the maximum value of π_{jk} equals $\pi_{jk}^{\max} = .6$, and $\sigma_{jk}^{\max} = .24$. The more univariate proportions differ, the more the difference deflates the maximum covariance attains. For example, let $\pi_j = .6$ and let $\pi_k = .5, .4, .3, .2, .1$; then, $\pi_{jk}^{\max} = \min(\pi_j, \pi_k) = \pi_k$, and $\sigma_{jk}^{\max} = .20, .16, .12, .08, .04$, respectively. Results for polytomous items and continuous variables show similar trends but are less dramatic, and for the product–moment correlation, one finds comparable results.

The scalability coefficient for two items, denoted H_{jk}, corrects the inter-item covariance for the deflating effect that different item-score distributions have, and equals

$$H_{jk} = \frac{\sigma_{jk}}{\sigma_{jk}^{\max}}. \tag{3.25}$$

The scalability coefficient for item j with respect to the other $J-1$ items in the test or questionnaire, denoted H_j, compares the sum of the covariances of item j with each of the other $J-1$ items, with the sum of the corresponding maximum possible covariances, so that

$$H_j = \frac{\sum_{k:k \neq j} \sigma_{jk}}{\sum_{k:k \neq j} \sigma_{jk}^{\max}}. \tag{3.26}$$

Using the properties of the covariance of linear combinations of variables, and consistent with Chapter 2, we define rest score, $R_{(j)} = \sum_{k:k \neq j} X_j$, and write Equation (3.26) as a conjecture based on several numerical results as

$$H_j = \frac{\sigma\left(X_j, \sum_{k:k \neq j} X_k\right)}{\sigma^{\max}\left(X_j, \sum_{k:k \neq j} X_k\right)} = \frac{\sigma\left(X_j, R_{(j)}\right)}{\sigma^{\max}\left(X_j, R_{(j)}\right)}. \tag{3.27}$$

The scalability coefficient for the set of J items is defined as (Mokken, 1971, pp. 148–153)

$$H = \frac{\sum\sum_{j<k}\sigma_{jk}}{\sum\sum_{j<k}\sigma_{jk}^{max}}.$$ (3.28)

Doubling the covariances in numerator and denominator, one can derive that

$$H = \frac{\sum_{j=1}^{J}\sigma(X_j, R_{(j)})}{\sum\sum_{j\neq k}\sigma_{jk}^{max}}.$$ (3.29)

Clearly, the scalability coefficients are (sums of) normed item-test covariances, corrected for the deflating influence of the marginal distributions of the item sores, and attain values different from, for example, the corrected item-total correlations discussed in Chapter 2. This is clear if one divides the numerator and the denominator of coefficient H_j in Equation (3.27) by the product $\sigma_{X_j}\sigma_{R_{(j)}}$ and obtains another conjecture by implication,

$$H_j = \frac{\rho(X_j, R_{(j)})}{\rho^{max}(X_j, R_{(j)})},$$ (3.30)

showing that $H_j \geq \rho(X_j, R_{(j)})$, but how much H_j exceeds $\rho(X_j, R_{(j)})$ depends on the maximum correlation in the denominator. For example, suppose, for item j, one finds $\rho(X_j, R_{(j)}) = .5$ and $\rho^{max}(X_j, R_{(j)}) = .6$, so that $H_j = \frac{.5}{.6} \approx .83$, and for item k one finds $\rho(X_k, R_{(k)}) = .6$ and $\rho^{max}(X_k, R_{(k)}) = .8$, implying $H_k = .75$. The example shows that the corrected item-total correlation and the item scalability coefficient can be reversely ordered, a result also found in real data.

According to Equation (3.23), inter-item covariances are non-negative. Maximum possible inter-item covariances given item-score distributions are positive (zero only if $\pi_j = 0$ or $\pi_k = 0$ or $\pi_j = \pi_k = 0$, which represent extreme circumstances). Hence, for Equation (3.25) we have bounds

$$0 \leq H_{jk} \leq 1, \quad \text{for all } j, k = 1, \dots, J; j \neq k;$$ (3.31)

for Equation (3.26), we have bounds

$$0 \leq H_j \leq 1, \quad \text{for all } j = 1, \dots, J;$$ (3.32)

and for Equation (3.28), we have bounds

$$0 \leq H \leq 1.$$ (3.33)

From Equations (3.31), (3.32), and (3.33) follows that negative values of the scalability coefficients contradict the monotone homogeneity model and that positive values tend to support the model. As scalability values are smaller, multidimensionality, local dependence of the data, or non-monotone ISRFs cannot be excluded (e.g., Mokken, Lewis, & Sijtsma, 1986; Smits, Timmerman, & Meijer, 2012).

The three scalability coefficients have the following relationships (Sijtsma & Molenaar, 2002, p. 58):

$$\min_{k}\left(H_{jk}\right) \le H_j \le \max_{k}\left(H_{jk}\right); \tag{3.34}$$

$$\min_{j}\left(H_j\right) \le H \le \max_{j}\left(H_j\right); \tag{3.35}$$

and

$$\min_{j,k}\left(H_{jk}\right) \le H \le \max_{j,k}\left(H_{jk}\right). \tag{3.36}$$

Mokken (1971, p. 184; also, see Sijtsma & Molenaar, 2002, pp. 67–69) defined a scale using coefficients H_{jk} and H_j.

Definition: A set of items constitutes a scale if, for a suitable chosen positive constant c,

a. For inter-item scalability coefficients,

$$H_{jk} > 0, \quad \text{all } j,k = 1,\ldots,J; j \ne k; \tag{3.37}$$

and

b. For item scalability coefficients,

$$H_j \ge c > 0, \quad \text{all } j = 1,\ldots,J. \tag{3.38}$$

Two remarks are in order. First, in part a. of the definition, Mokken (1971, p. 184) did not use inter-item scalability coefficients but demanded that for each item pair, product–moment correlation $\rho_{jk} > 0$. Because both H_{jk} and ρ_{jk} have a numerator equal to covariance, σ_{jk}, and both have positive denominators, $\sigma_{jk}^{\max}$ and $\sigma_j\sigma_k$, respectively, they have the same sign behavior; hence, for the definition of a scale,

$$H_{jk} > 0 \Leftrightarrow \rho_{jk} > 0. \tag{3.39}$$

Second, because from part b. of the definition of a scale we know that the smallest H_j value in a scale equals at least $c > 0$, that is, $H_j \ge c$, and from Equation (3.35) we know that $H \ge \min_j\left(H_j\right)$, the definition of a scale implies that $H \ge c$. Combined with Equation (3.33), we have the result that

$$c \le H \le 1. \tag{3.40}$$

Now that we know a lot about scalability coefficients and the definition of a scale, the question is what their relation with the model of monotone homogeneity is. We focus on Equation (3.23), which implies that as a borderline case all inter-item covariances can be 0, that is, $\sigma_{jk}=0$, implying that all $H_{jk}=0$ [Equation (3.25)] and all $H_j=0$ [Equation (3.26)]. The definition of a scale permits these values when we allow non-negativity to replace strict positivity, but clearly the range of items that satisfy the definition using $c=0$ or c small positive can be great. What one needs in a practically useful scale are items that have ISRFs (or IRFs) that have a positive slope and preferably one that is rather steep where the distribution of sum score X is located, because then the item clearly distinguishes low and high X values. The model of monotone homogeneity addresses IRF monotonicity but not IRF steepness related to the test-score distribution. Expressing the strength of the relationship between the item score and the rest score, corrected for the influence of marginal item-score distributions, scalability coefficient H_j fulfils precisely this role by relating IRF steepness to the test-score distribution. The use of constant c in the definition of a scale [Equation (3.38)] provides the researcher with the opportunity to manipulate the quality of the items she accepts in her scale. One may notice that constant c bears directly on the item's practical usefulness (Mokken et al., 1986), but the model of monotone homogeneity only implies $c \ge 0$, not a particular value of c. The next subsection comments on the zero lower bound.

For practical purposes, the following rules of thumb for the interpretation of the scale's strength are used (Mokken, 1971, p. 185): $.3 \le H < .4$ constitutes a weak scale; $.4 \le H < .5$ a medium scale; and $H \ge .5$ a strong scale. A set of items for which $H < .3$ is considered unscalable, but one may notice that such low values are consistent with the model of monotone homogeneity. The definition of a scale and the rules of thumb for interpreting results are used for assessing a priori defined sets of items and in an automated procedure that selects items in one or more scales while discarding unscalable items. Lower bound $c=.3$ is used as default in scale assessment and item selection. Van der Ark, Croon, and Sijtsma (2008) proposed marginal modelling (Bergsma, Croon, & Hagenaars, 2009) as a framework to derive the standard errors for the scaling coefficients and test hypotheses about the scaling coefficients.

Modified Scalability Bounds

In this section, we replace inter-item covariances, which are the basis of Mokken's scalability coefficients, with inter-item correlations, ρ_{jk}. Consider three random variables, X, Y, and Z, and let $\rho_{XY}=-1$. Then, if we know that $\rho_{XZ}=1$, it follows that $\rho_{YZ}=1$. That is, the first two correlations completely determine the third. This is an extreme example, but for dichotomous items, Ellis (2014) used the dependence between correlations on a set of variables to derive restrictions on the correlation between two items from the correlations each has with each of the other $J-2$ items. One may consult Van Bork, Grasman, and Waldorp (2018) for related developments in the context of the one-factor model.

First, we have to recall the concept of partial covariance. This is the covariance between the residuals of two variables X and Y that remains when both variables' linear regression on a third variable Z is subtracted from X and Y. Let $\beta_{XZ} = \rho_{XZ}\dfrac{\sigma_X}{\sigma_Z}$ and $\beta_{YZ} = \rho_{YZ}\dfrac{\sigma_Y}{\sigma_Z}$ be the regression coefficients, then the partial covariance equals

$$\sigma_{XY.Z} = \sigma\left(X - \rho_{XZ}\frac{\sigma_X}{\sigma_Z}Z, Y - \rho_{YZ}\frac{\sigma_Y}{\sigma_Z}Z\right).$$

The residual standard deviations of X and Y equal

$$\sigma_{X.Z} = \sigma_X\left(1-\rho_{XZ}^2\right)^{1/2} \text{ and } \sigma_{Y.Z} = \sigma_Y\left(1-\rho_{YZ}^2\right)^{1/2},$$

and together with the partial covariance, they can be used to obtain the partial correlation,

$$\begin{aligned}
\rho_{XY.Z} &= \frac{\rho_{XY} - \rho_{XZ}\rho_{YZ}}{\left[\left(1-\rho_{XZ}^2\right)\left(1-\rho_{YZ}^2\right)\right]^{1/2}} \\
&= \frac{\sigma_{XY} - \sigma_{XZ}\beta_{YZ}}{\left[\sigma_X^2\left(1-\rho_{XZ}^2\right)\sigma_Y^2\left(1-\rho_{YZ}^2\right)\right]^{1/2}}.
\end{aligned}$$

(3.41)

Ellis (2014, p. 306, Theorem 1) used the numerator in the second fraction in Equation (3.41) to prove the next result. Replace X, Y, and Z with item scores X_j, X_k, and X_l. Assume that the J items are dichotomous with 0,1 scores, and have positive variances and positive correlations smaller than 1; then, if the items are conditionally associated [Equation (3.22)], their partial correlations are non-negative,

$$\rho_{jk.l} \geq 0, \text{ all item triplets } j,k,l.$$

(3.42)

From the numerator in the first fraction of Equation (3.41) we then have (Ellis, 2014, p. 306, Corollary 1)

$$\rho_{jk} \geq \rho_{jl}\rho_{kl}, \text{ all item triplets } j,k,l.$$

(3.43)

Hence, the correlations items j and k have with item l restrict their mutual correlation by the inequality in Equation (3.43). Ellis (2014) provided a toy example in which $\rho_{jk} = .33$ and $\rho_{jl} = \rho_{kl} = .67$, thus all three correlations are positive [Equation (3.23)], but $\rho_{jk.l} = -.20$, which contradicts the monotone homogeneity model by Equation (3.42). The message is that one cannot conclude from positive inter-item correlations alone that they support the monotone homogeneity model for the data.

Equation (3.43) enables the derivation of lower and upper bounds for each of the $1/2\,J\,(J-1)$ correlations in the test. The lower and upper bounds usually are different from 0 and 1, respectively, and vary across item pairs. Given the monotone homogeneity model, for item pair (i, j) one obtains a lower bound,

$$\rho_{jk} \geq \max_{l}\left\{\rho_{jl}\rho_{kl}\,|\,l:l \neq j,k\right\}, \qquad (3.44)$$

and an upper bound

$$\rho_{jk} \leq \min_{l}\left\{\left.\frac{\rho_{jl}}{\rho_{kl}},\frac{\rho_{kl}}{\rho_{jl}}\right|\,l:l \neq j,k\right\}. \qquad (3.45)$$

Because the lower bound implies the upper bound, Ellis (2014) limited his attention to two lower bounds for inter-item correlations. The first lower bound is Equation (3.44). When the number of items, J, is large, the number of products for each inter-item correlation to choose from equals $J-2$, and chance capitalization (Chapter 2) is likely to occur. The second lower bound is the 90th percentile of the distribution of the products $\rho_{jl}\rho_{kl}$, but Ellis (2014) noticed that this choice also comes with drawbacks. We refrain from discussing the problems both lower bounds may involve, but see Ellis (2014).

To obtain a lower bound for inter-item scalability coefficient H_{jk} [Equation (3.25)], we first notice that

$$H_{jk} = \frac{\sigma_{jk}}{\sigma_{jk}^{\max}} = \frac{\sigma_{jk}/\sigma_j\sigma_k}{\sigma_{jk}^{\max}/\sigma_j\sigma_k} = \frac{\rho_{jk}}{\rho_{jk}^{\max}}. \qquad (3.46)$$

The lower bound for H_{jk} is obtained by inserting the right-hand side of Equation (3.44) in the numerator of H_{jk}, so that

$$H_{jk} \geq \max_{l}\left\{\left.\frac{\rho_{jl}\rho_{kl}}{\sigma_j\sigma_k}\right|\,l:l \neq j,k\right\}. \qquad (3.47)$$

Ellis (2014) recommended testing negative partial correlations for significance by means of the Mantel–Haenszel statistic (e.g., Rosenbaum, 1984). One can also transform lower bound results to the other scalability coefficients, but then one should notice that item scalability H_j [Equation (3.26)] and the scalability coefficient for J items [Equation (3.28)] compare weighted sums of covariances. For example, one can readily check that

$$H_j = \frac{\sum_{k:k\neq j}\sigma_j\sigma_k\rho_{jk}}{\sum_{k:k\neq j}\sigma_j\sigma_k\rho_{jk}^{\max}}. \tag{3.48}$$

One can use the results from this section to obtain an item-dependent lower bound for each item-scalability coefficient H_j. Because little experience has been accumulated with respect to the use of the modified lower bound result, we refrain from further treatment.

Mokken's Automated Item Selection Procedure

Depending on the composition of the item set, the automated item selection procedure (Mokken, 1971, Chap. 5; Sijtsma & Molenaar, 2002, Chap. 5) divides the start set of J_{start} items into G scales, indexed $g = 1,\ldots,G$. Each of these scales contains J_g items. One may use the algorithm in two ways. The exploratory approach entails feeding an experimental item set to the algorithm. The degree to which the researcher is certain about the composition of the item set depends on the strength of the theoretical foundation of the attribute and the operationalization. A relatively strong foundation may still leave uncertainty about the degree to which individual items discriminate sum scores, expressed by varying H_j values, and depending on the value of lower bound c, weakly discriminating items having $H_j < c$ are not admitted to a scale. Given c, such items contribute too little to a reliable person ordering. Another feature of an initial item set may be that it assesses different aspects or sub-attributes of an attribute and perhaps one, a few, or a whole subset of items that misrepresent the attribute. Items may fail to represent the attribute, because they were badly formulated or do not assess the target attribute. Such items usually come out of the automated item selection procedure as unscalable, and their number is denoted J_{unsc}. The researcher may further manipulate the item selection by defining a kernel of items that she considers key to measurement of the attribute, and then letting the automated item selection procedure find out whether the kernel can be extended with one or more of the remaining items that do not belong to the kernel. The confirmatory approach entails a user-specified definition of the scale or the subscales, and for each the scalability coefficients are estimated without any item selection interfering.

The automated item selection procedure is a bottom-up algorithm that selects two items for a start, or starts with a user-defined kernel of items, and adds items one by one until a selection criterion no longer is satisfied. As long as the criterion is satisfied, the procedure continues selecting items, if available, and thus aims at selecting the largest subset from the entire item set. If the criterion is dissatisfied, the first selection round is completed. If at least two items remain unselected, from the remaining items the procedure starts selecting a second item subset following the same procedure and if possible, a third, a fourth, and so on, until only unscalable items or no items remain. The item subsets the procedure selects are each consistent with the definition of a scale. Different subscales contain unique subsets of items and thus are exclusive. A scale must contain as many sufficient-quality items as possible, and the researcher controls the definition of item quality by choosing the lower bound value c in the definition of a scale. Different choices of c can produce different subscales (Hemker, Sijtsma, & Molenaar, 1995).

The division of the item set into several subscales and possibly unscalable items can be represented as

$$J_{\text{start}} = \sum_{g=1}^{G} J_g + J_{\text{unsc}}. \tag{3.49}$$

In its simplest form, the automated item selection procedure produces one scale including all items, so that $G=1$ and $J_{\text{start}}=J_1=J$. The algorithm contains the following steps (Sijtsma & Molenaar, 2002, pp. 71–72), based on the definition of a scale.

1. Select item pair (j, k_1) that has the largest item-pair scalability coefficient H_{jk_1} of all item pairs that have values significantly greater than 0, and exceeds lower bound c. If the researcher defined a kernel containing at least one item (Molenaar & Sijtsma, 2000, p. 40), this kernel replaces the item pair. From then on, the procedure is identical for both versions of a selected set. We discuss the procedure for item pair (j, k_1).

2. Select from the set of remaining $J_{\text{start}} - 2$ items, item k_2, which

   i. Correlates positively with the items already selected, which are j and k_1 or, similarly, for $l=j, k_1$, has a positive item-pair scalability coefficient H_{lk_2} [Equation (3.37)];

   ii. Has an H_{k_2} value [Equation (3.38)] with respect to the already selected items, j and k_1, that is significantly greater than 0 and also exceeds lower bound c; and

   iii. Maximizes the H value [Equation (3.28)] for the subset of items selected thus far, including item k_2.

   Several items can simultaneously satisfy requirements i. and ii. Requirement iii. in most cases produces a unique choice.

3. Select item k_3 from the remaining $J - 3$ items which

   i. Correlates positively with items j, k_1, and k_2 (or, $H_{lk_3} > 0$, with $l = j, k_1, k_2$);

   ii. Has an H_{k_3} value with respect to items j, k_1, and k_2, significantly greater than 0 and $H_{k_3} > c$; and

   iii. Maximizes H for items j, k_1, k_2, and k_3.

In the next steps, for each next item, the procedure outlined in steps 2 and 3 is repeated until none of the remaining items

  i. Correlates positively with all items already admitted to the selected item subset; and
 ii. Has an item scalability value with respect to the already selected items that is significantly positive and in excess of lower bound c.

Decision rules are effective for special cases. For example, in substep iii, two or more items that satisfied substeps i and ii may produce the same common H value, and a decision on how to proceed is needed. Furthermore, in each item-selection step, all for remaining items, one tests whether their H_i value is significantly greater than 0, and to control for chance capitalization, at any phase in the procedure one uses the Bonferroni correction for this test, taking into account the number of tests performed in the previous rounds and the present round.

After an item subset has been selected and at least two items remain, the procedure starts again, selecting a second item subset, if possible, from the items that remained unselected, and then a third subset, and so on. Items that remain unselected are unscalable. These are items, for example, correlating negatively with one or more items in any of the subsets. One can easily think of different variations on the procedure. For example, one may statistically test whether in step ii, the item scalability value is significantly greater than lower bound c, using Wald confidence intervals and a correction for chance capitalization. Statistically, this would be an improvement of the procedure, but the consequence would be that in smaller samples, items more likely are not admitted to a scale. The question is whether one can afford losing items and how important it is to know that $H_i > c$ is significant. The decision depends on many considerations that depend on the situation and involves whether the item is key to the attribute despite low scalability, the representativeness of the item for the attribute, the number of items that cover the attribute, the reliability of the sum score, and so on. Another modification of the automated item selection procedure might be to use the lower bound values Ellis (2014) derived for each item pair; see Equation (3.47). Again, using these lower bounds that take dependencies between item-pair associations into account, although theoretically defensible, might further limit possibilities for scale construction, and questions asked with significance testing for $H_i > c$ also apply here (Van der Ark et al., 2008). Nevertheless, experience has to accumulate using these new insights and improved item selection procedures may result from this.

Thus, ignoring Ellis' (2014) lower-bound results for the moment, we notice that default $c = .3$ has been proven useful in real-data analysis, but Hemker et al. (1995) found that the choice of c strongly influences item subset composition. For exploratory work, the authors recommended running the automated item selection procedure 12 times, consecutively using lower bounds $c = 0, .05, .10, \ldots, .55$. The resulting 12 item-selection results often produce one of two typical outcome patterns:

1. In unidimensional data, as c increases, one subsequently finds (a) most or all items in one scale; (b) one smaller scale; and (c) one or a few small scales and several unscalable items. Take the result in stage (a) as final.

2. In multidimensional data, as c increases, one subsequently finds (a) most or all items in one scale; (b) two or more scales; and (c) two or more smaller scales and several unscalable items. Take the result in stage (b) as final.

The difference between the two outcome patterns resides with results (b), where in the case of unidimensional data one expects to find one scale and in the case of multidimensional data at least two scales. These are ideal outcome patterns, but other, less well interpretable patterns are also possible, whereas real data's messy structure (Chapter 1) may obscure an ideal outcome pattern. The absence of a recognizable outcome pattern surely complicates the interpretation of one's item selection and unless item content clearly subdivides items already without psychometric analysis, provides a plea for confirmatory approaches.

Because the automated item selection procedure is sequential, once selected, an item no longer is a candidate for selection as the procedure continues. At the moment of selection, an item's scalability with respect to the items thus far selected exceeds c, hence, $H_l > c$, but when the item subset is completed, because of the items selected later, the scalability value H_l may have dropped below lower bound c (Mokken, 1971, p. 193). This means that, following the definition of a scale, in hindsight the item does not belong to the scale, but because it was never considered for selection in subsets selected later, we do not know whether it might better belong in one of those item subsets. A remedy is to take the second and next scales subsequently as the start set for re-selection rounds, in which only items from the first to the $(g-1)$st item subsets are candidates for selection into the gth subset (Molenaar & Sijtsma, 2000, pp. 39, 48; also, see Mokken, 1971, pp. 196, 199, 311–312). Straat, Van der Ark, and Sijtsma (2013) noticed that the first item subset selected sometimes contains fewer items than the next item subsets, which runs counter to the goal of having as many items in the first subset as possible, followed by the second subset, and so on. The problems we discuss here typically result from the procedure running into a local maximum. A definitive solution that avoids local maxima is to redefine the item selection algorithm, such that the first item subset is the longest, following by the second item subset, and so on. We discuss this possibility in the next subsection.

Modified Procedure to Produce Maximum-Length Scales

Straat et al. (2013) noticed that an item set consisting of J items can be partitioned in several ways into item subsets that all satisfy the definition of a scale. Let $\mathcal{M}$ be the set of such partitionings. Let F denote the number of partitionings in $\mathcal{M}$, indexed $f = 1, \ldots, F$. The number of item subsets in a partitioning equals G_f, and are indexed, as before, $g = 1, \ldots, G_f$. The number of items in item subset g in partitioning f equals J_{fg}. Vector $\mathbf{v}_f = (v_1, \ldots, v_J)$ informs one about the item selection in partitioning f, $v_j = 0$ meaning that item j was unscalable, $v_j = 1$ that item j was selected in the longest subset $g = 1$, $v_j = 2$ that item j was selected in the next-longest subset $g = 2$, and so on.

Given the definition of a scale, the first item subset must contain the largest number of items; the second item subset must contain the largest number of items from the complementary subset not selected in the first subset but no more items than the first subset; and so on, until no more subsets consistent with the scale definition are possible. Hence, the objective function governing the item selection should value having an additional item in subset g more than having any number of additional items in the next, smaller item subsets, $g+1, \ldots, G_f$. The objective function is a sum across item subsets in partitioning f of products that weigh the number of items in subset g by J^{-g}; that is,

$$O\left(\mathbf{v}_f\right) = \sum_{g=1}^{G_f} J^{-g} \times J_{fg}, \quad f = 1, \ldots, F. \tag{3.50}$$

Given that each next item subset contains at most as many items as the previous item subset, one knows that $J_{f1} \geq J_{f2} \geq \ldots \geq J_{fG_f}$, and furthermore that $J^{-1} \geq J^{-2} \geq \ldots \geq J^{-G_f}$, and thus the first product in Equation (3.50) is the largest, followed by the second, and so on. The more items accumulate in the first item subset, the larger $J^{-1} \times J_{f1}$, and the smaller at least one of the next ratios; then, the more items from the remaining items accumulate in the second subset, the larger $J^{-2} \times J_{f2}$, while the first ratio remains unchanged and at least one of the next ratios decreases; and so on. Hence, the higher the value of $O(\mathbf{v}_f)$.

The range of values for $O(\mathbf{v}_f)$ is $0 < O(\mathbf{v}_f) \leq 1$, with $O(\mathbf{v}_f) = 0$ when all items are unscalable and $O(\mathbf{v}_f) = 1$ when all items are in one scale. We use a few examples to show how Equation (3.50) works.

BOX 3.1   Numerical Examples of Equation (3.50)

To illustrate how one computes $O(\mathbf{v}_f)$, we consider a numerical example involving a set of ten items ($J = 10$) that is partitioned into four subsets. The first subset contains four items ($J_{f1} = 4$), the second subset contains three items ($J_{f2} = 3$), and the third subset contains two items ($J_{f3} = 2$), whereas the tenth item is unscalable. Hence, $v_f = (1, 1, 1, 1, 2, 2, 2, 3, 3, 0)$. From Equation (3.50), it follows that $O\left(\mathbf{v}_f\right) = 10^{-1} \times 4 + 10^{-2} \times 3 + 10^{-3} \times 2 = .4 + .03 + .002 = .432.$

Next, we consider how $O(\mathbf{v}_f)$ works in general. First, when all items are in the same subset, so that $G_f=1$ and $\mathbf{v}_f = (1,\ldots,1)$, then $O(\mathbf{v}_f) = J^{-1} \times J = 1$. When not all items are in one subset, given that the definition of a scale must hold and given a fixed number of items, a second subset must contain at least two items, so that $\mathbf{v}_f = (1,\ldots,1,2,2)$ represents the item-subset partitioning having the shortest second subscale, hence, the longest first subscale. Then, the first product of $O(\mathbf{v}_f)$ equals $J^{-1} \times J_1 = (J-2)/J$, so that compared to the all-items-in-one-set case at the start of this example, this ratio decreased by $\frac{2}{J}$. Because the second subset contributes ratio $J^{-2} \times J_2 = (2/J^2)$ to $O(\mathbf{v}_f)$, and $\frac{2}{J} > \frac{2}{J^2}$, we find that $O(\mathbf{v}_f)$ decreases by $\frac{2}{J} - \frac{2}{J^2} = \frac{2(J-1)}{J^2} > 0$; consequently,

$$O(\mathbf{v}_f) = 1 - \frac{2(J-1)}{J^2} < 1. \tag{3.51}$$

In general, moving more items from the first subset to the second subset as long as $J_1 \geq J_2$, with each next item $O(\mathbf{v}_f)$ diminishes with another quantity $(J-1)/J^2$. Hence, for $G_f=2$, the best solution has $J-2$ items in the first subset and two items in the second subset. A third subset again must contain at least two items. We consider $\mathbf{v}_f = (1,\ldots,1,2,2,3,3)$, and determine the $O(\mathbf{v}_f)$ value, which equals

$$O(\mathbf{v}_f) = 1 - \frac{2J(2J-1)-2}{J^3}. \tag{3.52}$$

Then, we subtract Equation (3.52) for $\mathbf{v}_f = (1,\ldots,1,2,2,3,3)$ from Equation (3.51) for $\mathbf{v}_f = (1,\ldots,1,2,2)$, which produces difference

$$\frac{2(J^2-1)}{J^3} > 0. \tag{3.53}$$

Because Equation (3.53) is positive, it demonstrates that adding the smallest possible third subscale decreases $O(\mathbf{v}_f)$. Further moving items from the first subset to any of the smaller subsets always decreases $O(\mathbf{v}_f)$; hence, moving items from smaller to larger subsets increases $O(\mathbf{v}_f)$. Thus, for $G_f=3$, selection $\mathbf{v}_f = (1,\ldots,1,2,2,3,3)$ produces the largest $O(\mathbf{v}_f)$, but the selection is only accepted as final provided the three subsets satisfy the definition of a scale and there is not a solution for $G_f=2$ that is consistent with the scale definition and has a higher $O(\mathbf{v}_f)$ value.

The partitioning f for which $O(\mathbf{v}_f)$ is maximal and satisfies the definition of a scale is denoted $\operatorname{argmax}_f\left(O(\mathbf{v}_f)\right)$, and the partitioning for which this maximum is attained is denoted $\mathbf{v}_*$. This is the final partitioning, and it may differ from the partitioning that Mokken's automated item selection procedure produced. The final partitioning may not have the greatest H values possible given the number of selected items, and other subsets of equal length may have a higher H value. This result is due to giving priority to realizing the longest scales possible. Straat et al. (2013) implemented a genetic algorithm to find the optimal solution, $\mathbf{v}_*$. The authors compared their method using the genetic algorithm to find $\operatorname{argmax}_f\left(O(\mathbf{v}_f)\right)$ and hence $\mathbf{v}_*$, with Mokken's automated item selection procedure and a modified procedure. The modified procedure allows item l to leave the item subset selected thus far when selection of a next item in the subset causes $H_l < c$. Item l then is candidate for selection in one of the scales, including the scale it left when an item selected after item l left the scale rendered $H_l > c$, and item l may re-enter the item subset. The authors used simulated data having a known dimensional structure and found that the genetic algorithm maximizing Equation (3.50) produced the best item partitioning representing the true dimensionality more often than the two competing algorithms. The automated item selection procedure and the modified version usually found the same partitionings, but the partitionings did not always represent the true dimensionality. The genetic algorithm improved upon its competitors especially when the H_j values were close to lower bound c. Because the procedure Hemker et al. (1995) promoted, running the automated item selection procedure consecutively for $c = 0, .05, .10, \ldots, .55$, likely contains a lower bound c to which one or several H_j values are close, selection or rejection of an item from a subset is a close call. The genetic algorithm optimizing Equation (3.50) does not allow $H_j < c$.

Sample Size and Concluding Remarks

Straat et al. (2014) noticed that, being highly multivariate procedures aimed at optimization, the automated item selection procedure and the genetic algorithm both tend to capitalize on chance (also, see Chapter 2), and investigated the minimum sample sizes needed to obtain replicable scales. Using simulated, one- and two-dimensional, five-point polytomous-item data, the authors varied sample size in 16 steps from $N = 50$ to $N = 3,500$ and replicated a design with factors test length (J), correlation between latent variables θ_1 and θ_2, and item-scalability coefficient H_j with 100 replicated data sets in each design cell. The automated item selection procedure and the genetic algorithm used default lower bound $c = .3$. The baseline item clusters were theoretically defined and confirmed using simulated data based on $N = 1,000,000$. The Per Element Accuracy (PEA; Hogarty, Hines, Kromrey, Ferron, & Mumford, 2005) expressed the proportion of correctly classified items.

In general, requiring that $PEA \geq p$, where p is a user-defined lower bound for PEA that the researcher finds acceptable, distance $\left(\hat{H}_j - c\right)$

predominantly determined minimally required sample size. Other design factors, such as test length, had little effect. When one knows little about the test or the questionnaire and the population, Straat et al. (2014) recommended to always use $N > 1,500$. When using the procedure that Hemker et al. (1995) proposed, increasing c with steps of .05, one of the c values will produce $\left(\hat{H}_j - c\right)$ to be close to 0, which is the value where the greatest sample is needed to obtain results such that $PEA \geq .8$. If possible, one should try to construct items with high H_j, such that $H_j \gg c$. Already for $N = 250$, this situation leads to high PEA values. When $\rho_{\theta_1\theta_2} = .6$ and items discriminate well, the genetic algorithm incorrectly selected all items in one scale $(PEA = .5)$, suggesting we need to accumulate more experience with respect to this procedure.

Van der Ark (2007, 2012) included the automated item selection procedure and the genetic algorithm in the software package mokken. The automated item selection procedure sometimes finds a better partitioning, expressed by a higher $O(\mathbf{v}_f)$ value, than the genetic algorithm. However, if the genetic algorithm uses the partitioning the automated item selection procedure produced as a startset, the genetic algorithm surely produces a partitioning that has an objective function value $O(\mathbf{v}_f)$ that is at least as large as that of the partitioning that the automated item selection procedures produced. The disadvantage of the genetic algorithm is its long computing time, especially when the number of items is large. The resulting solution may be investigated next with respect to local independence [Equation (3.2)] and monotonicity [Equation (3.8)].

Van Abswoude, Vermunt, Hemker, and Van der Ark (2004) and Brusco, Köhn, and Steinley (2015) suggested alternative algorithms for item selection that are interesting. We do not intend to provide a comprehensive treatment of all possibilities and prefer to stay close to the original proposal that Mokken (1971) made and which has become popular among researchers constructing scales (e.g., Sijtsma & Van der Ark, 2017). Thus, we limit attention to Mokken's automated item selection procedure, which is not perfect but functions well as a fast and steady approximation to the objective function in Equation (3.50), which explicitly aims at improving Mokken's procedure.

Local Independence

In nonparametric IRT, methods for investigating local independence are rare and in most cases need more research to establish their usefulness in goodness-of-fit research. We concentrate on the property of conditional association (Holland & Rosenbaum, 1986; Rosenbaum, 1984; Straat et al., 2016) to explore local independence in a set of J dichotomous or polytomous items. For this purpose, we repeat Equation (3.22),

$$\sigma[f_1(\mathbf{Y}), f_2(\mathbf{Y}) \mid h(\mathbf{Z}) = \mathbf{z}] \geq 0, \tag{3.22}$$

and discuss three of its special cases that we use to investigate local independence. For this purpose, we need the following item indices: We use a, b, c, and d to identify items that are in $\mathbf{Y}$, $\mathbf{Z}$, or neither of the two sets. Specifically,

- Items in $\mathbf{Y}$ have indices a and b; hence, we write X_a and X_b;
- Items in $\mathbf{Z}$ have index c; hence, we write X_c; and
- Items neither in $\mathbf{Y}$ nor $\mathbf{Z}$ have index d; hence, we write X_d.

If we indicate items in a broader context, we use indices j, k, and l, as in the present and the other chapters of the monograph. Following Straat et al. (2016), we make the following choices:

- $\mathbf{Y} = (X_a, X_b)$, and $f_1(\mathbf{Y}) = X_a$ and $f_2(\mathbf{Y}) = X_b$.

For $\mathbf{Z}$, we choose

- $\mathbf{Z} = \varnothing$, meaning vector $\mathbf{Z}$ has dimension 0, and $h(\mathbf{Z})$ is undefined; this is Case 1;
- $\mathbf{Z} = (X_c)$, and $h(\mathbf{Z}) = X_c$; this is Case 2; and
- $\mathbf{Z} = (X_c)_{c:c \neq a,b}$, meaning $\mathbf{Z}$ contains $J-2$ item scores but not X_a and X_b, for example, $X_a = X_2$, $X_b = X_3$, so that $\mathbf{Z} = (X_1, X_4, \ldots, X_J)$; define rest score $R_{(ab)} = \sum_{c:c \neq a,b} X_c$, for example, $R_{(2,3)} = X - X_2 - X_3$, and $h(\mathbf{Z}) = R_{(ab)}$; this is Case 3.

Straat et al. (2016) worked out the Cases 1, 2, and 3 as follows:

Case 1: Ignore $h(\mathbf{Z})$ (Rosenbaum, 1984), that is, $\mathbf{Z} = \varnothing$; then, conditional association [Equation (3.22)] reduces to

$$\sigma(X_a, X_b) \geq 0. \tag{3.54}$$

Equation (3.54) says that all inter-item covariances are non-negative; also, see Equation (3.23). One has to check $\frac{1}{2}J(J-1)$ signs. For example, for $J = 20$, one has to check the signs of 190 covariances.

Case 2: Let $h(\mathbf{Z}) = X_c$, then conditional association reduces to

$$\sigma(X_a, X_b \mid X_c = x) \equiv \sigma_{ab|c(x)} \geq 0. \tag{3.55}$$

That is, in each group of people who have the same score x on item c, the inter-item covariances must be non-negative. The number of signs one must check becomes quite large as J grows; that is, for each value of x, one has to check all item triplets, meaning $(M+1)\binom{J}{3}$ trip-

lets in total. For $M+1=5$ and $J=10$, one has to check the signs of 600 inter-item covariances, and for $J=20$, one has to check the signs of 5,700 inter-item covariances.

Case 3: Let $h(\mathbf{Z}) = R_{(ab)}$, then conditional association reduces to

$$\sigma\left(X_a, X_b \mid R_{(ab)} = r\right) \equiv \sigma_{ab|R(r)} \geq 0. \tag{3.56}$$

In each group of people having the same rest score r, the inter-item covariance must be non-negative. The number of signs to check grows further; that is, given $M+1$ different item scores, for each value of r, $r = 0, \ldots, M(J-2)$, one has to check the sign of the conditional covariances in Equation (3.56) for item pair (a,b), and repeat

this for each item pair, meaning $M(J-2)\binom{J}{2}$ pairs in total. For $J=10$,

one has to check the signs of 1,440 covariances, and for $J=20$, the number equals 13,680.

Together, the signs of the large numbers of covariances provide a strong case for or against the model of monotone homogeneity but also present the problem of how to combine so much information to reach one intelligible conclusion. The problem becomes even greater when one assesses other versions of Equation (3.22). For example, one may choose to define functions f_1 and f_2 on subsets of items and define h on a cut score based on another subset. The number of possibilities is enormous and so are the numbers of signs to check, especially as J grows. Thus, it is advisable to make defendable choices and not just try possibilities. Because they are so dominant in measurement, considering inter-item covariances and conditioning on other item scores and rest scores seem reasonable choices.

The CA Method

Conditional association is an implication of the monotone homogeneity model, not of a particular assumption of the model, and Straat et al. (2016) investigated whether negative signs in Equations (3.54), (3.55), or (3.56) provide information about violations of particular assumptions of the monotone homogeneity model. They found that negative conditional covariances did not identify violations of monotonicity. Only specific choices of negative conditional covariances identified weak positive local dependence, $\sigma_{jk|\theta} > 0$, and weak negative local dependence, $\sigma_{jk|\theta} < 0$; see Equation (3.1). We continue using item indices a, b, and c to preserve the relation with the results for conditional association just derived. Reiterating, we know then that $\mathbf{Y} = \left(X_a, X_b\right)$, and $\mathbf{Z} = \left(X_1, \ldots, X_{a-1}, X_{a+1}, \ldots, X_{b-1}, X_{b+1}, \ldots, X_J\right)$; $F_1(\mathbf{Y}) = X_a$ and $F_2(\mathbf{Y}) = X_b$; and

$h_1(\mathbf{Z}) = \varnothing$, $h_2(\mathbf{Z}) = X_c$, and $h_3(\mathbf{Z}) = R_{(ab)}$. Referring to Equations (3.54), (3.55), or (3.56), only the next three conditional covariances proved useful for identifying local dependence (Straat et al., 2016). Using simulated data, these authors showed that if items a and b are negative locally dependent (i.e., $\sigma_{ab|\theta} < 0$), then in samples, one has a probability of approximately .65 to find that $S_{ab|c(x)} < 0$ and .70 that $S_{ab|R(r)} < 0$. These results render $\sigma_{ab|c(x)}$ and $\sigma_{ab|R(r)}$ useful indicators of negative local dependence. Straat et al. (2016) also found that if either item a or item b is in a positive locally dependent item pair (i.e., either $\sigma_{aj|\theta} > 0$ or $\sigma_{bj|\theta} > 0$; $j \neq a, b$), then in samples, one has a probability of approximately .30 to find that $S_{ab|c(x)} < 0$ and .30 that $S_{ab|R(r)} < 0$. Hence, $\sigma_{ab|c(x)}$ and $\sigma_{ab|R(r)}$ are also indicators of positive local dependence.

Straat et al. (2016) defined three indices that quantified the degree to which individual items are likely to be in a pair of locally dependent items. An automated procedure named Conditional Association (CA) identified in each step the item that was involved most frequently in locally dependent item pairs and removed this item from the item set. The result of the procedure is a locally independent set of items, provided it exists for the item set of interest. A study using simulated data showed that the procedures' specificity—the percentage of replicated data sets in which procedure CA correctly identified a locally independent item—was 89.5%. Type I sensitivity was the percentage of replicated data sets in which procedure CA correctly identified a locally dependent item and removed it. Type II sensitivity (caution: This terminology does not refer to the Type II error in hypothesis testing) was the percentage of replicated data sets in which procedure CA correctly removed a locally dependent item pair in each step. Compared to Type II sensitivity, Type I sensitivity aims at retaining as many items as possible and by definition is smaller than Type II sensitivity. The authors found that Type I sensitivity ranged from 24% to 66%, and Type II sensitivity ranged from 42% to nearly 100%. An application of procedure CA to a real-data set proved useful but also suggested that more experience needs to be accumulated. Procedure CA has been implemented in the R package mokken (Van der Ark, 2007, 2012).

The DETECT Method

Weak local independence, $\sigma_{jk|\theta} \geq 0$, all $j, k; j \neq k$ [Equation (3.1)], is the basis of an item selection procedure named DETECT (Stout et al., 1996; Zhang & Stout, 1999a), aimed at finding dimensionally distinct item clusters. Let $S(.,.)$ denote sample covariance. Based on consistency results for dichotomous-item tests (Stout, 1987) and for polytomous-item tests (Junker, 1991), they estimated $\sigma_{jk|\theta}$ twice, using $\hat{\theta} = R_{(jk)}$ (producing a negatively biased covariance) and $\hat{\theta} = X$ (producing a positively biased covariance), yielding $S_{jk|R(jk)}$ and $S_{jk|X}$, and then taking the mean of the two estimates, thus producing smaller bias. Estimated conditional covariances close to 0 suggest weak local independence. We distinguish unidimensionality and multidimensionality.

For unidimensionality, conditional association implies that $\sigma_{jk|R(jk)} \geq 0$. Junker (1993, p. 1370) showed that for the Rasch model (Chapter 4) and for independently and identically distributed (i.i.d.) coin flips producing scores 0/1, we have that $\sigma_{jk|X} < 0$. For the monotone homogeneity model, with the assumptions of unidimensionality, local independence, and monotonicity, as far as we know, the sign of conditional covariance $\sigma_{jk|X}$ is unknown. The magnitude of the conditional covariance is also unknown. Together, these circumstances render $\sigma_{jk|X}$ somewhat problematic for investigating local independence (Sijtsma & Meijer, 2007).

Multidimensionality occurs when, for example, different subsets of items predominantly measure different latent variables. Then, it is reasonable to assume one latent variable, θ_α, that is a linear combination of Q latent variables, θ_q. Total scores such as $R_{(jk)}$ and X summarize test performance but ignore multidimensionality. Zhang and Stout (1999b) showed that, approximately, if $\sigma_{jk|\theta_\alpha}$ and its estimate, which averages the two inter-item covariances conditional on $R_{(jk)}$ and X, are positive, the two items measure the same latent variable, and if the covariances are negative, the items measure different latent variables.

Assume that different subsets of items predominantly measure different latent variables. Then, the DETECT algorithm uses the different signs of $\sigma_{jk|\theta_\alpha}$ to divide an item set into item clusters that together approach weak local independence as well as possible, given all other compositions of item clusters. Let $\wp$ denote an arbitrary partitioning of the J-item set into disjoint item subsets, let $d_{jk}(\wp) = 1$ if items j and k are in the same cluster, and let $d_{jk}(\wp) = -1$ if the items are in different clusters. DETECT uses a genetic algorithm (Zhang & Stout, 1999a) to find the partitioning, $\wp^*$, that maximizes the loss function

$$D_\alpha(\wp) = \frac{2}{J(J-1)} \sum_{j=1}^{J-1} \sum_{k=j+1}^{J} d_{jk}(\wp)\mathcal{E}\left(\sigma_{jk|\theta_\alpha}\right). \tag{3.57}$$

Partitioning $\wp^*$ is considered the best description of the dimensionality of the J-item set.

Comparative Research

Van Abswoude, Van der Ark, and Sijtsma (2004) compared Mokken's automated item selection procedure that selects items satisfying the scale definition in Equations (3.37) and (3.38) with the DETECT procedure. Based on simulated data generated from a model representing the true dimensionality, the authors found that, compared to the automated item selection procedure, DETECT was better capable of retrieving the true dimensionality but that it required a greater sample size to arrive at reliable results and was more vulnerable to chance capitalization when the sample size was small. DETECT assesses weak local independence using conditional covariances, and the automated item selection procedure assesses IRF and ISRF steepness using scalability coefficients, hence the authors recommended combining both approaches' merits into one procedure.

Monotonicity

Many methods are available to estimate IRFs and ISRFs. We discuss only methods that are used in the context of investigating the model of monotone homogeneity.

Binning

The first method assesses an observable consequence of the monotonicity assumption [Equation (3.8)]. First, we consider the case of J dichotomous items. If one assumes the monotonicity assumption to represent latent monotonicity, then manifest monotonicity is

$$P\left(X_j = 1 \mid R_{(j)} = r\right) \text{ monotone non-decreasing in } r,$$

$$\text{for } j = 1, \ldots, J, \text{ and } r = 1, \ldots, J-1. \tag{3.58}$$

As in Chapter 2, rest score $R_{(j)}$ is a sum score based on $J-1$ items, comparable to sum score X, which is based on J items. Given the assumptions of unidimensionality, local independence, and monotonicity, like sum score X, rest score $R_{(j)}$ stochastically orders latent variable θ [Equation (3.17)]. Based on statistical results that we do not discuss here (but see Hemker et al., 1997; Junker, 1991; Stout, 1987, 1990), one estimates latent probability $P\left(X_j = 1 \mid \theta\right)$ by means of $P\left(X_j = 1 \mid R_{(j)} = r\right)$. Let n_r denote the sample frequency that has a score of 1 on item j, and let n_{jr} denote the sample frequency that has both a 1 score on item j and a rest score r, then

$$\hat{P}\left(X_j = 1 \mid R_{(j)} = r\right) = \frac{n_{jr}}{n_r}. \tag{3.59}$$

Junker (1993) showed that for dichotomous items, unidimensionality, local independence, and monotonicity together imply manifest monotonicity [Equation (3.58)], where manifest probabilities are estimated by means of Equation (3.59). Junker and Sijtsma (2000) showed that conditioning on sum score X thus including the target item j, $P\left(X_j = 1 \mid X = x\right)$, does not imply a manifest monotonicity property, hence is uninformative of the fit of the model of monotone homogeneity to the data.

Equations (3.58) and (3.59) suggest a simple method to investigate monotonicity. For increasing values of rest score r, one uses Equation (3.59) to compute conditional probabilities and plots the resulting fractions as a function of r. This is a simple nonparametric regression method called binning (Fox, 2000, pp. 8–13). Manifest monotonicity means that the plot consists of function values that increase monotonically while adjacent values can be equal. Local decreases of fractions as r increases suggest violations of manifest monotonicity, hence of latent monotonicity [Equation (3.8)], and can be tested statistically by means of a normal approximation of the binomial test (Molenaar, 1970; Molenaar & Sijtsma, 2000). Suppose the fractions [Equation (3.59)] decrease between rest scores r and $r+a$, $a=2,3,\ldots$, then statistical testing gains power relative to testing each decrease between two adjacent scale values, when assessing the difference

$$\hat{P}(X_j = 1 \mid R_{(j)} = r) - \hat{P}(X_j = 1 \mid R_{(j)} = r+a), \qquad (3.60)$$

so that the number of statistical tests is reduced. Nonsignificant differences can be ignored, and significant differences provide evidence against manifest monotonicity, hence latent monotonicity and thus the model of monotone homogeneity.

The number of fractions is determined by the number of rest scores, which equals J. When the sample size is small, in particular in the tails of the distribution the rest-score groups—that is, groups consisting of people who have the same rests score—contain few observations, rendering statistical power modest if not negligible. One may merge adjacent rest-score groups to obtain bigger groups to increase power. That is,

$$\hat{P}(X_j = 1 \mid R_{(j)} \in (r,\ldots,r+a)) - \hat{P}(X_j = 1 \mid R_{(j)} \in (r+b,\ldots,r+c)), \quad (3.61)$$

is one of the possibilities, with $a \geq 1$, $b = a+1$, and $c > b$. Van der Ark (2014) discussed estimating confidence envelopes for the manifest IRF defined in Equation (3.58). One can use the confidence envelopes for visually inspecting the curves for decreases.

Order-Restricted Likelihood Ratio Test

Tijmstra, Hessen, Van der Heijden, and Sijtsma (2013) proposed a likelihood ratio test to find out whether for dichotomous item j manifest monotonicity can be rejected. The likelihood ratio test assesses the complete IRF at once. We use notation $P_{jr} = P(X_j = 1 \mid R_{(j)} = r)$ and consider the hypotheses

$$H_0 : P_{j0} \leq P_{j1} \leq \ldots \leq P_{j,J-1}$$

against

$$H_A : H_0 \text{ does not hold.}$$

For the likelihood ratio test, one needs three vectors of maximum likelihood estimates of the conditional response probabilities P_{jr}. First, vector $\hat{\mathbf{P}}_e = (\hat{P}_j,\ldots,\hat{P}_j)$ contains for each of the J elements the mean score on item j

irrespective of the rest score; that is, let n_j be the sample frequency of 1 scores on item j, then

$$\hat{P}_j = \frac{n_j}{N}. \tag{3.62}$$

Vector $\hat{\mathbf{P}}_e$ serves as borderline case for manifest monotonicity. Second, vector $\hat{\mathbf{P}}_u = \left(\hat{P}_{j0}, \ldots, \hat{P}_{j,J-1} \right)$ contains the J observed conditional response probabilities that are computed from Equation (3.59) and need not be consistent with the null hypothesis H_0. Vector $\hat{\mathbf{P}}_m = \left(\tilde{P}_{j0}, \ldots, \tilde{P}_{j,J-1} \right)$ contains conditional response probabilities that are consistent with the null hypothesis H_0 and that are derived by means of the pool-adjacent-violators algorithm from the observed conditional response probabilities in $\hat{\mathbf{P}}_u$.

Let us consider a toy example. For $J = 3$ and $N = 100$, consider target item 1. Based on items 2 and 3, let the frequency distribution for $R_{(1)} = 0, 1, 2$ be $n_r = 25$, 50, 25. Say, for $n_{jr} = n_{1r} = 15, 20, 20$, using Equation (3.62), one finds

$$\hat{P}_1 = \frac{15 + 20 + 20}{25 + 50 + 25} = .55; \text{ hence, } \hat{\mathbf{P}}_e = (.55, .55, .55). \tag{3.63}$$

(Note: If the mean was computed using sum score X, the same result was obtained). Using Equation (3.59), one finds

$$\hat{P}_{j0} = \frac{15}{25} = .60, \, \hat{P}_{j1} = \frac{20}{50} = .40, \text{ and}$$

$$\hat{P}_{j2} = \frac{20}{25} = .80; \text{ hence } \hat{\mathbf{P}}_u = (.60, .40, .80). \tag{3.64}$$

Because $\hat{P}_{j0} > \hat{P}_{j1}$, vector $\hat{\mathbf{P}}_u$ contains one violation of manifest monotonicity. The pool-adjacent-violator algorithm is used to obtain a maximum-likelihood estimate that is consistent with manifest monotonicity and is closest to $\hat{\mathbf{P}}_u$ in a least-squares sense; this is vector $\hat{\mathbf{P}}_m$. To obtain $\hat{\mathbf{P}}_m$, one replaces probabilities $\hat{P}_{j0}$ and $\hat{P}_{j1}$ with

$$\tilde{P}_{j0} = \tilde{P}_{j1} = \frac{n_{j0} + n_{j1}}{n_0 + n_1} = \frac{15 + 20}{25 + 50} \approx .47, \text{ so that } \hat{\mathbf{P}}_m = (.47, .47, .80). \tag{3.65}$$

If $\hat{P}_{j2}$ were smaller than .47, we would have averaged $\tilde{P}_{j0}$, $\tilde{P}_{j1}$, and $\hat{P}_{j2}$, and this would have produced three equal probabilities and, because all three probabilities were involved, $\hat{\mathbf{P}}_m = \hat{\mathbf{P}}_e$.

Vectors $\hat{\mathbf{P}}_m$ and $\hat{\mathbf{P}}_u$ are the basis of a likelihood ratio statistic, based on binomial models for each count n_{jr}, and parameters P_{jr} and n_r, for $r = 0, \ldots, J-1$. Tijmstra et al. (2013) showed that the log-likelihood equals

$$\ell\left(P_{jr} \mid n_{jr}, n_r\right) = n_{jr} \log P_{jr} + \left(n_r - n_{jr}\right) \log\left(1 - P_{jr}\right) + \log\left(\frac{n_r!}{n_{jr}!\left(n_r - n_{jr}\right)!}\right). \qquad (3.66)$$

Because for different rest-score groups, the counts n_{jr} are independent, one can add log-likelihoods. Let $\mathbf{n}_j = \left(n_{j0}, \ldots, n_{j,J-1}\right)$ and $\mathbf{n} = \left(n_0, \ldots, n_{J-1}\right)$; then, the log-likelihood for the full range of rest scores is

$$\ell\left(\mathbf{P} \mid \mathbf{n}_j, \mathbf{n}\right) = \sum_{r=0}^{J-1} \ell\left(P_{jr} \mid n_{jr}, n_r\right). \qquad (3.67)$$

We use the unconstrained $\hat{\mathbf{P}}_u$ to represent the alternative hypothesis H_A; if $\hat{\mathbf{P}}_u$ does not contain violations of manifest monotonicity, then $\hat{\mathbf{P}}_u = \hat{\mathbf{P}}_m$ and H_0 is not rejected. Because $\hat{\mathbf{P}}_u$ uses unconstrained estimates, its log-likelihood, $\ell\left(\mathbf{P}_u \mid \mathbf{n}_j, \mathbf{n}\right)$, is greater than the log-likelihood of the constrained estimates, $\ell\left(\mathbf{P}_m \mid \mathbf{n}_j, \mathbf{n}\right)$. Hence, we define the likelihood ratio as

$$T = -2\left[\ell\left(\hat{\mathbf{P}}_m \mid \mathbf{n}_j, \mathbf{n}\right) - \ell\left(\hat{\mathbf{P}}_u \mid \mathbf{n}_j, \mathbf{n}\right)\right]. \qquad (3.68)$$

Null hypothesis H_0 allows a great variety of orderings $\hat{\mathbf{P}}_m$, hence finding an appropriate null distribution for T—under H_0 we rather speak of the T_0-distribution—is complex, and different $\hat{\mathbf{P}}_m$s yield different probabilities of exceedance, $P\left(T_0 \geq T\right)$. This problem is solved by using the least favorable null distribution, which is the distribution of T_0 that maximizes $P\left(T_0 \geq T\right)$. This choice avoids inflated Type I errors, meaning that the actual significance level cannot exceed the nominal significance level. The value of $\hat{\mathbf{P}}_m$ that produces the least favorable null distribution is $\hat{\mathbf{P}}_e$, which is at the boundary of the space of orderings that are consistent with H_0. Tijmstra et al. (2013) generated under H_0 a large number of data sets using a binomial distribution with $\hat{\mathbf{P}}_e$ and $\mathbf{n}$ based on simulated, horizontal IRFs having no relationship with latent variable θ and, using the null distribution, found that the Type I error did not exceed the significance level. The test's power to detect local decreases of monotonicity was at least .75 for all test lengths when $N > 300$. Because probabilities close to or equal to 0 or 1 are problematic for computing the log-likelihood in Equation (3.66), the authors suggested merging a small-frequency rest-score group with one or more neighbor rest-score groups to obtain a common probability unequal to 0 or 1.

Binning assesses one item at a time by estimating its IRF and evaluating, either visually or statistically, the monotonicity property. Karabatsos and Sheu (2004) used a Bayesian approach including Markov Chain Monte Carlo simulation to assess the monotonicity assumption [Equation (3.8)] for all J items simultaneously. The procedure also allows the assessment of the fit of individual items. Tijmstra, Hoijtink, and Sijtsma (2015) proposed a Bayesian approach to assessing monotonicity that provides an alternative to the frequentist approach geared toward falsification of monotonicity, thus providing only indirect support in favor of manifest monotonicity. Bayes factors quantify the degree of support available in the data in favor of manifest monotonicity or against manifest monotonicity. Informative hypotheses can be used to determine the support for manifest monotonicity over substantively or statistically relevant alternatives to manifest monotonicity.

Kernel Smoothing

Ramsay (1991, 1997) suggested kernel smoothing for estimating IRFs and assessing monotonicity. Kernel smoothing aims at smoothing an irregular curve by means of local averaging in such a way that the largest weight is assigned to each score considered at that moment and smaller weights to their neighborhood scores (Fox, 1997, pp. 27–30; 2000, pp. 17–19). We concentrate on kernel smoothing to estimate IRFs, because it is simple and effective. Ramsay (1991) defined four steps, which he implemented in his software program *Testgraf98* (Ramsay, 2000). Here, it suffices to discuss three steps and leave out one, which mainly has computational relevance:

1. *Rank.* Sum scores $X = x_i$ are replaced with N rank scores. Because for J dichotomous items, only $J+1$ different sum-score groups can be distinguished, within sum-score groups people are randomly ordered, so that in total N different rank scores, denoted $U = u_i$, emerge. Different person scores producing fewer ties may replace sum score X, for example, a total score that weighs items by their capacity to distinguish the people having their sum scores in the lower and upper quartiles of X (Ramsay, 1991).

2. *Enumerate.* Rank scores U are replaced with N consecutive quantiles, denoted $Q = q_i$, covering areas of size $(N+1)^{-1}$ of a distribution that the researcher considers adequate for her test application. Differently, for person i with quantile score q_i, the area to the left of this value equals $i/(N+1)$, $i = 1,\ldots,N$. The standard normal distribution is a popular choice. Compared with binning and the order-restrained likelihood ratio approach, with the present approach, N quantiles, each corresponding to one person, replace a smaller number of J rest-score groups.

3. *Smooth.* For item j, one estimates probabilities $P_j(\theta)$ for increasing values of θ, which is approximated ordinally using target $q_i = \theta$ and several other quantile scores in the neighborhood of θ, such that each target quantile score is replaced with a weighted average of all scores in its neighborhood using a kernel function to specify the weights. For consecutive values of θ, one plots the estimates $\hat{P}_j(\theta)$ and corresponding confidence envelopes that allow the assessment of local decreases.

Kernel smoothing has the following properties that we reiterate briefly. We start by noting that one obtains estimate $\widehat{P}_j(\theta)$ by means of a weighted average of all data for item j, so that

$$\widehat{P}_j(\theta) = \sum_{i=1}^{N} w_i(\theta) X_{ij}. \tag{3.69}$$

Weights $w_i(\theta)$ are greatest for the target $q_i = \theta$ and decrease to 0 as $|q_i - \theta|$ increases. The degree to which the researcher wishes neighboring quantiles to affect the estimate $\widehat{P}_j(\theta)$ is manipulated by means of bandwidth h. Together, these choices define a weight proportional with kernel function K,

$$w_i(\theta) \propto K\left(\frac{|q_i - \theta|}{h}\right). \tag{3.70}$$

Smaller bandwidth h has the effect that fewer observations close to the target quantile are relevant for estimating $P_j(\theta)$, and the choice of h manipulates the tradeoff between bias and precision. That is, if only the target quantile and its nearest neighbors are used, then estimation occurs with little bias but is imprecise. Larger h involves more quantiles, hence produces greater precision but more bias.

The constraints $w_i(\theta) \geq 0$ and $\sum w_i(\theta) = 1$ are needed to guarantee $0 \leq \widehat{P}_j(\theta) \leq 1$. Positive kernel functions and the Nadaraya–Watson weights

$$w_i(\theta) = \frac{K\left(|q_i - \theta|/h\right)}{\sum_s K\left(|q_s - \theta|/h\right)}, \tag{3.71}$$

together attain this goal. Inserting Equation (3.71) in Equation (3.69), one obtains

$$\widehat{P}_j(\theta) = \frac{\sum_{i=1}^{N} K\left(\dfrac{|q_i - \theta|}{h}\right) X_{ij}}{\sum_s K\left(\dfrac{|q_s - \theta|}{h}\right)}. \tag{3.72}$$

Different kernel functions $K(.)$ may be chosen for this purpose, and the standard normal has precisely the property of assigning most weight to the target quantile score, which replaces the mean of the normal density, and decreasing weight as quantile scores are located further away from the target observation. Inserting the standard normal in Equation (3.72), and realizing that the target quantile now equals $z_i = 0$, yields

$$\hat{P}_j(\theta) = \frac{\sum_{i=1}^{N}\left(\frac{1}{\sqrt{2\pi}}e^{-\frac{1}{2}\left(\frac{|q_i-\theta|}{h}\right)}\right)X_{ij}}{\sum_s\left(\frac{1}{\sqrt{2\pi}}e^{-\frac{1}{2}\left(\frac{|q_s-\theta|}{h}\right)}\right)}. \tag{3.73}$$

When the target quantile is closer to the tails of the distribution, quantile scores on either side of the target score are unevenly distributed. This imbalance causes the smoothed function to flatten at the extremes of the scale, but Ramsay (1991) recommended not using corrections because IRFs well targeted with respect to the person distribution flatten out on either side anyway. Confidence envelopes can be estimated by means of

$$\hat{P}_j(\theta) \pm z_{(1/2)\alpha}\left[\sum_{i=1}^{N}w_i(\theta)^2\hat{P}_j(q_i)\left(1-\hat{P}_j(q_i)\right)\right]^{\frac{1}{2}}, \tag{3.74}$$

where $z_{(1/2)\alpha}$ is the deviate under the standard normal that corresponds to a $(1-\alpha) \times 100\%$ confidence interval for $P_j(\theta)$.

Kernel smoothing has been used regularly for assessing IRFs (Meijer & Baneke, 2004; Ramsay, 1991, 2000), assessing IRF fit of parametric IRT models by comparing them to nonparametric IRT models (Douglas & Cohen, 2001), and assessing local independence in item pairs (Habing, 2001). Emons, Sijtsma, and Meijer (2004) used kernel smoothing to estimate person response functions for assessing person fit. Nonparametric regression methods other than binning and local averaging by means of kernel smoothing have been proposed and applied. An example is spline regression (Marsh & Cormier, 2002; Ramsay & Silverman, 1997). Rather than taking the unweighted mean of the observations in a bin, as in binning, spline regression estimates a polynomial regression model in each bin. The regression curves must be such that between bins they connect smoothly to produce one smooth curve across the bins. The transition from one bin to the next must not be abrupt such that jumpy areas in the curve emerge. This guarantees smooth first and second derivatives. Green and Silverman (1994) discuss methods that control the balance between bias and precision and prevent curves from becoming jumpy.

Polytomous-Item Monotonicity

Similar to the impossibility of generalizing the stochastic ordering property in Equation (3.17) from dichotomous-item tests to polytomous-item tests, it proves impossible to generalize manifest monotonicity [Equation (3.58)] to polytomous-item tests. That is, replacing latent variable θ in $P(X_j \geq x \mid \theta)$ with rest score $R_{(j)}$ obtained from adding scores on a set of $J-1$ polytomous items not including polytomous item j yields a manifest probability $P\left(X_j \geq x \mid R_{(j)}\right)$, which is not necessarily non-decreasing in the rest score when latent monotonicity is true (result due to Hemker; cited in Junker & Sijtsma, 2000). Because for polytomous-item tests, the monotone homogeneity model does not imply manifest monotonicity [Equation (3.58)], violations of monotonicity in a sequence of manifest probabilities estimated from the data do not necessarily provide evidence against non-decreasing ISRFs. We discuss a few methods for assessing monotonicity in polytomous-item tests that do assess "manifest monotonicity" but only provide heuristic evidence for or against latent monotonicity.

First, for item j, for each value r of $R_{(j)}$, probabilities $P\left(X_j \geq x \mid R_{(j)}\right)$ can be estimated for $x = 1,\dots,M$. Let n_{jxr} be the sample frequency who have a score x on item j and also a rest score r. Then, in rest-score group r, one computes fraction

$$\hat{P}\left(X_j \geq x \mid R_{(j)} = r\right) = \sum_{y=x}^{M} \hat{P}\left(X_j = x \mid R_{(j)} = r\right) = \frac{\sum_{y=x}^{M} n_{jyr}}{n_r}. \tag{3.75}$$

For increasing values of r, manifest monotonicity can be assessed graphically and local violations of manifest monotonicity can be tested for significance. Estimated ISRFs can provide interesting heuristic information about the item's functioning, but if one considers this kind of information too detailed, an alternative is to study for each item the mean of its M ISRFs. We call this mean the IRF and define it as

$$M^{-1}\mathcal{E}\left(X_j \mid \theta\right) = M^{-1}\sum_{x=1}^{M} P\left(X_j \geq x \mid \theta\right). \tag{3.76}$$

Replacing latent variable θ with rest score $R_{(j)}$, for each r, one can estimate $M^{-1}\mathcal{E}\left(X_j \mid R_{(j)} = r\right)$ as the mean item score for the people who have a rest score r, graphically inspect monotonicity, statistically test for violations of monotonicity, and interpret results heuristically. One can use kernel smoothing to estimate a nonparametric regression curve for the ISRFs and the IRFs and their confidence envelopes; for example, see Sijtsma and Van der Ark (2017).

Junker (1996; Sijtsma & Van der Ark, 2001) showed that the model of monotone homogeneity implies the next "manifest monotonicity" property. Let D_j be a dichotomization of polytomous item score X_j at score x, so that

$$D_j = 0 \text{ if } X_j < x \text{ and } D_j = 1 \text{ if } X_j \geq x, \text{ for } j = 1,\dots,J. \tag{3.77}$$

Define a rest score based on the dichotomized item scores,

$$D_{(j)} = \sum_{k:k \neq j} D_k. \tag{3.78}$$

Given the monotone homogeneity model, one can prove that the following "manifest monotonicity" result is true for polytomous-item tests,

$$P\left(X_j \geq x \mid D_{(j)} = d\right) \text{ monotone non-decreasing in } d,$$

$$\text{for } d = 1, \ldots, J-1, \text{ and } x = 1, \ldots, M. \qquad (3.79)$$

The result in Equation (3.79) is also true if one dichotomizes different items at different values of x. To investigate monotonicity, for each item one may consider Equation (3.79) for each of the M different item scores x and for different versions of rest score $D_{(j)}$. As far as we know, little experience has been accumulated using this method.

Data Example: The Type D Scale14 (DS14) Revisited Using Nonparametric IRT

In Chapter 2, we analyzed data from the Type D Scale-14 (DS14; Denollet, 2005) by means of CTT and factor analysis methods, and here, for the same data, we investigated the scalability and the assumptions of monotonicity and local dependence. Table 2.7 provides the polytomously scored DS14 items, the structure of the questionnaire, and descriptive item statistics. See Chapter 2 for more information about DS14. We ran the DETECT analyses by means of the R package sirt (Robitzsch, 2018). We did all other analyses using the R package mokken (Van der Ark, 2007, 2012). A scale analysis directed at identifying one or more scales that are consistent with the model of monotone homogeneity, and sometimes also the model of double monotonicity, is called a Mokken scale analysis (Molenaar & Sijtsma, 2000; Sijtsma & Van der Ark, 2017). It is sometimes believed that the terminology of Mokken scale analysis refers to the models one assesses, but it does not; it rather refers to the activity of item analysis directed at identifying scales.

First, we did a confirmatory analysis to check whether the seven-item Social inhibition (Si) scale and the seven-item Negative affectivity (Na) scale could be considered two separate scales. Second, we used the automated item selection procedure to investigate whether a procedure blind to the items' content and the researcher's intentions also identified the Si and Na scales as separate. As an aside, we notice that the automated item selection procedure is often used as part of an exploratory strategy, when one lets the data decide on the composition of the scale and the number of different scales. Our use here, however, is confirmatory, aimed at finding support for the hypothesized structure of DS14.

For the Si scale, all H_{jk} values [Equation (3.25)] were positive, and all H_j values [Equations (3.26) and (3.27)] exceeded .4 (Table 3.1, first numerical column). The value of $H_{Si3^*, Si6}$ was just smaller than Ellis' (2014) lower bound [Equation (3.47)]. Because Ellis derived the lower bound for dichotomous items and the DS14 items are polytomous, and because the method's specificity may be low, it is not clear whether we should take this violation seriously. For the Si scale, we found scalability for all items together [Equations (3.28) and (3.29)] equal to $H = .519$ ($SE = .022$) and concluded that H is not significantly greater than .5. Hence, we preferred to label scale Si as medium (i.e., $.4 \leq H < .5$) rather than strong (i.e., $H \geq .5$).

For the Na scale, all H_{jk} values were positive and exceeded Ellis' second lower bound. All H_j values exceeded .48 (Table 3.1, third numerical column). For the Na scale, we found $H = .547$ ($SE = .022$) and concluded that H is greater than .5; thus, we labeled Na a strong scale.

For the entire 14-item scale, item Si3* had negative H_{jk} values with items Na2 and Na5, and several other item pairs had H_{jk} values that were not significantly greater than 0. Hence, the 14 items together do not meet the criteria for a scale [Equations (3.37) and (3.38)]. In addition, the H_{jk} values involving item Si1* or item Si3 and any of the Na items were often smaller than Ellis' lower bound [Equation (3.47)]. Table 3.1 (fifth numerical column) shows the H_j values and scalability H for all 14 items together.

Next, we used the automated item selection procedure to find out whether we could confirm the two-scale structure of DS14. For consecutive lower bounds $c = .00, .05, .10, \ldots, .80$, the automated item selection procedure produced the following results: For $0 < c \leq .30$, except item Si3*, all other 13 items were selected into the same scale. Item Si3* was left out because it had a few small negative inter-item correlations, which is inconsistent with the

TABLE 3.1

Item and Total Scalability Coefficients (Last Row) for the Social Inhibition Scale (Si), the Negative Affectivity Scale (Na), and the Total Scale (DS14). Standard Errors in Parentheses

Item	Si H_j	SE	Na H_j	SE	DS14 H_j	SE
Si1*	.562	(.026)			.343	(.025)
Si3*	.448	(.031)			.224	(.029)
Si6	.493	(.031)			.436	(.024)
Si8	.571	(.025)			.409	(.024)
Si10	.549	(.024)			.373	(.024)
Si11	.493	(.029)			.337	(.026)
Si14	.516	(.028)			.375	(.024)
Na2			.485	(.028)	.281	(.028)
Na4			.561	(.030)	.404	(.026)
Na5			.502	(.028)	.311	(.027)
Na7			.592	(.026)	.433	(.024)
Na9			.514	(.029)	.354	(.026)
Na12			.562	(.025)	.367	(.023)
Na13			.616	(.024)	.423	(.024)
H	.519	(.022)	.547	(.022)	.361	(.018)

Note: SE = standard error for H_j; * = negatively worded item

definition of a scale [Equation (3.37); $\rho_{jk} \geq 0$ is exchangeable with $H_{jk} \geq 0$]. For $.30 < c \leq .40$, the items were partitioned into two clusters. For $c = .40$, all items were correctly classified into an Si cluster and an Na cluster; for $c = .35$, item Si6 was misclassified into the Na cluster. For $.40 < c \leq .70$, the complete set of 14 items fell apart into several small scales and a few unscalable items. For $c > .70$, all items were unscalable. Hence, if at least a weak scale $(.3 \leq H < .4)$ is required, Mokken scale analysis suggests that Social inhibition and Negative affectivity should be measured separately.

We investigated manifest monotonicity [Equation (3.58)] separately for the Si scale and the Na scale. Based on the default values of joining adjacent rest-score groups used by software package mokken, we found only one significant violation:

$$P\left(X_{\text{Si6}} \geq 3 \mid R_{(\text{Si6})} \in \{8,9\}\right) = .108 > .081 = P\left(X_{\text{Si6}} \geq 3 \mid R_{(\text{Si6})} \in \{10,11\}\right).$$

Visual inspection of the ISRFs of item Si6 and the 95% confidence envelopes (Figure 3.6a) suggested that the violation was minor. For bandwidth $h \geq 1.1$, ISRFs from the same item estimated by means of spline regression did not intersect. Therefore, bandwidth $h = 1.1$ was selected for visual inspection of monotonicity of item Si6 using spline regression. We did not find violations of monotonicity (Figure 3.6b). The order-constrained likelihood ratio statistic [Equation (3.68)] applied to the four ISRFs of item Si6 also suggested absence of violation of ISRF monotonicity (Table 3.2).

We investigated local independence for the Si scale and the Na scale. CA procedure did not detect violations for the Na scale, but for the Si scale, items Si3* and Si6 were flagged for negative local dependence. Based on the DETECT procedure $(D = 21.96)$, the optimal partitioning for locally independent item scores assigned the items to the Si cluster and the Na cluster. This result corroborates the result from the automated item selection procedure that the DS14 data are two-dimensional.

The results suggest that DS14 consists of the Si scale and the Na scale and that people measured using DS14 should receive scores on both scales separately. When used separately, there is little evidence against the model of monotone homogeneity, thus implying that the sum score is a robust measure for ordering people with respect to social inhibition and negative affectivity. One may notice that the sum score is not mathematically justified as in SOL [Equation (3.17)], because the items are polytomously scored. However, item Si3* (I often talk to strangers) and item Si6 (I often feel inhibited in social interactions) may be negative locally dependent, and item Si3* has small negative correlations with other items. This result may suggest that including two counter-indicative items in the 14-item DS14 may have had the effect of confusing some of the subjects in the sample, thus rendering the two items problematic.

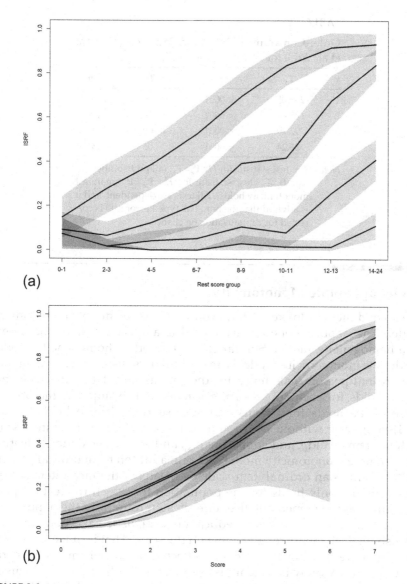

FIGURE 3.6
(a) Visual inspection of the ISRFs and the confidence envelopes of item Si6 based on rest-score groups and (b) spline smoothing.

TABLE 3.2

Order-Constrained Likelihood Ratio Test (Tijmstra
et al., 2013) for $P(X_{Si6} \geq x)$ for $x = 1,2,3,4$

x	T_A	p	T_B	p
1	251.05	<.0001	.00	>.9999
2	185.84	<.0001	.00	.9999
3	97.79	<.0001	.00	.9969
4	36.24	<.0001	.05	.8131

Note: T_A=test statistic for H_0: $\mathbf{P}_e = \mathbf{P}_u$, (all probabilities are
equal), T_B=test statistic for H_0: $\mathbf{P}_m = \mathbf{P}_u$ (manifest
monotonicity holds). All tests are dependent, and it
is advised that the nominal Type I error rate is corrected for capitalization on chance.

Model of Double Monotonicity

The second model Mokken (1971) proposed was the double monotonicity
model. The double monotonicity model is a special case of the monotone homogeneity model. Similar to the monotone homogeneity model,
the double monotonicity model is an ordinal person-measurement model.
The ordinal person-scale holds for the J items that define the scale, but
it also holds for each subset of the J items. For example, if the ordinal
person-scale holds for J items and sum score X, it also holds for the
$(J-1)$-item scale without item j and sum score $R_{(j)}$, the $(J-2)$-item scale
without items j and k and sum score $R_{(jk)}$, and so on. The defining feature
of the double monotonicity model is that in addition to an ordinal person-
scale, it defines an ordinal item-scale by means of the mean item scores.
This scale not only holds for the population for which the test or questionnaire was developed but also for each subgroup from the population,
for example, defined by gender, education level, or social economic class,
but also for each measurement value on the scale. That is, if we know that
item j is more difficult than item k, reflected by their item-mean scores
$\mu_j < \mu_k$, then we know that item j is the most difficult of the two items in
each subgroup and for each measurement value. We emphasize that this
result only concerns the *ordering* of the items by their mean scores, not the
numerical values of the mean scores. That is, people having lower θ values have lower probabilities $P_j(\theta)$ than people having higher θ values, and
we expect the mean item score in the lower-θ subgroup, L, to be smaller
than in the higher-θ group, H, and likewise for item k. However, across
different groups we have $\mu_{j|(\theta)_L} < \mu_{k|(\theta)_L}$ and $\mu_{j|(\theta)_H} < \mu_{k|(\theta)_H}$.

The invariant ordering of items across different θ values is important, for
example, in intelligence testing. In these tests, one often presents items to

respondents in order of descending mean score, so that respondents start with the easiest items while each next item is a little more difficult than its predecessor is. This way, respondents can get used to the unusual assessment situation in which they find themselves and are not immediately discouraged by first problems that are very difficult, while other respondents have some time to overcome possible test anxiety or just nervousness induced by a high-stakes testing occasion. In general, items having an invariant ordering facilitate the interpretation of the scale, because by indexing (not presentation ordering) for each measurement value Item 1 is the most difficult (maximum performance) or the least popular (typical behavior), followed by Item 2, and so on, until Item *J*, which is the easiest or the most popular. Being able to relate item performance to sum score *X* lends the sum score more meaning.

Only a few IRT models imply an invariant item ordering (Sijtsma & Hemker, 1998), among them the double monotonicity model. The model assumes unidimensionality, local independence, and monotonicity, implying an ordinal person-scale, and adds the fourth assumption of non-intersecting IRFs. Different from Equation (3.76), we conveniently define the IRF as the sum rather than the mean of the *M* ISRFs of item *j*,

$$\mu_{j|\theta} = \mathcal{E}\left(X_j \mid \theta\right) = \sum_{x=1}^{M} P\left(X_j \geq x \mid \theta\right). \tag{3.80}$$

For dichotomous items, $M=1$ and thus $\mu_{j|\theta} = \mathcal{E}\left(X_j \mid \theta\right) = P_j(\theta)$. The double monotonicity model assumes that the IRFs of the *J* items do not intersect. Following Sijtsma and Hemker (1998), this means the following. A set of items with $M+1$ ordered item scores has an invariant item ordering if the items can be ordered and numbered accordingly, such that

$$\mathcal{E}\left(X_1 \mid \theta\right) \leq \mathcal{E}\left(X_2 \mid \theta\right) \leq \ldots \leq \mathcal{E}\left(X_J \mid \theta\right); \tag{3.81}$$

Equation (3.81) allows for the possibility that ties exist for particular θ values or intervals of θ, and IRFs may even coincide completely. Using item means, we have

$$\mu_{1|\theta} \leq \mu_{2|\theta} \leq \cdots \leq \mu_{J|\theta}. \tag{3.82}$$

Sijtsma and Junker (1996) studied and defined an invariant item ordering for dichotomous items,

$$P_1(\theta) \leq P_2(\theta) \leq \cdots \leq P_J(\theta), \text{ for all } \theta. \tag{3.83}$$

For three dichotomous items, Figure 3.7a shows the non-intersecting IRFs, $\mu_{j|\theta} = \mathcal{E}\left(X_j \mid \theta\right) = P_j(\theta)$. For three polytomous items, each having four ordered scores, Figure 3.7b shows the three ISRFs of each item, $P(X_j \geq x \mid \theta)$, $x = 1, 2, 3$, and Figure 3.7c shows the sum of the three ISRFs, which is $\mu_{j|\theta}$ [Equation (3.80)]. The ISRFs of the same item cannot intersect by definition, because from Equation (3.9), for two adjacent ISRFs,

$$P(X_j \geq x \mid \theta) - P(X_j \geq x+1 \mid \theta) = P(X_j = x \mid \theta) \geq 0, \tag{3.84}$$

and because Equation (3.84) is true for all M pairs of adjacent ISRFs, it is true for all pairs of ISRFs.

Molenaar (1997) originally defined the polytomous-item double monotonicity model for nonintersecting ISRFs rather than IRFs and required the ISRFs of different items not to intersect. The original double monotonicity model thus required a total of $J \times M$ ISRFs not to intersect. One may notice that, by definition, the M ISRFs of one item cannot intersect but that the sets of M ISRFs of different items can intersect numerously (Figure 3.7b), and that there is no reason to expect the number of intersections is moderate in real data. This presents a practical argument to consider IRFs rather than ISRFs, meanwhile acknowledging that J IRFs also may show numerous intersections. A theoretical argument against using ISRFs is that a set of $J \times M$ nonintersecting ISRFs does not imply an invariant item ordering [Equation (3.81)]. Hence, requiring that $J \times M$ ISRFs do not intersect presents one with an extremely demanding requirement that real data will likely not satisfy that, if satisfied, does not imply a useful measurement property and, consequently, does not serve a useful purpose. By definition, non-intersection of J IRFs implies an invariant item ordering. Thus, acknowledging real data also will not readily satisfy non-intersection of IRFs, we nevertheless focus on IRFs rather than the lower-aggregate level ISRFs.

Goodness-of-Fit Methods

Several methods exist to investigate whether a set of items is invariantly ordered. We briefly discuss the most important methods. First, we define the mean score of item j in a population with cumulative distribution $G(\theta)$ as

$$\mu_j = \mathcal{E}(X_j) = \int \mathcal{E}(X_j \mid \theta) dG(\theta). \tag{3.85}$$

Based on a standard normal θ, the three dichotomous items in Figure 3.7a have means equal to $\mu_1 = .28$ (solid IRF), $\mu_2 = .75$ (dashed IRF), and $\mu_3 = .91$ (dotted IRF), and the three polytomous items in Figure 3.7b and Figure 3.7c have means equal to $\mu_1 = 0.72$ (dotted IRF), $\mu_2 = 1.14$ (dashed IRF), and $\mu_3 = 1.40$ (solid IRF).

For all plots in Figure 3.7, item 1 has the smallest mean and item 3 the greatest mean, but in real data one uses the sample mean to order the J items

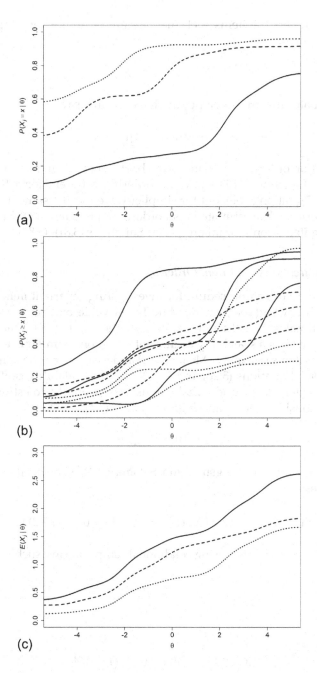

FIGURE 3.7
(a) Non-intersecting IRFs of three dichotomous items; (b) ISRFs of three polytomous items, each having four ordered response categories, ISRFs intersect; (c) non-intersecting IRFs of the three polytomous items in panel (b).

and then numbers the items such that, assuming no ties for simplicity, for example,

$$\bar{X}_1 < \bar{X}_2 < \ldots < \bar{X}_J, \tag{3.86}$$

which estimates the unknown population ordering, say,

$$\mu_1 < \mu_2 < \ldots < \mu_J. \tag{3.87}$$

The population ordering of the items can be different from the sample ordering, so that Equation (3.87) shows an ordering different form the sample ordering in Equation (3.86), but for simplicity, we use the same item indices in both orderings. The estimated item ordering in Equation (3.86) is the basis for ordering items conditional on θ, as in Equations (3.81)–(3.83).

Method Manifest Invariant Item Ordering

We discuss a two-step procedure for investigating invariant item ordering, based on work by Ligtvoet, Van der Ark, Te Marvelde, and Sijtsma (2010). The first step estimates the IRFs and assesses whether a set of IRFs intersect. The second step defines a scalability coefficient based on a similar coefficient for dichotomous items (Sijtsma & Meijer, 1992) and inspired by the item scalability coefficient [Equations (3.26) and (3.27)] and the scalability coefficient for J items [Equations (3.28) and (3.29)]. We use rest score $R_{(jk)}$ to estimate latent variable θ. Consider items j and k, and assume $\mu_j < \mu_k$, so that

$$\mathcal{E}(X_j \mid \theta) \leq \mathcal{E}(X_k \mid \theta). \tag{3.88}$$

Ligtvoet, Van der Ark, Bergsma, and Sijtsma (2011) showed that Equation (3.88) implies

$$\mathcal{E}\left(X_j \mid R_{(jk)} = r\right) \leq \mathcal{E}\left(X_k \mid R_{(jk)} = r\right), \text{ all } r = 0, \ldots, (J-2)M. \tag{3.89}$$

Equation (3.89) is estimated using conditional sample means, such as for item j,

$$\bar{X}_{j|r} = \frac{\sum_{x=0}^{M} n_{jrx}}{n_r}, \tag{3.90}$$

so that

$$\bar{X}_{j|r} \leq \bar{X}_{k|r}, \quad \text{all } r = 0, \ldots, (J-2)M \tag{3.91}$$

provides evidence of Equation (3.88), and

$$\bar{X}_{j|r} > \bar{X}_{k|r}, \quad \text{all } r = 0, \ldots, (J-2)M \tag{3.92}$$

provides evidence against Equation (3.88). Ligtvoet et al. (2010) used a one-sided, one-sample t-test for testing

$$H_0 : \mathcal{E}\left(X_j \mid R_{(jk)} = r\right) = \mathcal{E}\left(X_k \mid R_{(jk)} = r\right)$$

against

$$H_A : \mathcal{E}\left(X_j \mid R_{(jk)} = r\right) > \mathcal{E}\left(X_k \mid R_{(jk)} = r\right), \quad \text{all } r.$$

Rejection of the null hypothesis for at least one rest-score group rejects invariant item ordering for items j and k. If the frequency of observations in a rest-score group is small, to increase statistical power, adjacent rest-score groups can be merged until a preset group size is reached. To avoid testing small and unimportant reversals, one may choose testing differences only when $\bar{X}_{j|r} - \bar{X}_{k|r} > \Delta$, $\Delta > 0$, and a reasonable choice based on experience is $\Delta = .03$ for dichotomous items and based a simulation study (Ligtvoet et al., 2010), $M\Delta = M \times .03$ for polytomous items.

Method manifest invariant item ordering consists of several steps, aimed at selecting a subset, if available, of items having an invariant ordering. A study using simulated data from models either implying or not implying an invariant item ordering showed that the method's sensitivity—the probability of correctly identifying invariant item ordering—and the specificity—the probability of correctly rejecting invariant item ordering—benefit from higher item discrimination, more item scores, and larger sample sizes. The sensitivity is worse for short tests, but then the specificity is better. The steps of method manifest invariant item ordering are the following:

1. For each of the $J-1$ item pairs that includes item j, count the frequency that a reversal [Equation (3.92)] at least equal to $M\Delta$ is significant, and add counts across the $J-1$ item pairs. This produces the total frequency that item j is involved in a rejection of H_0. In step s of the procedure, denote the total frequency U_{js}, $j = 1, \ldots, J$;

2. If for all items $U_{j1} = 0$, then the set of J items are invariantly ordered; otherwise, reject the item having the highest U_{j1} value, and repeat step 1 for the remaining $J-1$ items. Determine the U_{j2} values. If in step s, K items (K at most equal to the number of available items) have the same U_{js} value, reject the $K-1$ items having the smallest item-scalability coefficients, H_j [Equations (3.26) and (3.27)];

3. Reject the item having the highest U_{j2} value and so on.

4. Stop when only items having $U_{js} = 0$ remain.

5. For the remaining item set, compute coefficient H^T.

Coefficient H^T is a measure for the accuracy by which N respondents order J items. To define coefficient H^T, take the transposed data matrix, $\mathbf{X}^T$, thus reversing the roles of persons and items, and for person v, define the vector containing her J item scores, $\mathbf{X}_v = \left(X_{1v}, \ldots, X_{Jv}\right)$, and assume $\sigma(\mathbf{X}_v) > 0$, for $v = 1, \ldots, N$. Let $\sigma\left(\mathbf{X}_v, \mathbf{X}_w\right)$ denote the covariance between the scores of persons v and w. Let $\sigma^{\max}_{(\mathbf{X}_v, \mathbf{X}_w)}$ be the maximum covariance, given the marginal distributions of the scores of persons v and w. Further, we define the sum score on item j as $S_j = \sum_{v=1}^{N} X_{vj}$ and the sum score minus the score of person v as $S_{j(v)} = S_j - X_{vj}$. The sum scores for J items are collected in vectors $\mathbf{S} = \left(S_1, \ldots, S_J\right)$ and $\mathbf{S}_{(v)} = \left(S_{1(v)}, \ldots, S_{J(v)}\right)$. Using this notation, analogous to Equations (3.26) and (3.27), we define a person scalability coefficient,

$$H_v^{\mathrm{T}} = \frac{\sum_{w: w \neq v} \sigma\left(\mathbf{X}_v, \mathbf{X}_w\right)}{\sum_{w: w \neq v} \sigma^{\max}_{(\mathbf{X}_v, \mathbf{X}_w)}} = \frac{\sigma\left(\mathbf{X}_v, \mathbf{S}_{(v)}\right)}{\sigma^{\max}_{\left(\mathbf{X}_v, \mathbf{S}_{(v)}\right)}}. \tag{3.93}$$

Coefficient H_v^{T} expresses the weighted normalized covariance between the item scores of person v and the corrected sum scores on the J items, and this shows the degree to which person v answers items in a manner consistent with the difficulty ordering of the items (except the scores of the person of interest, but without much effect, because even for small N, $\mathbf{S} \approx \mathbf{S}_{(v)}$). Following Equations (3.28) and (3.29), coefficient H^{T} summarizes the information provided by N H_v^{T}s and is defined as

$$H^{\mathrm{T}} = \frac{\sum_v \sigma\left(\mathbf{X}_v, \mathbf{S}_{(v)}\right)}{\sum_v \sigma^{\max}_{(\mathbf{X}_v, \mathbf{S}_{(v)})}}. \tag{3.94}$$

In samples, coefficient $\hat{H}_v^{\mathrm{T}}$ is informative about an invariant item ordering in the following way. Given that one has established that a set of items has an invariant ordering, then, when for most θs or preferably all θs the IRFs are further apart, $\hat{H}_v^{\mathrm{T}}$ will attain a higher value, expressing that the ordering of the item scores is more stable and better in agreement with the item totals. Provided an invariant item ordering holds, many high $\hat{H}_v^{\mathrm{T}}$s provide support for a relatively large spread of the IRFs. As non-intersecting IRFs are further apart, the sample $\hat{H}^{\mathrm{T}}$ value tends to be higher.

For invariantly ordered items, and given local independence, $0 \leq H_v^{\mathrm{T}} \leq 1$ and $0 \leq H^{\mathrm{T}} \leq 1$. Values $H_v^{\mathrm{T}} = 0$ and $H^{\mathrm{T}} = 0$ are obtained when all J IRFs coincide and $\sigma\left(\mathbf{X}_v, \mathbf{X}_w\right) = 0$, all $v \neq w$. Value $H_v^{\mathrm{T}} = 1$ is obtained if $\sigma\left(\mathbf{X}_v, \mathbf{S}_{(v)}\right) = \sigma^{\max}_{(\mathbf{X}_v, \mathbf{S}_{(v)})}$, and value $H^{\mathrm{T}} = 1$ if $\sigma\left(\mathbf{X}_v, \mathbf{S}_{(v)}\right) = \sigma^{\max}_{(\mathbf{X}_v, \mathbf{S}_{(v)})}$, $v = 1, \ldots, N$. A computational study confirmed that H^{T} increased as IRF locations were further apart and IRF slopes were steeper, both IRF characteristics having the effect that non-intersecting IRFs were more widely spaced. Ligtvoet et al. (2010) tentatively suggested that $H^{\mathrm{T}} < .3$ indicates that IRFs are so close together than an invariant item ordering is too inaccurate to be useful, and $H^T \geq .3$, in different gradations, gives evidence of increasing accuracy of an invariant item ordering. The authors noticed that more experience using real data is needed to establish or fine-tune the usefulness of the combination of method manifest invariant item ordering and coefficient H^T.

Sometimes researchers incorrectly assume that Mokken's automated item selection procedure selects scales that have an invariant item ordering. It does not. Sijtsma, Meijer, and Van der Ark (2011) argued that sets of intersecting and non-intersecting IRFs can have the same H values [Equations (3.28) and (3.29)]. Thus, different H values do not distinguish sets of intersecting IRFs from sets of non-intersecting IRFs and are ineffective as selection criteria in item selection aimed at identifying scales that have an invariant item ordering.

Other Methods for Investigating Manifest Invariant Item Ordering

For dichotomous items, Sijtsma and Molenaar (2002, pp. 98–99) called Equation (3.89) the *rest-score method* and recommended investigating invariant item ordering for a set of items pair-wise and for each item pair across all rest-score values, $r = 0, \ldots, J - 2$. The conclusion on whether the set of items is invariantly ordered is based on combining the results with respect to the significant violations found and finally depends on judgment and experience. The authors discussed two variations on the rest-score method. Like the rest-score method, both methods condition on an observable variable that stochastically orders groups on the latent variable θ and thus are ordinal estimates of the ordering on θ.

The *item-splitting method* (Sijtsma & Junker, 1996) is defined as

$$\mathcal{E}\left(X_j \mid X_l = x\right) \leq \mathcal{E}\left(X_k \mid X_l = x\right), \text{ with } x = 0, 1, \text{ and } l \neq j, k. \tag{3.95}$$

Thus, conditioning on a third item l, in both score groups and across each of the other $J - 2$ items, the ordering of the mean scores on items j and k must be the same, except for possible ties. Some conditioning items may produce subgroups that obscure violations of invariant item ordering, but using $J - 2$ conditioning items having varying item means splitting the sum-score distribution at different scale values, one will probably find violations, if any exist. Sijtsma and Molenaar (2002, pp. 104–105) show that the item-splitting method is mathematically equivalent with the $\mathbf{P}(++)/\mathbf{P}(--)$ method (Mokken, 1971, pp. 132–135). We return to this method in the section on reliability based on the non-intersection of IRFs.

The *rest-score splitting method* (Sijtsma & Junker, 1996; also, see Rosenbaum, 1987a) is defined as

$$\mathcal{E}\left(X_j \mid R_{(jk)} \leq r\right) \leq \mathcal{E}\left(X_k \mid R_{(jk)} \leq r\right), \tag{3.96}$$

and

$$\mathcal{E}\left(X_j \mid R_{(jk)} > r\right) \le \mathcal{E}\left(X_k \mid R_{(jk)} > r\right), \tag{3.97}$$

all $r = 0,\ldots,(J-3)$. The rest-score splitting method combines the merits of the other two methods. That is, conditioning on rest score $R_{(jk)}$ provides a more reliable subgrouping than conditioning on just one item score X_l, and by dichotomizing the group for each value of r, one often obtains groups larger than single rest-score groups that have enough power to detect violations of invariant item ordering.

The rest-score method was already generalized to IRFs for polytomous items (section on method manifest invariant item ordering), and the item-score method and the rest-score splitting method may likewise be generalized. The three methods allow the researcher to do a fine-grained analysis of invariant item ordering or lack thereof, in the latter situation learning a lot of detail, such as the region on the scale where IRFs intersect and what this may mean. As with many methods allowing detailed data analysis, the researcher needs to make many decisions that are often arbitrary, and she receives many local results that are difficult to combine into one decision. Sijtsma and Molenaar (2002, Chap. 6) provide a detailed discussion about the pros and cons of the methods. Sijtsma and Junker (1996; also, Ligtvoet et al., 2011; Rosenbaum, 1987a, b) provided more statistical background.

Tijmstra et al. (2011) proved for dichotomous items that using sum score X including item scores X_j and X_k, invariant item ordering [Equation (3.81)] implies

$$\mathcal{E}\left(X_j \mid X = x\right) \le \mathcal{E}\left(X_k \mid X = x\right), \text{ all } x = 1,\ldots,J-1. \tag{3.98}$$

Sum scores $x = 0$ and $x = J$ imply expectation equal to 0 and 1, respectively, and thus do not contain information about the item ordering. Equation (3.98) may be labeled manifest invariant item ordering across the sum score. The result in Equation (3.98) is remarkable, given that conditioning on sum score X is inappropriate for investigating manifest monotonicity and must be replaced by conditioning on a variable that does not involve the target item, such as rest score $R_{(j)}$ [Equation (3.58)]. The rest-score method [Equation (3.89)] and other methods avoiding conditioning on a manifest variable that includes the target items necessarily have to investigate invariant item ordering for pairs of items and then combine the numerous results into one conclusion about invariant item ordering for the whole J-item set. Using item triplets, conditioning on $R_{(jkl)}$, item quartets, and so on, is possible but further complicates the research. Conditioning on the sum score has the great advantage that one can actually study the whole set of J items at once. The steps are to first estimate the overall item ordering from the data and number the items accordingly; see Equation (3.86), and then check

$$\overline{X}_{1|x} \le \overline{X}_{2|x} \le \ldots \le \overline{X}_{J|x}, \text{ all } x = 1, \ldots, J-1. \tag{3.99}$$

Tijmstra et al. (2011) proposed using Kendall's W (Kendall & Babington Smith, 1939), with $0 \le W \le 1$, to express the degree to which J means are ordered similarly for different values of x. $W=0$ means absence of a common ordering, and $W=1$ means perfect ordinal association. Briefly, W is defined as follows. For each $x = 1, \ldots, J-1$, replace the item-mean orderings with rank numbers, R_{jx}, with $R_{j1} = 1$ for the smallest mean, and so on. Next, for each item, add the rank numbers across the different x values to produce sum $S_j = \sum_{x=1}^{J-1} R_{jx}$, and determine $S = \sum_{j=1}^{J-1}(S_j - \overline{S})^2$. Then, define

$$W = \frac{12S}{(J-1)^2(J^3-J)}. \tag{3.100}$$

A version that weighs test-score groups by their size, n_x, gives more prevalent scores greater weight, so that

$$S_j^* = \frac{J-1}{\sum_{x=1}^{J-1} n_x} \sum_{x=1}^{J-1} n_x R_{jx}, \tag{3.101}$$

and the sum of squared differences of the quantities S_{j,n_x}^* to their mean is substituted in Equation (3.100) for S.

The sampling distribution is only known for $W=0$, and can be used to test whether a common ordering is absent. However, this is the opposite of what one wants to know, which is whether the data support an invariant item ordering. This leaves Kendall's W a descriptive measure for all practical purposes. For dichotomous items, Karabatsos and Shue (2004; also, see Tijmstra et al., 2011) discussed a Bayesian approach to finding a posterior p-value that informs one whether the data support a manifest invariant item ordering across the sum score; henceforth, in this subsection, manifest invariant item ordering, for short. We only outline the approach. Let $\mathbf{F} = ((f_{jx}))$ be a matrix containing fractions $f_{jx} = \overline{X}_{j|x}$ [Equation (3.99)], $j = 1, \ldots, J$ and $x = 1, \ldots, J-1$. Matrix $\Delta = ((P_{jx}))$ contains the item-sum-score regressions $P_{jx} = \mathcal{E}(X_j \mid X)$ [Equation (3.98)], $j = 1, \ldots, J$ and $x = 1, \ldots, J-1$. The likelihood of the data $\mathbf{F}$ given Δ is the product of $J(J-1)$ independent variates, n_{jx}, that are binomially distributed, $\mathcal{B}(n_x, P_{jx})$, such that

$$L(\mathbf{F} \mid \Delta) = \prod_{j=1}^{J} \prod_{x=1}^{J-1} \binom{n_x}{n_{jx}} P_{jx}^{n_{jx}} (1 - P_{jx})^{n_x - n_{jx}}. \tag{3.102}$$

The prior distribution of Δ is denoted $\pi(\Delta)$, which is restricted such that all the matrices that are consistent with manifest invariant item ordering have equal positive probability, p [i.e., $\pi(\Delta) > p$] whereas all other matrices that are inconsistent with manifest invariant item ordering have probability 0 [i.e., $\pi(\Delta) = 0$]. Let Ω denote the subset of matrices Δ that is consistent with manifest invariant item ordering, so that $\Omega \subset \Delta$. Combining the likelihood [Equation (3.102)] and the prior produces the well-known order-constrained posterior,

$$\pi(\mathbf{F} \mid \Delta) = \frac{L(\mathbf{F} \mid \Delta)\pi(\Delta)}{\int_\Omega L(\mathbf{F} \mid \Delta)\pi(\Delta)d\Delta}. \tag{3.103}$$

The solution of Equation (3.103) is unfeasible using analytical methods, and one thus tackles the problem through another route, using the Gibbs sampler to mimic a posterior distribution and a chi-square discrepancy measure to obtain a posterior predictive p-value. This p-value shows how likely it is to find data as extreme as $\mathbf{F}$, given that manifest invariant item ordering holds. Tijmstra et al. (2011) discussed a real-data example in which the posterior predictive p-value proved useful in distinguishing the two instances in which manifest invariant item ordering was and was not tenable, and this corresponded to relatively high and low values of Kendall's W, with and without weighing for group size.

Reliability

Meijer, Sijtsma, and Molenaar (1995) argued that the reliability of a single dichotomous item provides information about the slope of the item's IRF. Given that in the nonparametric IRT context one refrains from parametric IRF definitions allowing the estimation of a slope parameter, item reliability thus provides an alternative, a possibility Tucker (1946) already acknowledged. Following the logic of Chapter 2 for reliability definition, let $\pi_{X_j X_j'} = \pi_{jj}$ denote the proportion of people from the target population who obtained a 1 score both on an administration of item j and an independent second administration of item j. In addition, one may notice that the proportions of 1 scores on the two item replications are equal, so that $\pi_j = \pi_j'$. The reliability of item j is the product-moment correlation between item scores X_j and X_j', denoted by $\rho_{jj'}$, and Mokken (1971, p. 143) derived item reliability as

$$\rho_{jj'} = 1 - \frac{\pi_j - \pi_{jj}}{\pi_j\left(1 - \pi_j\right)} = \frac{\pi_{jj} - \pi_j^2}{\pi_j\left(1 - \pi_j\right)}. \tag{3.104}$$

After having read Chapter 2, the reader will already have anticipated the problem of the unavailability of independent replications of the same measurements, meaning that one needs to approximate bivariate proportion π_{jj} based on a single test administration. In the context of nonparametric IRT, one does these using properties of IRFs.

Given J invariantly ordered IRFs, assuming that item j is not the first or the last item in the ordering, find the two nearest neighbors of item j, and index them items $j-1$ and $j+1$, so that

$$P_{j-1}(\theta) < P_j(\theta) < P_{j+1}(\theta). \tag{3.105}$$

Thus, item $j-1$ is the nearest neighbor of item j that is uniformly more difficult, and item $j+1$ is the nearest neighbor that is uniformly easier. Mokken (1971, pp. 142–147) used the two nearest neighbors based on the idea that of all other $J-1$ items, they are the most similar to item j and thus can approximate independent replications of item j the best. For an arbitrary item j, we write

$$\pi_{jj} = \int P_j(\theta)P_j'(\theta)dG(\theta), \tag{3.106}$$

notice that $P_j(\theta) = P_j'(\theta)$, and replace one of the IRFs by a linear function of one of the two neighboring IRFs (i.e., Mokken's Method 1) or both neighboring IRFs (Mokken's Method 2); that is, for scalars a, b, and c, we approximate the IRF of item j by

$$\tilde{P}_j(\theta) = a + b\,P_{j-1}(\theta) + c\,P_{j+1}(\theta). \tag{3.107}$$

Mokken (1971, pp. 142–147) discussed the choice of the scalars, which uses the logic that the IRF closest to the IRF of item j receives all weight (e.g., if this is item $j-1$, then $b>0$ and $c=0$; Method 1) and that the IRF closest to the IRF of item j receives the greatest weight (e.g., if this is item $j-1$, then $b>c>0$; Method 2). For the first and the last items in the ordering, only one neighbor is available as a best approximation to item j and Method 1 is the only possibility. For both methods, scalars $a=0$. Substitution of Equation (3.107) for one of the conditional probabilities in Equation (3.106) and integrating yields

$$\tilde{\pi}_{jj} = a\pi_j + b\pi_{j-1,j} + c\pi_{j,j+1}. \tag{3.108}$$

Bivariate proportion $\tilde{\pi}_{jj}$ can be estimated from Equation (3.108) when sample fractions are inserted for proportions π_j, $\pi_{j-1,j}$, and $\pi_{j,j+1}$, and inserting the result $\hat{\pi}_{jj}$ in Equation (3.104) yields the estimate of item reliability. Sijtsma and Molenaar (1987) proposed a third method for approximating the bivariate proportion $\tilde{\pi}_{jj}$, denoted method MS. For simulated data, Meijer et al. (1995) found that $\hat{\pi}_{jj}$ based on method MS produced an estimated item reliability almost without bias and a standard error acceptably small when sample size was at least 300. For more results, see Zijlmans, Van der Ark, Tijmstra, and Sijtsma (2018) who compared results with those for three different item-reliability methods.

Mokken (1971) used his Methods 1 and 2, and Sijtsma and Molenaar (1987) used their method MS to estimate the reliability of the sum score, defined by

$$\rho_{XX'} = 1 - \frac{\sum_{j=1}^{J}\left(\pi_j - \pi_{jj}\right)}{\sigma_X^2}. \tag{3.109}$$

The bivariate proportions $\hat{\pi}_{jj}$ estimated by means of one of the three approximation methods can be inserted in Equation (3.109) together with the sample fractions $\hat{\pi}_j$ and the sample variance S_X^2, and this yields estimate $\hat{\rho}_{XX'}$. It is unknown whether estimate $\hat{\rho}_{XX'}$ underestimates or overestimates reliability $\rho_{XX'}$, but based on simulated data, Sijtsma and Molenaar (1987) found that the bias was negligible for all three approximation methods, whereas the standard error was comparable to the standard errors of methods $\lambda_3(\alpha)$ and λ_2 (Chapter 2). Molenaar and Sijtsma (1988) generalized method MS for sumscore reliability estimation for polytomous items; also, see Van der Ark (2010).

Data Example: The Type D Scale14 (DS14) Continued

For the Si subscale of the DS14 (Denollet, 2005), using the default values discussed previously, only item pair (Si10, Si11) showed a signification violation of invariant item ordering (Figure 3.8). Because $H_{\text{Si10}} = .55 > .49 = H_{\text{Si11}}$, item Si11 is a candidate for removal. We removed item Si11, and for the remaining

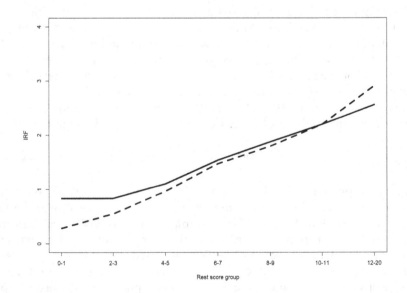

FIGURE 3.8
Mean item scores for items Si10 (dashed curve) and Si11 (solid curve) conditional on combined rest scores $r = \{0,1\}$, $\{2,3\}$, $\{4,5\}$, $\{6,7\}$, $\{8,9\}$, $\{10,11\}$, $\{12,13, \ldots, 20\}$ as estimates of $\mathcal{E}(X_{\text{Si10}} \mid \theta)$ and $\mathcal{E}(X_{\text{Si11}} \mid \theta)$, respectively, show a violation of manifest invariant item ordering.

six items, we found $H^T = .12$. Visual inspection of the IRFs estimated by means of kernel smoothing revealed that the different IRFs were rather similar, implying they were difficult to distinguish and did not allow a clear-cut ordering (Figure 3.9b). The low H^T value is consistent with IRFs being close together and expresses the near impossibility of ordering the items involved. With item Si11, the reliability (method MS) of the sum score of all seven Si

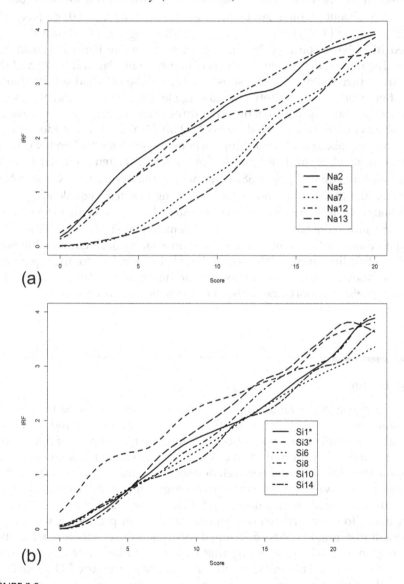

FIGURE 3.9
(a) The estimated IRFs of the five Na items that satisfy manifest invariant item ordering and (b) the six Si items that satisfy manifest invariant item ordering. The estimated IRFs were obtained using kernel smoothing.

items equaled $\hat{\rho}_{XX'}^{MS} = .87$, and without item Si11, the reliability of sum score based on the remaining six items equaled $\hat{\rho}_{XX'}^{MS} = .86$.

For the Na subscale, using the default values discussed previously, the four item pairs (Na4, Na7), (Na4, Na9), (Na7, Na9), and (Na9, Na13) showed a significant violation of an invariant item ordering. After having removed items Na4 and Na9, violations of manifest invariant item ordering were absent, which supports the ordering $\mathcal{E}(X_{Na13}|\theta) < \mathcal{E}(X_{Na7}|\theta) < \mathcal{E}(X_{Na5}|\theta) < \mathcal{E}(X_{Na12}|\theta) < \mathcal{E}(X_{Na2}|\theta)$ for all θ. Figure 3.9a shows the estimated IRFs. Based on $H^T = .38$, the ordering of the five items is considered weak. The figure suggests two subsets of items: Items Na2 and Na12, and the remaining three items. Within a subset, we found the IRFs had similar shapes and often overlapped, but between subsets, clearly, items Na2 and Na12 were invariantly more popular than the other three items. Hence, one may consider the subsets invariantly ordered. Items Na2 and Na12 concern almost obsessive worrying about irrelevant things, whereas items Na5, Na7, and Na13 concern a consistent pessimistic attitude. For the seven Na items, the reliability of the sum score equaled $\hat{\rho}_{XX'}^{MS} = .88$. After having removed items Na4 and Na9, the reliability of the sum score of the remaining five items equaled $\hat{\rho}_{XX'}^{MS} = .84$.

Although of little interest in this data example, we notice that the reliability of the sum score based on all 14 DS14 items equaled $\hat{\rho}_{XX'}^{MS} = .88$. This value was based on a set of items that were not invariantly ordered, possibly biasing the reliability estimate. Moreover, the reliability values for the separate Si and Na scales are almost as high as that for the complete DS14 scale, and this result probably supports the division of the scale in two subscales.

Discussion

Nonparametric IRT is an attempt to construct an ordinal scale for persons and sometimes also for items, putting fewer constraints on the data than several other models do, in an effort to retain a large number of items from the experimental version of the measurement instrument in areas in which attribute theories have not been well developed. In this sense, nonparametric IRT agrees well with social science, psychology, and health science, in which most theories for attributes do not yet have a fine-grained and well-tested level that allows for trustworthy operationalization producing valid measurement instruments. Well-developed theory supports only few attributes, resulting in an attribute structure that requires a tailor-made measurement model (Chapter 1). This model need not be a nonparametric IRT model but rather, depending on the attribute structure, a model of which the mathematical structure aligns with the structure and the dynamics of the target attribute.

For an IRT model, both nonparametric and parametric, to hold in a population, the basic question is whether the target attribute is unidimensional and item responses relate monotonically to the attribute, so that evidence of a person's attribute level accumulates as she solves more items correctly or endorses more items with greater conviction. IRT models vary with respect to the definition of the monotonicity assumption: Nonparametric or parametric, intersecting or non-intersecting, and for parametric definitions, the kind of response probability and the number of item parameters. In addition, parametric IRT models may decompose item parameters into basic parameters, include covariates, or define a multidimensional latent variable, which provide possibilities to quantify solution processes, integrate group structure in the measurement, or represent the complex structure of the attribute, respectively (Chapter 4).

Different IRT models were developed to adapt to items being dichotomously or polytomously scored, multiple-choice items providing an opportunity for guessing for the correct answers, items to have different discrimination power, data being unidimensional or multidimensional, and relations between items to be locally independent or dependent. Each of these characteristics alludes to particular response properties, often strongly related to the intricacies of the items: Number of response categories, constructed response or open response, and different items nested within the same problem setting, as in an exam consisting of testlets (Wainer, Bradlow, & Wang, 2007). Thus, the wealth of IRT models proposed originates from tailoring models to stimulus characteristics and acknowledging that responses may be steered by varying psychological processes (e.g., guessing) requiring models that provide more data flexibility while preserving particular measurement characteristics. All of this makes clear that the genesis of the IRT family is not so much psychological but rather (albeit not exclusively) technical. In Chapter 4, we will see that several researchers proposing IRT models also were driven by motives to better connect measurement to the typical characteristics of psychological attributes. Mokken motivated his nonparametric IRT approach to measurement by his conviction that the social sciences were not ready for more demanding parametric IRT models. This is a fine example of a decision driven by the idiosyncrasies of the substantive field, and we will meet other examples.

All too often, researchers consider nonparametric and parametric IRT different approaches to measurement rather than variations of the same problem that are different with respect to the statistics of model fitting rather than the assumptions that define the models. In Chapter 4, based on work of Hemker et al. (1996, 1997), Hemker et al. (2001), and Van der Ark (2001, 2005), we discuss a hierarchy of IRT models in which nonparametric IRT models lay the foundation and parametric IRT models are the specializations of such models. The hierarchy clarifies that nonparametric and parametric IRT models constitute one family of strongly related measurement models. We also discuss the connection of different IRT models to the structure of particular psychological attributes.

4

Parametric Item Response Theory and Structural Extensions

... the power of this line [the cycloid] to measure time

—**Christiaan Huygens (Yoder, 2005)**

Introduction

Compared to Chapter 3, about nonparametric IRT, formally, the main difference that the reader encounters in Chapter 4 is that now the item response function (IRF) is parametrically defined. This is typical of parametric IRT. First, we give an example and then we provide some historical background before we discuss parametric IRT models in more detail. Since the 1970s, parametric IRT is the dominant measurement model in psychometric studies. We discuss the best-known models for dichotomous and polytomous items and provide taxonomies that reveal the models' hierarchical relations and measurement properties. We also discuss some models that relate IRT to mechanisms explaining the responses that people give to items, using formal aids such as basic parameters explaining item difficulty, covariates explaining the latent variable, and multiple latent variables replacing a single latent variable. The Rasch model is a famous IRT model that receives extra attention despite the model's restrictiveness with respect to the data. However, its theory was extremely well developed and a misfitting Rasch model can be highly informative about one's items. Moreover, the theory of the Rasch model has great didactical value.

One of the first examples of a parametric IRT model was discussed extensively by Lord (1952, p. 5), who defined the IRF, then called the item characteristic curve (Tucker, 1946) or trace line (e.g., Lazarsfeld, 1950), by means of the cumulative normal distribution function of latent variable θ with mean b_j and variance a_j^{-2},

$$P_j(\theta) = \frac{1}{\sqrt{2\pi}} \int_{-\infty}^{z_j} \exp(-z^2/2) dz; \quad z_j = a_j(\theta - b_j), a_j > 0. \tag{4.1}$$

Equation (4.1) is the normal-ogive model. Figure 4.1 shows two IRFs based on Equation (4.1). The ordinate that corresponds with $\theta = b_j$ is $P_j(\theta) = .5$. Hence, for each IRF, response probability $P_j(\theta) = .5$ corresponds with parameter b_j, and this correspondence provides one with a basis for interpreting b_j as the location parameter of item j. The inverse of parameter a_j, which is a_j^{-1}, is the standard deviation of a normal distribution, so that a_j represents the inverse standard deviation: The smaller the standard deviation, the higher a_j. Lord and Novick (1968, p. 368) showed that at $\theta = b_j$, which is the inflexion point of the normal ogive, $(b_j, .5)$, the first derivative of $P_j(\theta)$ equals $a_j / \sqrt{2\pi}$, which is proportional to a_j. Thus, in the neighborhood of $(b_j, .5)$, the rate of change of the IRF is proportional to a_j. Hence, parameter a_j is called the slope parameter.

In the context of measurement, b_j and a_j often have interpretations different from location and slope parameters, respectively. Parameter b_j is interpreted as the item difficulty parameter, because, given a fixed value for slope parameter a_j, a greater b_j value corresponds with an IRF that shifts to the right on the scale, and one needs a higher θ value to obtain a 1 score with the same probability. Theoretically, $-\infty \le b_j \le \infty$, but if the distribution of θ is normed to be, for example, standard normal, one will find almost all values of b_j between −3 and 3. Likewise, parameter a_j is interpreted as the discrimination parameter, because a higher a_j value corresponds with a steeper IRF slope and as a_j increases, probabilities $P_j(\theta)$ to the left of $\theta = b_j$ grow smaller and probabilities $P_j(\theta)$ to the right grow larger. Hence, the item better distinguishes or discriminates people left and right of location b_j as a_j increases. Theoretically, $-\infty \le a_j \le \infty$, but obviously, for most item types and attributes

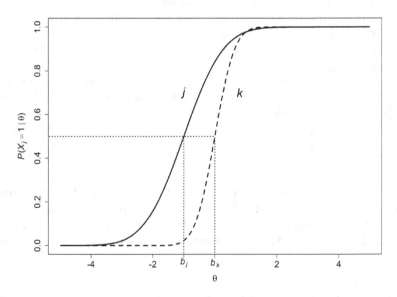

FIGURE 4.1
Two normal-ogive IRFs. For the solid IRF, $a_j = 1$ and $b_j = -1$; for the dashed IRF, $a_k = 2$ and $b_k = 0$.

negative slopes are meaningless. If they appear, negative signs usually are the result of incorrect item coding and can be rendered positive by re-coding. Given a standard normal θ, one will find many values of a_j between .5 and 2.5.

A reader unfamiliar with IRT but acquainted with statistics will understand that Equation (4.1) is only one of the many monotone functions that one could select to replace the nonparametric monotonicity assumption discussed in Chapter 3. One could add person parameters and thus increase the dimensionality of the model and define different item parameters, but one could also choose different function families. All of this happens in parametric IRT, as we will see in this chapter. Each choice defines a new model that assesses particular aspects of the person and the item and their interaction.

We will say a little bit more about the history of parametric IRT before we discuss the most prominent IRT models. Several authors discussed predecessors of modern IRT prior to Lord (1952), such as Richardson (1936), Ferguson (1942), Lawley (1943), Finney (1944), Brogden (1946), and Cronbach and Warrington (1952). Binet (cited in Cronbach, 1949, p. 108) already estimated response functions from real data, plotting for an item the proportion of correct responses as a function of age. In a way, one may frame Thurstone's model of pairwise comparison (Thurstone, 1927a) as a precursor of the normal-ogive model; see Chapter 6. In his Psychometric Monograph, which was his doctoral dissertation at Princeton University a year previously, Lord (1952, p. v) noticed that the theory of mental test scores he presented "is more appropriate and more powerful in the area for which it is intended than is a direct application of the classical theory of errors". One of the key features of the new theory was to investigate "the relation of these variables (i.e., test scores, true scores, and errors of measurement; *the authors*) to the 'ability' involved in taking the test". In our opinion, looking back, this is probably the greatest innovation of the new theory relative to CTT; also, see Tucker (1946). In Chapter 2, we explained that CTT is an error theory without any claims on the composition of the true score. The true score can be anything, ranging from the unidimensional measurement of some attribute of interest to the sum of a score on an arithmetic item, today's temperature at noon, and one's shoe size. IRT is an attempt to relate a test score to an attribute that different items have in common, excluding the ludicrous "arithmetic—temperature—shoe size" example, but also less extreme cases. Thus, unlike CTT, IRT directs attention to *what* the test measures, whereas CTT focused on reliability of *whatever* the test measured. A latent variable represents the common attribute, and empirically investigating the fit of an IRT model to the data entails singling out the items that are connected by the latent variable. We refer to Chapter 2, where we discussed the difference between CTT and factor analysis and noticed the practical necessity of involving factor analysis and similar methods to select items in a scale, in an effort to compensate for what CTT was missing. The missing link in CTT was a restriction on the true scores of different items allegedly measuring the same thing that causes the different items to correlate. An obvious example is a common latent variable that represents the commonality of the different items. Despite this missing link, one usually finds that, also in the CTT context, different items correlate positively. However, there is nothing in CTT that prescribes items to correlate positively, and the fact that in practice they often do is because test constructors understand very well that in a test for a particular attribute the different items represent different aspects of the same attribute. Thus, they construct their items to have a common core. Intriguingly, CTT does not prescribe this, and the absence of a common cause for different items in principle allows meaningless measurement.

Lord (1952, p. 2) pointed out the advantages of the IRT approach (again, the name of IRT was introduced much later; e.g., see Lord, 1980), for example, a standard error of measurement that varies across examinees' abilities and fine-grained possibilities for the selection of items to construct a test for a particular purpose. He also pointed out that the normal-ogive IRF is assumed whenever one computes a tetrachoric correlation between a dichotomous item and a dichotomized, normally distributed measure of ability. Prior to Lord (1952), Lawley (1943) discussed a more restricted version of what might be called early IRT, assuming that all inter-item correlations were equal, but like Lord (1952), he assumed a normal-ogive IRF. Brogden (1946) studied the relation of the distribution of the item difficulties and the correlation of a test (i.e., the sum score; *the authors*) with "a perfect measure of the characteristic the test is intended to measure". The quantification of this characteristic was the true score but defined differently from the true score in CTT; see Equation (2.5) (Lord & Novick, 1968, p. 30). For a set of items to measure the same characteristic to the same degree, a requirement important to his goal we choose to ignore here, Brogden (1946) proposed that their tetrachoric correlations were equal and that their biserial correlations with a continuous, normally distributed and "perfect measure of the common characteristic" were equal. Even though he did not mention it, these proposals imply a normal-ogive model for the item responses. Our main point is that psychometricians began to think about underlying attributes that one expected the items measured *in common*, and thus to compensate the shortcoming of CTT of not defining items' true scores to have a common core.

Tucker (1946) pointed out that CTT implies that perfectly reliable dichotomous items—perfectly discriminating items, in modern IRT parlance—having the same difficulty, divide the group of people who took the test into two subgroups, one subgroup having sum score 0 and another subgroup having sum score $X = J$. He studied correlations between sum score and scale score using a measurement model that assumes a monotone relationship between the ability scale, which was the true-score scale, and the probability of a correct answer. Tucker (1946) assumed a normal ogive for this purpose, with location parameter, s_j [b_j in Equation (4.1)], representing the value on the "scale of the ability" for which the probability of a correct response equaled

.5, and σ_j (i.e., a_j^{-1}), representing the "spread of the curve", hence, the inverse of the slope of discrimination parameter. Assuming a standard normal distribution for the ability scale in combination with Equation (4.1), he derived the correlation of the sum score and the scale score and investigated this correlation as a function of several item and test parameters, studying the conditions under which the sum score best represented the scale value, that is, correlated highest. Tucker (1946) expressed his amazement that items having modest discrimination measured the latent variable better than items having maximum discrimination. This tension between item discrimination and validity—low item discrimination produces a distribution of sum scores that is well-spread along the scale (i.e., interpreted as high validity), and high item discrimination produces a distribution that represents only the extreme sum scores (i.e., interpreted as low validity)—became known as the attenuation paradox (Loevinger, 1954). It was later discussed in the context of IRT as the Rasch paradox (Michell, 2000; Borsboom & Mellenbergh, 2004). We discuss these paradoxes later in this chapter.

A Taxonomy for IRT Models

In this chapter, we collect most of the well-known unidimensional parametric and nonparametric IRT models in two hierarchical taxonomies (Hemker et al., 1997; Hemker et al., 2001; Sijtsma & Hemker, 2000). The taxonomies clarify how the different models are related, which models have particular properties in common, and which models belong to conceptually different sets of models. For example, let us consider the monotone homogeneity model for dichotomous items (Chapter 3), based on the assumptions of unidimensionality, monotonicity, and local independence. The monotone homogeneity model implies the property of stochastic ordering of the latent variable by the observable sum score [SOL; Equation (3.16)]. Then, by logical implication all parametric IRT models for dichotomous items assuming unidimensionality, monotonicity, and local independence imply SOL as well. For example, the normal-ogive model [Equation (4.1)] is an example of a parametric IRT model, based on the three assumptions of the model of monotone homogeneity and adding the restriction of a normal-ogive IRF, which allows the estimation of two parameters for each item and for each person taking the test a position on the θ scale. However, irrespective of this additional restriction on the IRF, one can consider the normal-ogive model as a monotone homogeneity model; hence the SOL property holds also for this parametric version of the monotone homogeneity model. Such special cases, when they fit the data well, allow the ordering of persons by means of their sum scores, which provides an ordering along the scale of θ. The difference with the monotone homogeneity model is that in the latter model one does not have access to the θ estimates, whereas parametric models do facilitate access, which comes with particular advantages but at the expense of worse model fit, because of the additional restriction of a particular parametric function for the IRF.

Before we present the hierarchical taxonomies, we discuss a few basic and well-known IRT models one needs to know for understanding IRT in general and the hierarchical taxonomies in particular. Then, we discuss three different classes of IRT models for polytomous items. These classes are at the core of the taxonomy for polytomous-item IRT models, and all other polytomous IRT-models can be derived from these basic models. Thissen and Steinberg (1986) discussed three model types, two of which are in the hierarchical taxonomy for polytomous IRT models, whereas Hemker et al. (1997) argued that the third type describing multiple-choice responding with guessing did not fit well in their taxonomy and thus was left out. Mellenbergh (1995) and Hemker et al. (2001) discussed all three classes.

Some Basic IRT Models for Dichotomous Items

Guttman Model

Guttman (1944, 1950; see Walker, 1931, for a precursor) proposed a model for scale construction in the context of sociology, generally known as the Guttman scalogram. The basic idea was that if a person responds positively to a dichotomous item, she must also answer positively to items more popular than the first item. Popularity is quantified by means of the proportion of people from the group who answered an item positively. If we denote the popularity of item j by proportion π_j, then if for items j and k we find $\pi_j > \pi_k$, if person i provides a positive answer to item k, she must also provide a positive answer to item j, because item j is the more popular of the two items. Thus, for these two items the data must not show item-score patterns $(x_j, x_k) = (0,1)$, but the other three patterns are allowed, which are $(1,0)$, $(1,1)$, and $(0,0)$. Because exceptions are not allowed, Guttman's model is deterministic, meaning that one inconsistent item-score pattern observed with one person already rejects the model. Guttman did not use IRT concepts to explain his ideas, meaning that he did not use concepts such as dimensionality of measurement, local independence of item scores, and IRFs. His model entailed mainly the idea that someone responding positively to a particular item must not respond negatively to an item that was more popular.

None of the early IRT theorists, such as Lawley, Lord, or Tucker, referred to Guttman's ideas, but later, several IRT researchers did. In IRT notation, the Guttman scalogram can be written as

$$\theta - \delta_j \leq 0 \Leftrightarrow P_j(\theta) = 0 \qquad (4.2a)$$

and

$$\theta - \delta_j > 0 \Leftrightarrow P_j(\theta) = 1 \qquad (4.2b)$$

thus introducing a continuous latent variable and conditional response probabilities. Figure 4.2 shows that the IRFs have a peculiar form, often called step functions, because at item locations δ_j and δ_k the function values $P_j(\theta) = 0$ and $P_k(\theta) = 0$ jump to function values equal to 1. The vertical line segments actually should not be drawn, but then the IRFs would become invisible because the horizontal line segments coincide with the two horizontal axes drawn in the figure. Clearly, the 0 and 1 response probabilities show that the model does not tolerate exceptions. In Figure 4.2, persons v and w having latent variable values $\theta_v \neq \theta_w$ are in the interval between δ_j and δ_k, and the logic of the model implies that both persons have sum scores $X_v = X_w = 1$ with complete certainty. For person u, we have $\theta_u > \delta_k$, so that $X_u = 2$. Given the discussions of CTT and nonparametric IRT in the previous chapters, it is clear that deterministic models for item scores are unrealistic. However, they can provide great discussion tools, as we will see later.

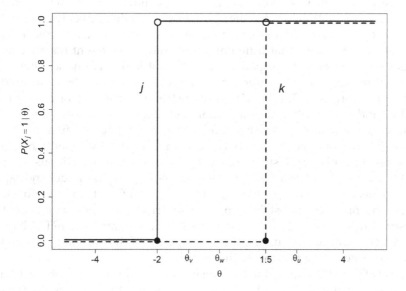

FIGURE 4.2
Two IRFs consistent with the Guttman model. For the solid IRF, $\delta_j = -2$; for the dashed IRF, $\delta_k = 1.5$. Three θ values are shown: $\theta_v = -1$, $\theta_w = 0$, and $\theta_u = 2.5$.

Normal-Ogive Models

Before we discuss the normal-ogive model in some more depth, to distinguish the model from the deterministic Guttman model, we show how it can be expressed as an error model. By doing this, we follow Sijtsma and Junker (2006) who noticed that one of the similarities between CTT and IRT is that both are error models, whereas the difference is that IRT produces a model for random error that opens possibilities for parameter estimation and goodness-of-fit assessment and CTT does not. The idea is simple.

We redefine the normal-ogive model [Equation (4.1)] as a model for comparing a person with an item, where the question is whether person i dominates item j, that is, whether $\theta_i > b_j$, and we assume that a standard normal error, denoted ε_{ij}, affects the comparison. Thus, as in CTT, a random error affects the response process, such that the outcome does not only depend on the difference between person parameter θ_i and item location b_j. We define a fallible response variable

$$y_{ij} = a_j\left(\theta_i - b_j\right) + \varepsilon_{ij}, \tag{4.3}$$

and we rewrite the normal-ogive model as

$$P_j(\theta) = P\left(y_{ij} > 0\right) = P\left[\varepsilon_{ij} > -a_j\left(\theta_i - b_j\right)\right] = \frac{1}{\sqrt{2\pi}} \int_{-\infty}^{a_j\left(\theta_i - b_j\right)} \exp\left(-y^2/2\right) dy. \tag{4.4}$$

With respect to the Guttman model, which excludes random measurement error, Equation (4.4) shows that the normal-ogive model explicitly models the error that governs the comparison between a person and an item, and a slope parameter that shows that different people having different measurement values are differentially responsive to different items. We notice that CTT does not model error as a function of the difference between a sum score and a true score, $X_i - T_i$. What CTT does instead is assuming for groups that $\mathcal{E}(E) = 0$ and $\rho_{EH} = 0$, where H is a random variable that does not include random measurement error, E. This choice renders goodness-of-fit assessment impossible, unless one collects data from an independent repetition of the same test or a parallel test. However, in the absence of repetitions or parallel tests, CTT is a tautological model. The normal-ogive model and other IRT models facilitate the assessment of goodness of fit, which is obviously a necessary condition for establishing the empirical usefulness of models for constructing scales. All of this also holds for nonparametric IRT (Chapter 3). We now continue discussing the assumptions of the normal-ogive model and some interesting features of the model.

Lord (1952, p. 62) enumerated the restrictions and the assumptions of the normal-ogive model. The restrictions are item scores that are dichotomous, a test score that is the sum of the item scores, and a sample size large enough to ignore inaccuracies due to sampling fluctuation. The three assumptions he mentioned are: (1) The matrix of tetrachoric inter-item correlations has rank 1 when appropriate communalities are inserted in the main diagonal. This assumption replaces the assumption of a unidimensional latent variable in IRT. (2) A measure of ability exists such that the probability of a correct answer to any item is a normal-ogive function. In modern IRT terminology, the IRF thus is defined by means of a monotone non-decreasing, normal-ogive function, which defines the scale of the attribute. (3) The ability distribution is Gaussian. For most derivations, the third assumption is unnecessary. Lord (1952, p. 8) discussed local independence without mentioning the term, but did not include it explicitly in his list of assumptions.

Lord and Novick (1968, p. 361) hypothesized that for a particular test one has chosen the dimension of vector $\boldsymbol{\theta}$ such that it spans the complete latent space. Hence, local independence [Equation (3.7)] holds, and Lord and Novick assumed this throughout their discussion of the normal-ogive model. They (ibid., p. 366) defined the normal-ogive model by means of two assumptions that are understandably more up-to-date than the ones Lord (1952) presented 16 years previously:

1. The latent-variable space is one-dimensional, meaning one needs one latent variable, θ.
2. The metric for θ can be chosen such that the regression of the score of each item in the test on θ is the normal-ogive function; see Equation (4.1).

A subtlety is that one can distinguish the situation in which one already knows that one needs only one latent variable, which implies local independence rendering the assumption superfluous, and the situation in which one hypothesizes unidimensionality, implying one needs to study whether local independence is satisfied in one's data (Chapter 3). We will assume, as we did in Chapter 3, that social, behavioral, and health data do not provide enough certainty to accept a hypothesized dimensionality as the truth. Thus, we propose to assume local independence in any IRT model definition.

Finally, as an alternative to simply adopting a set of assumptions as reasonable and justifying use of the normal ogive from that perspective, Lord and Novick (1968, pp. 370–371; also, see Baker, 1992, pp. 8–9) discussed conditions implying the normal-ogive model rather than assuming it. They assumed a latent, continuous item score, denoted Ω_j, with $-\infty \leq \Omega_j \leq \infty$, that they linearly regressed on latent variable θ. For a fixed value of θ, higher values of item score Ω_j mean a higher propensity to produce a positive response, hence, a 1 score. Figure 4.3 shows a linear regression function and several distributions of Ω_j conditional on θ. These conditional distributions play the role of distributions of residuals conditional on the independent variable in a linear regression model. Now, assume we dichotomize each of these continuous distributions at the same threshold, ω_j, so that, if $\Omega_j < \omega_j$, then $X_j = 0$, and if

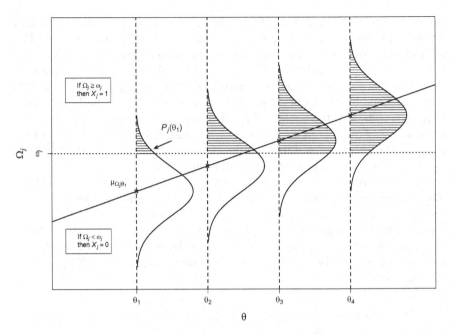

FIGURE 4.3

Linear regression of latent response variable Ω_j on latent variable θ, and four normal distributions of Ω_j conditional on θ with equal variances and varying conditional mean $\mu_{\Omega_j|\theta}$. When plotted as functions of θ, response probabilities $P_j(\theta)$ constitute a normal ogive; see Figure 4.1. The dotted horizontal line corresponds to $\omega_j = 0$; the regression line is $\Omega_j = .5 \times \theta$.

$\Omega_j \geq \omega_j$, then $X_j = 1$. Let the conditional distributions of Ω_j given θ have means $\mu_{\Omega_j|\theta}$ and variances $\sigma^2_{\Omega_j|\theta}$. The assumptions of the linear regression model that are relevant here are that (1) all conditional distributions of Ω_j given θ have equal variance, $\sigma^2_{\Omega_j|\theta} = \sigma^2$, (2) the conditional distribution of Ω_j given θ is normal, $\mathcal{N}\left(\mu_{\Omega_j|\theta}, \sigma^2\right)$, and (3) θ has no measurement error, hence it is fixed. From these assumptions, it follows that the regression function passes through the conditional means of the distributions of Ω_j, $\mu_{\Omega_j|\theta}$ (Lord & Novick, 1968, pp. 370–371), shown in Figure 4.3. The constant threshold ω_j cuts through the conditional distributions and separates the unshaded from the shaded areas. The shaded areas correspond to $P_j(\theta)$. When plotted, the conditional probabilities produce precisely the normal-ogive IRF.

Because the regression model implies the normal-ogive model, one might argue that checking the regression model for one's data provides a way to investigate the fit of the normal-ogive model to the data. However, this approach is unfeasible because in practice the latent response-variable Ω_j is unknown and cannot be regressed on the latent variable θ; that is, the model is purely hypothetical and serves as an aid to justify the use of a normal-ogive IRT model.

Birnbaum (1968, p. 404) proposed an extension of the normal-ogive model to a model with three item parameters. We denote the normal ogive in Equation (4.1) by $\Phi(z_j) = \Phi\left[a_j\left(\theta - b_j\right)\right]$; then, Birnbaum's model is defined as

$$P_j(\theta) = c_j + \left(1 - c_j\right)\Phi\left[a_j\left(\theta - b_j\right)\right]. \tag{4.5}$$

Parameter c_j is the lower asymptote of the IRF; that is,

$$c_j = \lim_{\theta \to -\infty} P_j(\theta). \tag{4.6}$$

The interpretation of the c parameter is relevant when items have the multiple-choice format. Then, people having little knowledge or a badly developed skill, that is, a low θ value, still have a chance greater than 0 of producing a positive answer. For example, when a multiple-choice item has four response options, one of which is correct, one could assume that $c_j = .25$, but it may be more realistic that many people with low θ values will not blindly guess for the correct answer. That is, $c_j = .25$ is not likely in real life, and this is what numerous data analyses have shown. For example, one of the three distractors (i.e., incorrect options) may be so unlikely that even people with low θ values are able to eliminate it and then guess from the remaining three options with higher success probability, that is, $c_j > .25$. This may not be true for all low-θ people, but this is something that one cannot know and definitely calls for c_j to be estimated from the data. Another possibility is that for low-θ people one or more distractors look attractive, thus drawing them to this distractor, reducing the success probability below blind-guessing level; that is, $c_j < .25$. The c parameter has been called the guessing parameter, but this name clearly is too restrictive.

A normal-ogive model with only a location parameter is easily defined when $a_j = a$, meaning equal discrimination for all items in the test, and arbitrarily fixing $a = 1$ yields

$$P_j(\theta) = \Phi\left(\theta - b_j\right). \tag{4.7}$$

This one-parameter normal-ogive model is more restrictive than the model with three parameters [Equation (4.5)] and may easily fail to fit the data when there is no reason to expect that all items will discriminate equally well.

Because it proved mathematically convenient, psychometricians such as Rasch (1960) and Birnbaum (1968; previously published in internal reports in the 1950s, unavailable to the authors; also, see Maxwell, 1959) replaced the normal-ogive model with exponential models that are at the core of this chapter. We use Latin letters to denote item parameters in the normal-ogive models and Greek letters to denote item parameters in exponential IRT models. The unidimensional, logistic IRT model with two item parameters that is widely used since the 1970s is the two-parameter logistic model (Birnbaum, 1968, pp. 399–402), which is defined as

$$P_j(\theta) = \frac{\exp\left[\alpha_j\left(\theta - \delta_j\right)\right]}{1 + \exp\left[\alpha_j\left(\theta - \delta_j\right)\right]}. \tag{4.8}$$

Item parameter δ_j represents the location or difficulty parameter, and item parameter α_j represents the slope or discrimination parameter. Like the normal ogive, the logistic function has its inflexion point at coordinates $(\delta_j, .5)$ and a slope parameter equal to the inverse of the standard deviation, in this case, of the logistic density. To amplify the differences, Figure 4.4 shows a normal-ogive IRF [Equation (4.1), solid curve] and a logistic IRF [Equation (4.6), dotted curve] with equal location parameters, $b_j = \delta_j = .5$, and equal slope parameters, $a_j = \alpha_j = 1.5$. The result is two curves that have the same location but different slopes (but notice that the slope parameters in different models are equal). The slopes are different because the logistic density has smaller variance than the normal density, which is reflected by the larger slope parameter that is the inverse of the standard deviation of the logistic. Hence, if the logistic IRF (dotted curve) has the same slope as the normal-ogive IRF (solid curve), its slope parameter a_j is greater than α_j. It has been established that if the logit, $\alpha_j\left(\theta - \delta_j\right)$, is multiplied by the constant $D = 1.702$, then the

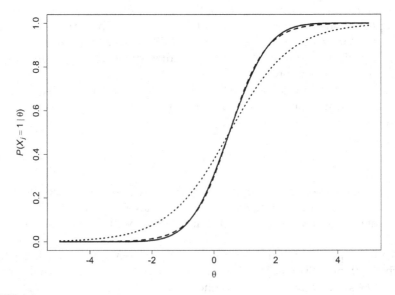

FIGURE 4.4
Two IRFs (solid and dotted curves) with the same item parameters. The solid curve is a normal-ogive IRF [Equation (4.1)] with $a_j = 1$ and $b_j = 0.5$. The dotted curve is a logistic IRF [Equation (4.8)] with $\alpha_j = 1$ and $\delta_j = 0.5$. The dashed curve is a logistic IRF with $\alpha_j = 1$, $\delta_j = 0.5$, and $D = 1.702$ [Equation (4.10)], which is barely distinguishable from the solid curve.

difference between the normal-ogive IRF and the logistic IRF never exceeds .01 (e.g., Camilli, 1994); that is, denoting the logistic function in Equation (4.8) by $\Psi\left[\alpha_j\left(\theta-\delta_j\right)\right]$, we have

$$\left|\Phi\left[a_j\left(\theta-b_j\right)\right]-\Psi\left[\alpha_j\left(\theta-\delta_j\right)\right]\right|<.01, \quad -\infty\leq\theta\leq\infty. \tag{4.9}$$

This small difference can be ignored for all practical purposes. In older writings, one regularly sees Equation (4.9) written as

$$P_j(\theta)=\frac{\exp\left[D\alpha_j\left(\theta-\delta_j\right)\right]}{1+\exp\left[D\alpha_j\left(\theta-\delta_j\right)\right]}=\frac{\exp\left[\alpha_j^*\left(\theta-\delta_j\right)\right]}{1+\exp\left[\alpha_j^*\left(\theta-\delta_j\right)\right]}. \tag{4.10}$$

The dashed curve in Figure 4.4 nearly coincides with the normal ogive in Equation (4.1).

When the normal-ogive was standard in IRT, Equation (4.10) was necessary to justify using logistic IRFs, but since the exponential family of equations has become dominant, few researchers include constant D anymore in the model equations and instead simply write α_j for the logistic slope parameter. Baker (1992, p. 18) mentions three advantages of using logistic IRFs. First, the use of a closed form logistic IRF frees the psychometrician from awkward numerical integration to arrive at a response probability. Second, the mathematics of the logistic density function is much simpler than that of the normal density. Third, the logistic approach facilitates handling polytomous ordered and unordered item scores. We add that, from a substantive perspective, there is no reason to think that one monotone response function must be preferred over another as long as enough flexibility to accommodate the uncertain nature of person responses to cognitive, personality, and attitude items is taken into account.

One-Parameter Logistic Model or Rasch Model

The Model, Separability of Parameters

The Rasch model (Fischer & Molenaar, 1995; Rasch, 1960, 1968; Von Davier, 2016), or one-parameter logistic model, is the simplest of the IRT models we discuss in this chapter, but its simplicity has not prevented many researchers from using the model to analyze their data. Psychometricians have devoted several books (also, see Andrich, 1988b; Bond & Fox, 2015; Fischer & Molenaar, 1995; Von Davier & Carstensen, 2007; Wright & Stone, 1979) and hundreds of articles to study the Rasch model's properties. Originally, there seemed to be a divide between European and Australian researchers on the one hand, several of whom were enthusiastic proponents of the Rasch model, and American researchers on the other hand, who were inclined to use more flexible IRT models to construct their scales, in particular, in educational measurement. One reason to use the Rasch model was that its simplicity rendered the estimation of its parameters and the goodness-of-fit research feasible, given the computational possibilities in the 1980s and 1990s. Nowadays, the availability of powerful computers facilitating computer-intensive statistics allows the estimation of models previously thought impossible and thus has eliminated this motivation. However, as we will see, the Rasch model has interesting properties justifying some attention.

Originally, some theorists assigned the Rasch model the allegedly unique merit that the model enables the measurement of persons independent of the difficulty of a set of items, but Irtel (1995) showed that the nonparametric model of monotone homogeneity and the two-parameter logistic model also have this property. We do not go into this property for the two-parameter logistic model and refer to Irtel (1995) instead, but recall that in Chapter 3, we discussed the SOL property of the model of monotone homogeneity, meaning that given the model, an ordering on sum score X implies (with random error) the ordering on latent variable θ, and noticed that if SOL holds for the set of J items, it also holds for any subset of these items. On the contrary, CTT confounds a person's scale value with the properties of the items used for measurement. For example, if one tests the same person twice, first using an easy arithmetic test and later a difficult arithmetic test, the person obtains high and low scale values for arithmetic, respectively, reflecting not only her arithmetic ability but also the varying difficulty level of the items. The Rasch model allows the determination of item properties independent of the group of tested people's distribution of scale values, whereas in CTT the same item is difficult in one low-level group and easy in another high-level group, and CTT is unable to separate item and person properties.

We first notice the properties of the latent variable scale of the Rasch model, and then we discuss the separability of the person and item parameters. The Rasch model is defined as

$$P_j(\theta) = \frac{\exp(\theta - \delta_j)}{1 + \exp(\theta - \delta_j)}. \tag{4.11}$$

In Equation (4.11), the probability of a 1 score is a monotone increasing function of latent variable θ. The Rasch model only has a location parameter to distinguish different items. Figure 4.5 shows two IRFs consistent with Equation (4.11). Items having different location parameters run parallel, and if $\delta_j < \delta_k$, then one can obtain one IRF from the other by shifting the IRF of item j a distance $|\delta_j - \delta_k|$ to the right or the IRF of item k along the same distance to the left. Translations of the kind $\theta^* = \theta + h$, h is a scalar, so that also $\delta_j^* = \delta_j + h$, leave the response probability in Equation (4.11) unchanged. One can see this by inserting θ^* and δ_j^* in Equation (4.11). This produces an exponent equal to $\theta^* - \delta_j^* = (\theta + h) - (\delta_j + h) = \theta - \delta_j$, and Equation (4.11) remains unchanged.

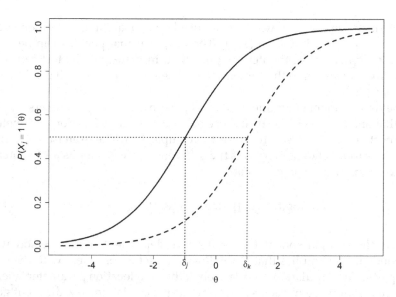

FIGURE 4.5
Two Rasch IRFs. For the solid IRF, $\delta_j = -1$, for the dashed IRF, $\delta_k = 1$.

A scale that allows transformations $\theta^* = \theta + h$ but no other transformation is a difference scale.

To show that person parameters can be evaluated separately from item parameters, we write the Rasch model as the logit of the response probability $P_j(\theta)$ [Equation (4.11)],

$$\text{logit}(P_j(\theta)) = \ln\left(\frac{P_j(\theta)}{1 - P_j(\theta)}\right) = \ln\left[\exp(\theta - \delta_j)\right] = \theta - \delta_j. \quad (4.12)$$

On the right-hand side, we have a linear IRF of latent variable θ with intercept equal to $-\delta_j$ and slope equal to 1. For two persons v and w with parameter values θ_v and θ_w, the difference of the logits is

$$\text{logit}(P_j(\theta_v)) - \text{logit}(P_j(\theta_w)) = \theta_v - \theta_w, \quad (4.13)$$

which does not depend on the item parameter δ_j. Hence, comparison of person measurement values on the logit scale is independent of the item used for the comparison. In addition, because the linear IRF has slope 1, a unit change on the latent variable scale, defined by $\theta_v - \theta_w = 1$, corresponds to a unit change on the logit scale. It follows that the unit of measurement in the Rasch model is the logit. However, the Rasch model does not impose a zero-point on the θ-scale. Thus, the scale has a fixed unit and an arbitrary zero-point, leaving the difference $\theta_v - \theta_w$ intact and allowing transformations

$\theta^* = \theta + h$, as we also saw when we focused on probabilities [Equation (4.11)] rather than logits [Equation (4.12)]. The property that persons can be compared independent of the item implies that, in principle, the test can have any difficulty level and the comparison of the same persons will produce the same result.

The Rasch model is symmetrical in its person and item parameters, implying that one can assess the difference between two item location parameters irrespective of the person used for the comparison. The exercise needed to show this independence requires the difference between logits for two items j and k, and one person v, so that

$$\text{logit}\big(P_j(\theta_v)\big) - \text{logit}\big(P_k(\theta_v)\big) = \delta_k - \delta_j. \tag{4.14}$$

Hence, the comparison of two items is independent of the person who responded. This implies that, if the Rasch model is consistent with the data in a particular population, comparison of the item location parameters yields the same difference in each subgroup of this population, say, different gender groups, different age groups, and different ability level groups.

A word of caution is in order. First, the results in Equations (4.13) and (4.14) are theoretical, but if one estimates person and item parameters from data collected by means of a random sample from the population of interest, one finds that the estimates' precision depends on the sample size (for item parameter estimates), test length (for person parameter estimates), and also on the position of the distribution of the latent variable relative to the item parameters. The greater the distance of the person parameter and the tests' item location parameters, the greater the standard error of the estimated person parameter, and similarly, the greater the distance between the item location parameter and the distribution of the latent variable, the greater the standard error of the estimated item location parameter. Thus, estimation precision of person and item parameters also depends on the distributions of item and person parameters, respectively. To better understand this, the concept of the information function is helpful, as we will see later in this chapter.

The property that people can be measured independent of the set of items used, that is, using any subset of items from the test or the larger item domain for which the Rasch model holds irrespective of their difficulties, is highly convenient in educational measurement when different cohorts from the same population were tested using partly different item sets for the same attribute. The different item sets need to contain some common items known as anchor items, because if one calibrated the scales to be, for example, standard normal, the item parameters may differ by a linear transformation. One needs the calibration of the same items on the different scales to find how the calibrations differ and thus to find the appropriate transformation to equate the scales.

Sufficiency and Estimation

Now that we have discussed some properties of the Rasch model, we enumerate its assumptions and then discuss the fourth assumption we have not encountered so far (Fischer, 1974, 1995a). The first three assumptions are those of the model of monotone homogeneity: unidimensional latent variable θ, local independence, and monotone IRF. The fourth and novel assumption is sufficiency of the sum score for persons, X_i, for the maximum likelihood estimation of latent variable θ_i, and the sufficiency of the sum score for items, S_j, for the maximum likelihood estimation of item location δ_j. We show next (Box 4.1) what sufficiency of statistics for the maximum likelihood estimation of parameters means.

BOX 4.1   Sufficiency of Total Scores for
Maximum Likelihood Estimation

For N persons with parameters $\theta = (\theta_1,...,\theta_N)$ and J dichotomous items with parameters $\delta = (\delta_1,...,\delta_J)$, of which the item scores x_{ij} are collected in data matrix $\mathbf{X}$, the likelihood of the Rasch model equals

$$L(\mathbf{X}\mid\theta,\delta) = \prod_{i=1}^{N}\prod_{j=1}^{J}P_j(\theta)^{x_{ij}}\left[1-P_j(\theta)\right]^{1-x_{ij}}. \qquad (4.15)$$

Inserting the Rasch model [Equation (4.11)] in Equation (4.15) yields

$$L(\mathbf{X}\mid\theta,\delta) = \prod_{i=1}^{N}\prod_{j=1}^{J}\frac{\exp(\theta_i-\delta_j)^{x_{ij}}}{1+\exp(\theta_i-\delta_j)} = \prod_{i=1}^{N}\prod_{j=1}^{J}\frac{\exp\left[x_{ij}(\theta_i-\delta_j)\right]}{1+\exp(\theta_i-\delta_j)}. \qquad (4.16)$$

We consider what happens when we multiply the terms in row i across items, $j = 1,...,J$. One may notice that in the product,

$$x_{ij} = 1 \text{ implies } P_j(\theta) = \frac{\exp(\theta_i-\delta_j)}{1+\exp(\theta_i-\delta_j)}, \text{ and}$$

$$x_{ij} = 0 \text{ implies } 1-P_j(\theta) = \frac{1}{1+\exp(\theta_i-\delta_j)}. \qquad (4.17)$$

Because vector $\mathbf{x}_i$ contains x_i 1 scores and $J-x_i$ 0 scores, parameter θ_i occurs x_i times in the product of all J terms. That is,

$$L(\mathbf{x}_i\mid\theta_i,\delta) = \prod_{j=1}^{J}\frac{\exp\left[x_{ij}(\theta_i-\delta_j)\right]}{1+\exp(\theta_i-\delta_j)} = \prod_{j=1}^{J}\frac{\exp(x_{ij}\theta_i)\exp(-x_{ij}\delta_j)}{1+\exp(\theta_i-\delta_j)},$$

and defining

$$C_{i,J}(\theta_i,\delta)=\left\{\prod_{j=1}^{J}\left[1+\exp(\theta_i-\delta_j)\right]\right\}^{-1},$$

we have

$$L(\mathbf{x}_i\mid\theta_i,\delta)=C_{i,J}(\theta_i,\delta)\exp(x_i\theta_i)\prod_{j=1}^{J}\exp(-x_{ij}\delta_j). \tag{4.18}$$

For parameter θ_i, the likelihood depends on the sum score x_i, that is, the number of items answered positively, but not on which of the J items were answered positively. In the next step, we take the product in Equation (4.18) across the rows of the data matrix,

$$L(\mathbf{X}\mid\theta,\delta)=\prod_{i=1}^{N}L(\mathbf{x}_i\mid\theta_i,\delta). \tag{4.19}$$

Then, defining

$$C_{N,J}(\theta,\delta)=\left\{\prod_{i=1}^{N}\prod_{j=1}^{J}\left[1+\exp(\theta_i-\delta_j)\right]\right\}^{-1}, \tag{4.20}$$

we write the likelihood as

$$L(\mathbf{X}\mid\theta,\delta)=\prod_{i=1}^{N}\left[\prod_{j=1}^{J}\frac{1}{1+\exp(\theta_i-\delta_j)}\right]\prod_{i=1}^{N}\left[\exp(x_i\theta_i)\prod_{j=1}^{J}\exp(-x_{ij}\delta_j)\right],$$

and re-arranging terms, we obtain

$$L(\mathbf{X}\mid\theta,\delta)=C_{N,J}(\theta,\delta)\exp\left[\sum_{i=1}^{N}\theta_i x_i-\sum_{j=1}^{J}\delta_j s_j\right]. \tag{4.21}$$

Now, it is clear that the likelihood for the complete data matrix $\mathbf{X}$ depends on the marginals of $\mathbf{X}$, that is, on $\mathbf{x}=(x_1,\dots,x_N)$ and $\mathbf{s}=(s_1,\dots,s_J)$, but not on the separate item scores, x_{ij}.

Equation (4.21) is a likelihood function from the exponential family (e.g., Molenaar, 1995), and the observable sum scores $\mathbf{x}=(x_1,\dots,x_N)$ and $\mathbf{s}=(s_1,\dots,s_J)$ are sufficient statistics for the estimation of parameters $\theta=(\theta_1,\dots,\theta_N)$ and $\delta=(\delta_1,\dots,\delta_J)$, respectively. Several estimation methods are available. For the Rasch model, of particular relevance is conditional maximum likelihood. We briefly outline its procedure and explain its relevance.

Let the item-score vector of person i be $\mathbf{x}_i$. The conditional likelihood is

$$L\left(\mathbf{x}_i \mid x_i, \theta_i, \delta\right) = \frac{P(\mathbf{x}_i \wedge x_i \mid \theta_i, \delta)}{P(x_i \mid \theta_i, \delta)} = \frac{P(\mathbf{x}_i \mid \theta_i, \delta)}{P(x_i \mid \theta_i, \delta)}. \tag{4.22}$$

In the numerator in the fraction on the right-hand side, probability $P(\mathbf{x}_i \mid \theta_i, \delta) = L\left(\mathbf{x}_i \mid \theta_i, \delta\right)$; see Equation (4.18). In the denominator, probability $P(x_i \mid \theta_i, \delta)$ is obtained as follows. We identify all item-score vectors for which $\sum_{j=1}^{J} x_{ij} = x_i$ and then add all corresponding probabilities $P(\mathbf{x}_i \mid \theta_i, \delta)$. This yields the desired probability $P(x_i \mid \theta_i, \delta)$. In mathematical notation,

$$P\left(x_i \mid \theta_i, \delta\right) = \sum_{\left\{x: \sum_j x_{ij} = x_i\right\}} P(\mathbf{x}_i \mid \theta_i, \delta). \tag{4.23}$$

In Equation (4.22), we replace the numerator with Equation (4.18) and the denominator with Equation (4.23), and this yields

$$L\left(\mathbf{x}_i \mid x_i, \theta_i, \delta\right) = \frac{\prod_{j=1}^{J} \exp\left(-x_{ij}\delta_j\right)}{\sum_{\left\{x: \sum_j x_{ij} = x_i\right\}} \prod_{j=1}^{J} \exp\left(-x_{ij}\delta_j\right)}. \tag{4.24}$$

One may notice that the conditional likelihood does not contain person parameter θ_i, hence $L\left(\mathbf{x}_i \mid x_i, \theta_i, \delta\right) = L\left(\mathbf{x}_i \mid x_i, \delta\right)$. If $\mathbf{x}_i = (0, \dots, 0) = \mathbf{0}_i$, or if $\mathbf{x}_i = (1, \dots, 1) = \mathbf{1}_i$, one may verify that in Equation (4.24), the denominator contains only one product, so that the numerator and the denominator are equal, hence, $L\left(\mathbf{x}_i \mid x_i, \delta\right) = P\left(\mathbf{0}_i \mid x_i, \delta\right) = P\left(\mathbf{1}_i \mid x_i, \delta\right) = 1$. For our purpose, it suffices to expand the likelihood this far.

The important result from Box 4.1 is that the likelihood does not contain the person parameter, thus allowing the estimation of the item locations or difficulties independent of the distribution of the person parameter. This result can be accomplished when one (1) expands the likelihood for the whole sample of people; (2) computes the natural logarithm of the likelihood; (3) takes the partial derivatives with respect to each of the item location parameters; and (4) sets the partial derivatives equal to 0 and solves for the values of the location parameters that maximize the function. Maximization can be done, for example, by means of the Newton–Raphson algorithm (Baker, 1992, p. 141).

One can also execute this procedure if the data matrix **X** is incomplete. Incompleteness can be caused by, for example, some people not answering some of the items (e.g., Van Ginkel, Sijtsma, Van der Ark, & Vermunt, 2010) or a structurally incomplete design that deliberately presents different but partly overlapping subsets of items from a larger item bank to different groups of examinees (Eggen & Verhelst, 2011; Mislevy & Sheenan, 1989; Verhelst & Glas, 1993). When data are missing not by design but due to other causes, the pattern of the missingness determines whether the item parameter estimates maintain their desirable statistical properties. When data are missing by design, if implemented correctly, the design satisfies the conditions that facilitate maximum likelihood estimation while maintaining desirable statistical properties. Desirable statistical properties are the following.

The maximum likelihood item-location parameter estimates, $\hat{\delta}_j$, $j=1,\ldots,J$, have the statistical property of *consistency*, meaning that, loosely speaking, as the sample size grows, the estimator $\hat{\delta}_j$ converges to parameter δ_j, finally landing on the spot as N continues to grow. Intuitively, if one collects more data, one collects more information about the parameter of interest, and finally, one knows all about the parameter that there is to know. Several formal definitions of consistency exist, but here we consider the well-known definition of *convergence in probability* (Casella & Berger, 1990, pp. 213–214, 323). Let sample size N grow, and estimate parameter δ_j for each next sample size. Then, estimator $\hat{\delta}_N$ is consistent if, for all $\varepsilon > 0$,

$$\lim_{N\to\infty} P\left(|\hat{\delta}_N - \delta| > \varepsilon\right) = 0. \tag{4.25}$$

No matter how small one chooses ε, as the sample continues growing, Equation (4.25) will be satisfied and the estimator will converge to the parameter.

Another statistical property is *unbiasedness* (e.g., Casella & Berger, 1990, pp. 303, 307). The conditional maximum likelihood estimator is not entirely without bias (Verhelst, Glas, & Van der Sluis, 1984). Suppose it would be an unbiased estimator, then the maximum likelihood estimator $\hat{\delta}_j$ obtained in an infinite number of independent samples of the same, finite size, has a mean equal to the parameter it attempts to estimate. Formally, estimator $\hat{\delta}_j$ is unbiased when

$$\mathcal{E}\left(\hat{\delta}_j\right) - \delta_j = 0. \tag{4.26}$$

Unbiasedness means that in separate samples, one may miss the parameter regularly, but in the long run the mean of the estimates coincides with the target parameter.

Although consistency and unbiasedness appear to be similar concepts, one does not imply the other. For example, if an estimator uses only a sub-sample of constant size, say, n_0, to estimate the parameter, the fact that the total sample size continues growing does not lead to consistency in the sense of Equation (4.25). Yet the estimator may be unbiased, because that property is defined for each finite sample size. Also, a consistent and unbiased estimator, say, $\hat{\varsigma}$, can be transformed into a biased estimator by adding, for example, quantity N^{-1}, producing biased estimator $\hat{\varsigma}^* = \hat{\varsigma} + N^{-1}$, but it is consistent because, if $N \to \infty$, then $N^{-1} \to 0$, so that $\hat{\varsigma}^* \to \hat{\varsigma}$.

If the sample size is larger, the spread of the estimates, for example, expressed by the variance of the estimator, is another statistical property of an estimator. The variance of an estimator represents the estimator's *efficiency* (e.g., Casella & Berger, 1990, pp. 308–311). Large efficiency, hence small variance, is desirable, because researchers usually have just one sample, and individual estimates are, on average, closer to the parameter one wishes to know. The best unbiased estimator of a parameter has the smallest variance; this is the *uniform minimum variance unbiased estimator* of the parameter (e.g., Casella & Berger, 1990, p. 307).

We refrain from further discussion but notice that consistency, unbiasedness, and efficiency are desirable for an estimator. They reflect the desiderata that, in the long run, one wishes to know all about the parameter (consistency), and for a limited sample size, one wishes not to be far off target (efficiency) and on average precisely on target (unbiasedness).

To obtain maximum likelihood estimates of the latent variable values, $\theta = (\theta_1, \ldots, \theta_N)$, proves a little more complex. One may choose to maximize Equation (4.21) jointly for θ and δ, known as joint maximum likelihood estimation (Baker, 1992, Chap. 4). However, this procedure produces inconsistent parameter estimates, due to the number of latent variable parameters growing with sample size N (e.g., Neyman & Scott, 1948). Because better estimation procedures are available, joint maximum likelihood is not used regularly. Hoijtink and Boomsma (1995) discussed estimation of latent variable θ and summarized several problems related to estimation, such as the limited number of items available and the varying distributions of individual item scores that depend on the unknown item parameters. Instead, they proposed estimating θ assuming that the maximum likelihood estimates of the item parameters are the true parameters. Thus, the elements $\hat{\delta}_j$, $j = 1, \ldots, J$, in $\hat{\delta}$ are substituted in Equation (4.21) for the elements δ_j, $j = 1, \ldots, J$, in δ, and the likelihood is further developed so that the parameters in θ can be estimated. In particular, the estimates of the extreme θs are biased (Lord, 1983). Warm (1989) suggested corrections for most of the bias. Other methods for estimation of parameters are available, but we do not further discuss them; see Baker (1992; Baker & Kim, 2004) for a detailed anthology of estimation in IRT including the Rasch model.

Information Functions and Measurement Precision

Not only in the Rasch model but in all of parametric IRT, the Fisher information function expresses how well each of the items as well as the whole test estimate the latent variable as a function of θ. We will see that the information function is closely related to the standard error of the estimated latent variable, $\hat{\theta}$, given parameter θ, that is, the conditional standard error, $\sigma_{\hat{\theta}|\theta}$. The conditional standard error is the IRT analogue of the standard error of measurement we discussed in the context of CTT in Chapter 2 [Equation (2.72)]. We noticed there that the standard error of measurement expresses measurement precision for single test scores. The conditional standard error depends on the scale and thus reveals how well the items constituting the test measure a particular θ value. For the Rasch model, we will see next that the closer the item location parameters are to the target θ value, the more precise they measure the target value.

Let $P_j'(\theta)$ denote the first derivative of the IRF with respect to θ, that is, $P_j'(\theta) = \partial P_j(\theta) / \partial \theta$. For dichotomous items, for several models but not for all models, Fisher's information function for item j equals

$$I_j(\theta) = \frac{\left[P_j'(\theta)\right]^2}{P_j(\theta)\left[1 - P_j(\theta)\right]}.$$

(4.27)

In Equation (4.27), the first derivative of the IRF with respect to latent variable θ gives the slope of the IRF at each value of θ. The denominator provides the conditional variance of the dichotomous item score. For the Rasch model, $P_j'(\theta) = P_j(\theta)\left[1 - P_j(\theta)\right]$, so that the item information function simplifies to

$$I_j(\theta) = P_j(\theta)\left[1 - P_j(\theta)\right] = \frac{\exp\left(\theta - \delta_j\right)}{\left[1 + \exp\left(\theta - \delta_j\right)\right]^2}.$$

(4.28)

When local independence holds, Fisher's information for the whole test equals

$$I(\theta) = \sum_{j=1}^{J} I_j(\theta).$$

(4.29)

Figure 4.6a provides information functions for two items consistent with the Rasch model but differing in item location, so that $\delta_j < \delta_k$. Clearly, because the IRFs are parallel, the information functions are also parallel. The graphs show that items provide maximum information at $\theta = \delta$, and that information levels off symmetrically at both sides of the maximum.

A maximum likelihood estimator, such as $\hat{\theta}$, has a normal distribution with mean θ and variance $\sigma_{\hat{\theta}|\theta}^2 = I(\theta)^{-1}$, that is, $\hat{\theta} \sim \mathcal{N}\left(\theta, I(\theta)^{-1}\right)$. Thus, the standard

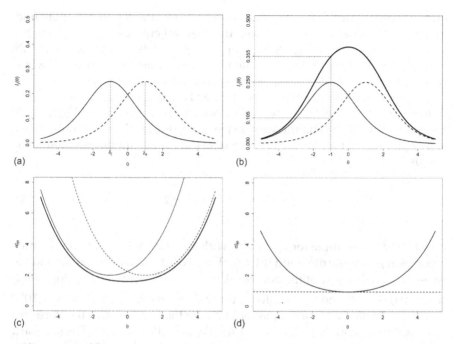

FIGURE 4.6
(a) Two item-information functions for Rasch items with $\delta_j = -1$ and $\delta_k = 1$, respectively. (b) The same item-information functions, showing for $\theta = -1$ that the test-information function is the sum of the item-information functions. (c) Standard error of $\hat{\theta}$ as a function of θ for the items j and k in (a) and (b) and for the two items combined (solid curve). (d) Standard error of $\hat{\theta}$ as a function of θ for five Rasch items with $\delta_j = -1.5 + 0.5 \times j (j = 1,...,5)$ (solid curve) plus the standard error of measurement (dashed curve; $\sigma_E \approx 9495$; $\rho_{XX'} \approx 0.5297$).

error equals $\sigma_{\hat{\theta}|\theta} = I(\theta)^{-1/2}$. This means that items measure θ most precisely at $\theta = \delta$ where information is maximal, whereas precision levels off as $|\theta - \delta|$ increases. Because maximum likelihood estimates are efficient, thus having minimum variance, using test scores other than maximum likelihood estimates reduces statistical information and measurement precision defined by the conditional standard error. A special merit of test information is that it is the simple sum of the item information contributions, thus allowing the assessment of unique item contributions to the test's measurement precision. In fact, one can build a test as if one were laying bricks on top of one another, and Figure 4.6b shows this metaphor for the item information functions that, when piled up, produce the test information function. This property facilitates the selection of items for particular measurement goals. For example, when one wishes to measure precisely at a cutoff score θ_0, the simple selection rule is to use items that have location parameters as close as possible to the cutoff value. The number of items one selects depends on the desired measurement precision, which in turn may be determined by the number of selection errors—false positives and false negatives—one is ready to tolerate. This is quite a complex area that is beyond the scope of this monograph. A more fine-grained categorization of people that uses several cutoff scores to divide the scale in several adjacent segments requires subsets of items that are maximally precise at each of the cutoffs. A scale that should diagnose across the whole range requires items that are spread at equal distances across the scale.

As we will see next, the use of the standard error of the items and the standard error of the whole test does not parallel the nice masonry of Figure 4.6b. Using Equation (4.29) and the knowledge that $I_j(\theta)$ attains its maximum value at $\theta = \delta_j$, it follows that for the Rasch model the maximum information value equals

$$I_j(\theta) = \frac{\exp(0)}{\left[1 + \exp(0)\right]^2} = .25. \tag{4.30}$$

We expand an example for two items with $\delta_j = -1$ and $\delta_k = 1$, and at $\theta = -1$ (cf. Figure 4.6b) compute the standard error based on the individual items and the two-item test. Because at the item level, $\sigma_{\hat{\theta}|\theta} = I_j(\theta)^{-1/2}$ and likewise for item k, we have $I_j(-1) = .25$ and $\sigma_{\hat{\theta}_j|\theta} = 2$ (for $\hat{\theta}_j$ that coincides with δ_j), and $I_k(-1) \approx .1050$ and $\sigma_{\hat{\theta}_k|\theta} \approx 3.086$. Following Equation (4.30), for the two-item test, the test information at $\theta = -1$ equals $I(-1) = I_j(-1) + I_k(-1) = .25 + .1050 = .3550$. The standard error for the two-item test at $\theta = -1$ equals $\sigma_{\hat{\theta}|\theta} = I(-1)^{-1/2} = .3350^{-1/2} \approx 1.6784$. Hence, at $\theta = -1$, measurement based on the two items together is more precise than measurement based on each of the individual items. Figure 4.6c shows the standard errors for the individual items and the two-item test as functions of θ.

Finally, for five items with $\delta_j = -1, -0.5, 0, 0.5, 1$, Figure 4.6d shows the conditional standard error based on these five items. For these items and $\theta \sim \mathcal{N}(0,1)$, we also computed the reliability, which equaled $\rho_{XX'} = \sigma_T^2 / \sigma_X^2 \approx .53$ [Equation (2.19)], and the standard error of measurement, which was $\sigma_E = \sigma_X \sqrt{1 - \rho_{XX'}} \approx .95$ [Equation (2.72)]. Figure 4.6d also shows the standard error of measurement as a horizontal dashed line. One sees right away that IRT, in this example, the Rasch model, facilitates measurement precision that is dependent on the scale, taking the location of the items in account, whereas CTT has more trouble incorporating such nuances (Chapter 2; Feldt et al., 1985).

Goodness-of-Fit Methods

Fischer (1974, Chap. 12; Fischer, 1995a) proved that the assumptions of unidimensionality, local independence, monotonicity, and sufficiency are identical to the Rasch model. To assess the goodness of fit of the Rasch model means that one has to inspect whether the data are consistent with each of the four assumptions. A drawback is that it proves impossible to assess all four assumptions at once, but only in subsets of two. Many goodness-of-fit methods for the Rasch model have been proposed. Glas and Verhelst (1995a) noticed that the plethora of goodness-of-fit methods needed ordering. They distinguished four classes of methods, two of which are particularly rele- vant for our discussion. The first class contains the methods that focus on the properties of the IRFs, being that they are strictly monotone increasing and parallel. Parallelism follows from the sufficiency of sum scores for model parameters (Fischer, 1974, pp. 193–203). Another consequence of sufficiency is parameter separation. This property implies that an item must have the same location parameter in different subgroups from the population in which the Rasch model holds for the test of interest. Subgroups can be different gender or age groups but also subgroups based on the sum score. The second class focuses on the assumptions of unidimensionality and local independence. We make another distinction within each group, which is between meth- ods focusing on the complete set of items and methods focusing on single items when assessing the IRFs, and methods focusing on pairs of items when assessing unidimensionality and local independence. We call such methods global when they assess sets of items and local when they assess individual items or item pairs. Combining the global and local methods for the assess- ment of the two sets of model features is a challenge, because a decision to leave out one or two items having non-monotone IRFs will affect the good- ness-of-fit results for the remaining items in the next analysis round. We will present a flowchart that we find convenient (Figure 4.7), which the researcher may use when assessing the fit of the Rasch model to the data.

The flowchart is based on the following procedure.

1. For the whole item set, test the null-hypothesis of monotonicity and sufficiency by means of a global test statistic.

2. If one rejects global monotonicity or sufficiency, use a local test statistic to identify the item or subset of items that violate these assumptions. Remove the violating item or the violating subsets of items from the item set, construct subsets of items, or execute a com- bination of these actions. Repeat 1. for the item set after removal of one or more items or for the newly formed item subset. Repeat the loop until one or more item sets remain for which 1. does not reject the null hypothesis.

3. For the item set or the item subsets resulting from 1. and 2., test the null hypothesis of unidimensionality and local independence.

4. If one rejects global unidimensionality or local independence, use a local statistic to identify pairs of items responsible for the misfit, and either remove the items, construct item subsets, or execute a combi- nation of these actions. Repeat 3. for the remaining item set of the newly formed item subsets. Repeat the loop until one or more item subsets remain for which one cannot refute unidimensionality and local independence. Go back to 1.

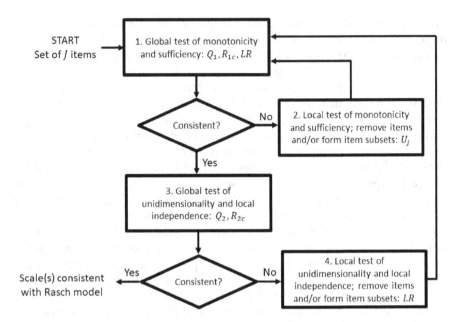

FIGURE 4.7
Flow chart for item analysis according to the Rasch model. Based on Figure 1, Sijtsma (1983).

Starting with monotonicity and sufficiency is rather arbitrary, and one might choose to start with steps 3 and 4, followed by steps 1 and 2. Our experience with either procedure is that it can be quite laborious, requiring many decisions as one moves forward, necessitating correction for the number of significance tests. Because the Rasch model is restrictive and because a full goodness-of-fit procedure is quite selective, the result can be the loss of many items or the breakdown of the item set in several item subsets. Although such a result can be discouraging when one expected a different result, the fact is that having one's item set dissected like this can be equally instructive as well.

We briefly discuss methods that one can use in the different stages of the flowchart, and use the numbering in the flowchart. All methods are consistent with conditional maximum likelihood estimation, but several can be adapted to other estimations contexts.

1—Monotonicity and Sufficiency, Global Methods. Several statistics assess observed and expected IRFs. Let N_{jx} be a random variable with realization n_{jx} for the count of the number of 1 scores on item j obtained by people who have a sum score x, $x = 1, \ldots, J-1$, on the item set of interest. For $x = 0$, one may notice that $n_{j0} = 0$. Let n_x be the frequency of sum score x, so that n_J denotes the number of people who have the maximum score of $x = J$; then, for $x = J$, we have that $n_{jJ} = n_J$. Hence, n_{j0} and n_{jJ} are uninformative. If some frequencies are small, then adjacent sum-score groups can be combined, such that sum scores are replaced with G adjacent, exhaustive and exclusive combined groups, enumerated by index $g = 1,...,G$, $G \leq J$; also, see Chapter 3, where joining low-frequency adjacent score groups was feasible to estimate IRFs. Then, we replace N_{jx} with N_{jg} and n_{jx} with n_{jg}. Observed frequency n_{jg} is compared with the corresponding frequency expected for the Rasch model, $\mathcal{E}(N_{jg} | n_g, \hat{\delta})$, given the frequency distribution of person sum scores and conditional maximum likelihood estimates for the item location parameters, collected in $\hat{\delta}$, so that

$$d_{jg}^* = n_{jg} - \mathcal{E}(N_{jg} | n_g, \hat{\delta}) \tag{4.31}$$

When the difference d_{jg}^* is divided by its estimated standard deviation, a standardized binomial variable, z_{jg}, results (Glas & Verhelst, 1995a). Summing across items and combined-score groups, one obtains

$$Q_1 = \frac{J-1}{J} \sum_{j=1}^{J} \sum_{g=1}^{G} z_{jg}^2 \tag{4.32}$$

(Van den Wollenberg, 1982a). Statistic Q_1 is approximately chi-squared distributed with $df = (G - 1)(J - 1)$. Glas (1988a) proposed statistic

$$R_{1c} = \sum_{g=1}^{G} \mathbf{d}_g' \mathbf{W}_g^{-1} \mathbf{d}_g. \tag{4.33}$$

In Equation (4.33), $\mathbf{d}_g$ is the vector with elements d_{jg}, with $d_{jg} = d_{jg}^*/\sqrt{N}$, N is the sample size. Matrix $\mathbf{W}_g$ is the covariance matrix of $\mathbf{d}_g$ and serves to incorporate the dependency between the scaled deviates in $\mathbf{d}_g$. This correction causes R_{1c} to have an asymptotic chi-squared distribution with $df = (G - 1)$ $(J - 1)$. Glas and Verhelst (1995a) showed how statistic R_{1c} can be accommodated to other estimation procedures, such as marginal maximum likelihood, and to structurally incomplete designs in which different subgroups of persons respond to different subsets of items.

2—Monotonicity and Sufficiency, Local Methods. Molenaar (1983) suggested the U_j statistic that assesses whether the observed IRF of item j is consistent with the function expected based on the Rasch model. Moreover, the statistic informs the researcher whether the observed IRF runs flatter or steeper than expected and at which scale locations the greatest deviations, if any, occur. Statistic U_j is defined as follows. First, one may notice that

$$n_{jx} - \mathcal{E}\left(N_{jg} | n_x, \hat{\delta}\right) > 0 \quad \text{for small } x \tag{4.34}$$

and

$$n_{jx} - \mathcal{E}\left(N_{jg} | n_x, \hat{\delta}\right) < 0 \quad \text{for large } x \tag{4.35}$$

suggests that the observed IRF is flatter than the expected IRF, and reversing the signs in the equations suggests a steeper sloped IRF. Combining this diagnostic information in one test statistic is done as follows. Let $n_m = N - n_1 - n_J$, then define a set L (L for low) of indices such that set L identifies as close to one-quarter of the people with the lowest sum scores as possible but no more than one-quarter; that is,

$$L = \{1, \ldots, n_l\} \text{ for } \sum_{x=1}^{n_l-1} n_x < \frac{n_m}{4} \leq \sum_{x=1}^{n_l} n_x. \tag{4.36}$$

Likewise, set H (H for high) contains indices such that set H identifies as close to one-quarter of the people with the highest sum scores as possible but no more than one-quarter; that is,

$$H = \{n_h, \ldots, n_{J-1}\} \text{ for } \sum_{x=n_h}^{J-1} n_x \geq \frac{n_m}{4} > \sum_{x=n_h+1}^{J-1} n_x. \tag{4.37}$$

Using these definitions and the standardized binomial variable z_{jx} (Glas & Verhelst, 1995a) that was used for defining global statistic Q_1 [Equation (4.32)], statistic U_j is defined as

$$U_j = \frac{\sum_L z_{jx} - \sum_H z_{jx}}{\left(n_l + J - n_h\right)^{1/2}}. \tag{4.38}$$

Statistic U_j is approximately standard normally distributed. Large positive U_j values are typical of an IRF that is flatter than expected, and large negative U_j values indicate an IRF that is steeper than expected. The reasons for a deviating U_j are usually unknown, but a large positive value suggests weak discrimination, which in itself may be a reason to remove an item from the test, more than a large negative value, which suggests strong discrimination. If a subset of items has large positive or large negative U_j values, this may be a sign that the subset measures an attribute different from the rest of the items, and different item subsets may be identified for further analysis. Removing individual items or selecting item subsets for further analysis as a rule has the effect of changing the U_j values of the remaining items or the U_j values in the newly formed item subsets, and items that initially looked good may have deviant U_j values later, and similarly for items that initially looked bad. When for some sum scores one finds low frequencies, adjacent low-frequency groups may be combined as with the global statistics. Two local statistics were defined that have asymptotic chi-squared distributions, because they incorporate the covariances between the deviates d_{jx} or d_{jg} (Glas & Verhelst, 1995a). To avoid too much detail, we refrain from discussing these alternative statistics.

3—*Unidimensionality and Local Independence, Global Methods.* Tests for uni-dimensionality and local independence focus on the latter assumption. If one latent variable is insufficient to let the covariance between items vanish if one conditions on an estimate of the latent variable, an explanation for the non-zero covariance is that at least one additional latent variable is needed; hence, the data are inconsistent with unidimensionality. Glas and Verhelst (1995a; Glas, 1988a) derived a test statistic for unidimensionality based on local independence. Let N_{2jk} be a random variable with realization n_{2jk} counting the number of people having a sum score x at least equal to 2, who have 1 scores on both items j and k. Sum scores equal to 0 and 1 cannot produce two 1 scores and are ignored. We define vector $\mathbf{n}$ that contains the frequencies n_x. Observed frequencies n_{2jk} are compared with the expectation, $\mathcal{E}(N_{2jk}|\mathbf{n},\hat{\delta})$, resulting in

$$d_{2jk}^* = n_{2jk} - \mathcal{E}(N_{2jk}|\mathbf{n},\hat{\delta}). \tag{4.39}$$

Vector $\mathbf{d}_2$ contains differences $d_{2jk} = d_{2jk}^*/\sqrt{N}$, and matrix $\mathbf{W}_2^{-1}$ is the inverse of the estimated covariance matrix for the differences d_{2jk}. To derive a statistic with known sampling properties, one also needs to consider vector $\mathbf{d}_1$, which contains for all items the difference between the number of people who have a sum score equal to 1 and a 1 score on the item of interest, and the expected value of this count, both terms divided by $\sqrt{N}$. For these differences, matrix $\mathbf{W}_1^{-1}$ is the inverse of the estimated covariance matrix. The test statistic is defined as

$$R_{2c} = \mathbf{d}_2'\mathbf{W}_2^{-1}\mathbf{d}_2 + \mathbf{d}_1'\mathbf{W}_1^{-1}\mathbf{d}_1, \tag{4.40}$$

which is asymptotically chi-squared distributed with $df = J(J-1)/2$.

We first discussed statistic R_{2c}, because it does not require choices for a sub-division of the group into subgroups when some sum-score groups have low frequencies, but the next statistic does. For statistic Q_2 (Van den Wollenberg, 1982a; we discuss the version Glas & Verhelst, 1995, proposed and which is com-putationally more efficient than the original version), the group is divided into G combined score groups, if necessary, but other groups based on covariates are also possible, and groups are enumerated $g = 1,...,G$, as before. Frequency n_{2jk} that was used for statistic R_{2c}, is further dissected into frequencies n_{2jkg} for similar counts within each of the subgroups, producing a difference (with $\hat{\omega}$ containing dummy quantities added for technical reasons)

$$d_{2jkg}^* = n_{2jkg} - \mathcal{E}(N_{2jkg}|\hat{\omega},\hat{\delta}). \tag{4.41}$$

Standardized differences z_{2jkg} are obtained when d_{2jkg}^* is divided by its esti-mated standard deviation. Then, statistic Q_2 equals

$$Q_2 = \frac{J-3}{J-1}\prod_{j=1}^{J-1}\prod_{k=j+1}^{J}\prod_{g=1}^{G}z_{2jkg}^2. \tag{4.42}$$

Statistic Q_2 has an approximate chi-squared distribution with $df = GJ(J-3)/2$, provided one estimates the item locations separately in each of the subgroups.

Before we discuss a local method to identify individual items or subsets of items deviating from the majority's dimensionality, we first discuss a likelihood ratio test for the Rasch model, proposed by Andersen (1973a). This *LR*-statistic takes full advantage of the Rasch model property that, when the model is consistent with the data in a particular population, its item location parameters for the whole population are also valid in each subpopulation, even if subpopulations are systematically different with respect to one or more covariates from the mother population. Thus, in subsamples from the greater sample, item parameter estimates must be the same except for sampling fluctuation. Let $L_{\text{TOT}}(\hat{\delta};\mathbf{X})$ be the likelihood for a random sample $\mathbf{X}$ drawn from the population that is maximized for obtaining item location estimates $\hat{\delta}$. In addition, let $L_g\left(\hat{\delta}_g;\mathbf{X}_g\right)$ be the maximized likelihood for subgroup g, $g = 1,...,G$, based on estimates $\hat{\delta}_g$ obtained from the data in subgroup g, denoted $\mathbf{X}_g$. The data submatrices are unique and together constitute the complete data. An essential step toward a likelihood ratio test is to understand that if one inserts the total-group estimates $\hat{\delta}$ in the likelihood of a subgroup, the next inequality,

$$L_g\left(\hat{\delta};\mathbf{X}_g\right) \le L_g\left(\hat{\delta}_g;\mathbf{X}_g\right) \tag{4.43}$$

is true, because the estimates $\hat{\delta}$ do not maximize the subgroup likelihood whereas the estimates $\hat{\delta}_g$ do. In addition, because for a fitting Rasch model

$$L_{\text{TOT}}\left(\hat{\delta};\mathbf{X}\right) \le \prod_{g=1}^{G} L_g\left(\hat{\delta}_g;\mathbf{X}_g\right), \tag{4.44}$$

the likelihood ratio

$$\lambda = \frac{L_{\text{TOT}}\left(\hat{\delta};\mathbf{X}\right)}{\prod_{g=1}^{G} L_g\left(\hat{\delta}_g;\mathbf{X}_g\right)} \le 1. \tag{4.45}$$

Applying the rules that $\ln\left(\dfrac{A}{B}\right) = \ln A - \ln B$ and $\ln\left(\prod_g B_g\right) = \sum_g \ln B_g$, one obtains

$$\ln\lambda = \ln L_{\text{TOT}}\left(\hat{\delta};\mathbf{X}\right) - \sum_{g=1}^{G} \ln L_g\left(\hat{\delta}_g;\mathbf{X}_g\right). \tag{4.46}$$

Taking $-2\ln\lambda$, one obtains

$$LR = 2\left[\sum_{g=1}^{G} \ln L_g\left(\hat{\boldsymbol{\delta}}_g; \mathbf{X}_g\right) - \ln L_{\text{TOT}}\left(\hat{\boldsymbol{\delta}}; \mathbf{X}\right)\right]. \qquad (4.47)$$

Statistic LR is asymptotically chi-squared distributed with $df = (J-1)(G-1)$. The LR test is also used when subgroups are identified by the same sum score, except values $x=0$ and $x=J$, so that $G=J-1$ and $df=(J-1)(J-2)$, a value one often encounters in the literature. Glas and Verhelst (1995a) discuss an adaptation of LR when not all item location parameters can be estimated in each subgroup.

Depending on the definition of subgroups, statistic LR is sensitive for different inconsistencies of the data and the Rasch model. Glas and Verhelst (1995a) mentioned the case when a low sum-score group produces smaller item parameter estimates than expected, suggesting items were too easy, possibly because of successful guessing for the correct answer. Van den Wollenberg (1979) suggested splitting the group into two subgroups, one defined by the smaller sum scores and the other by the larger sum scores. This setup was sensitive to IRFs that were non-monotone, had different slopes, or both (ibid., pp. 84–95). Van den Wollenberg (1982b) suggested a different subgrouping, rendering statistic LR sensitive to violations of unidimensionality and local independence, which we will discuss next.

4—*Unidimensionality and Local Independence, Local Methods.* Criteria for defining subgroups for the LR test often relate in the same degree to items measuring different attributes, provided they are available in the item set. For identifying multidimensionality, one needs to define subgroups based on a criterion related to the hypothesized multidimensionality. Suppose that one suspects that item subset A measures attribute A and that item subset B measures attribute B. Attributes A and B have a rather low correlation. Choose an item from subset A, say, item j_A, and form two groups, one having the so-called splitter item j_A correct—this is subgroup g_{1A}—and the other having the splitter item j_A incorrect—this is subgroup g_{0A}. One expects that in subgroup g_{1A} mean θ_A is larger than in subgroup g_{0A}, that is, $\bar{\theta}_{1A} > \bar{\theta}_{0A}$. For the other item subset to which the splitter item is not or only weakly related, one expects $\bar{\theta}_{1B} \approx \bar{\theta}_{0B}$. Hence, subgroup g_{1A} consists of people dispersed around the mean of attribute B and high on attribute A, and subgroup g_{0A} consists of people also dispersed around the mean of attribute B but low on attribute A.

Now, we simultaneously estimate all item parameters in each of these two groups. In subgroup g_{1A}, the items measuring attribute A are relatively easy relative to the items measuring attribute B, and in subgroup g_{0A}, the items measuring attribute A are relatively difficult. In a plot having the estimated item parameters in subgroups g_{0A} and g_{1A} on the abscissa and the ordinate, respectively, one expects all item parameter estimates for attribute B to lie on or close to the 45° line but the item parameter estimates for attribute A to lie below the 45° line. The latter result reflects item parameter estimates for attribute A to be relatively difficult in group g_{0A} and relatively easy in group g_{1A}; see Figure 4.8 for an example referring to fraction arithmetic items (Sijtsma, 1983). To obtain a clearer picture of the dimensionality of one's items, one should try several splitter items considered typical of attribute A and several typical of attribute B and look for persistent patterns in the item parameter plots. This may lead to the identification of item subsets measuring different attributes but also to item pairs or individual items that are candidates for removal from the test.

We reiterate the statistical methods we discussed also in Figure 4.7.

1. Statistics Q_1, R_{1c}, and LR based on subgroupings unrelated to possible multidimensionality are global methods for assessing monotonicity and sufficiency;

2. Statistic U_j is a local, item-oriented statistic for assessing monotonicity and sufficiency;

3. Statistics Q_2 and R_{2c} can be used to assess unidimensionality and local independence for sets of items;

4. Statistic LR based on splitter items can be used to identify subsets of items measuring different attributes.

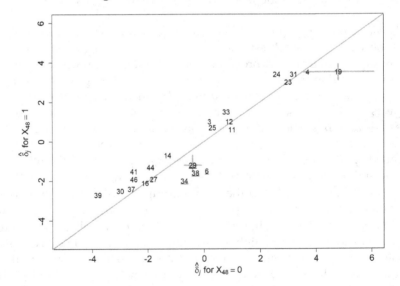

FIGURE 4.8

Graph showing $\hat{\delta}_j$ in groups for which $X_{48}=0$ (horizontal axis) and $X_{48}=1$ (vertical axis), and asymptotic 95% confidence intervals for items 19 and 29. Items concern different types of fraction arithmetic problems. Underlined item numbers refer to problems asking to find the common denominator of two fractions. The figure is based on Figure 2, Sijtsma (1983).

Several other goodness-of-fit methods have been proposed for the Rasch model, but we refrain from discussing all of them (but see Glas, 2016; Glas & Verhelst, 1995a).

The Rasch Paradox

The Rasch paradox resembles the attenuation paradox we encountered in the introduction to this chapter. Although it refers to a relatively small problem, we found discussing the problem challenging for thinking about IRT models and measurement and thus chose to discuss it here. The paradox originates from first observing that the deterministic Guttman model implies ordinal measurement and then noticing that if one replaces the Guttman model with a probabilistic model, such as the Rasch model, that only adds random error, one obtains a model that implies a metric scale for persons and items. The paradox is that adding error produces a more informative scale. It is not our intention to resolve the discussion (e.g., Borsboom & Mellenbergh, 2004; Humphry, 2013; Michell, 2000, 2014; Sijtsma, 2012; Sijtsma & Emons, 2013) definitively but to show instead how the Rasch model transforms into the Guttman model and what this means for the Rasch model's likelihood. We will also outline the relationship of this result with the attenuation paradox.

To clarify our line of reasoning, we write the Rasch model as

$$P_j(\theta) = \frac{\exp\left[\alpha\left(\theta - \delta_j\right)\right]}{1 + \exp\left[\alpha\left(\theta - \delta_j\right)\right]} = \left\{1 + \exp\left[-\alpha\left(\theta - \delta_j\right)\right]\right\}^{-1}. \qquad (4.48)$$

It is unusual to include a discrimination parameter other than $\alpha = 1$, but we notice that the latter value is the result of convention rather than necessity. Adaptation of the distribution of latent variable θ always allows one to insert a discrimination parameter equal to 1 in Equation (4.48). To investigate when the model transforms into the deterministic Guttman model, we choose α large, in fact, we let $\alpha \to \infty$ for three different cases, which are

$$\theta - \delta_j < 0 \Rightarrow P_j(\theta) = 0;$$

$$\theta - \delta_j = 0 \Rightarrow P_j(\theta) = .5;$$

$$\theta - \delta_j > 0 \Rightarrow P_j(\theta) = 1.$$

Figure 4.9 shows the resulting IRF, and one may notice that it differs only from a Guttman IRF in allowing response probability $P_j(\theta) = .5$ when $\theta - \delta_j = 0$. Thus, letting the discrimination parameter grow infinitely large positive,

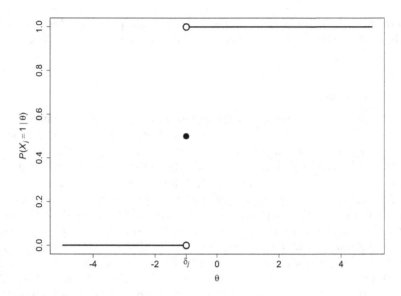

FIGURE 4.9
One-parameter logistic *IRF* with $\delta_j = -1$, when $\alpha \rightarrow \infty$.

the Rasch model reduces to a deterministic model, which resembles the Guttman model. What happens to the likelihood?

Let us assume four items with $\delta_1 < \delta_2 < \delta_3 < \delta_4$, and an item-score vector $\mathbf{X}_i$ = (1100). We use shorthand $P_j = P_j(\theta)$. The likelihood of the Rasch model for one item-score vector equals [based on Equation (4.18)]

$$L\left(\mathbf{X}_i \mid \theta_i, \delta\right) = \prod_{j=1}^{J} P_j^{x_{ij}} \left(1 - P_j\right)^{1 - x_{ij}}. \tag{4.49}$$

We consider the Rasch model for $\alpha \rightarrow \infty$ and evaluate three situations for $\mathbf{X}_i$, one in which θ is located between δ_2 and δ_3, one in which θ coincides with either δ_2 or δ_3, and one in which θ is located to the left of δ_2 or to the right of δ_3, and then notice the following results for the likelihood in Equation (4.49):

$$\delta_2 < \theta < \delta_3 \Rightarrow L = 1;$$

$$\theta = \delta_2 \vee \theta = \delta_3 \Rightarrow L = .5;$$

$$\theta < \delta_2 \vee \theta > \delta_3 \Rightarrow L = 0.$$

One may notice that in the interval $\delta_2 < \theta < \delta_3$, the likelihood is a plateau from which a maximum likelihood estimator cannot be derived. Moreover, values in the intervals $\theta < \delta_2$ and $\theta > \delta_3$ cannot produce item-score vector $\mathbf{X}_i$. Thus, one sees that in the limiting case of perfect discrimination, the Rasch model reduces to a deterministic model that does allow a person ordering, however, and that its statistical machinery breaks down. We do not think this is so much as a paradox but rather a limiting case of a statistical model, as there are more in statistical modeling. One may think of collinearity in multiple regression, when highly correlating independent variables produce highly inflated sampling error in the estimated regression coefficients (Fox, 1997, pp. 120–121) that tends to infinity when correlations tend to 1, whereas for correlations equal to 1 regression coefficients are no longer unique (ibid., pp. 212–213). Here, we see that for another statistical model, a limiting case can be identified that causes the model to break down as well.

The relation of the Rasch paradox, to maintain the phrase, to the older attenuation paradox is the following. Reading the articles devoted to the attenuation paradox (e.g., see Loevinger, 1954, for a review), what strikes one initially is the terminology that is different from modern psychometric parlance. However, it is clear that, at the time, people were puzzled by the phenomenon that as inter-item correlations grew beyond a particular boundary, the sum score was less informative about the attribute that the items measured in common. That is, high item reliability—in IRT jargon, high item discrimination—produced high inter-item correlations, and the resulting overrepresentation of extremely low or high sum scores and the underrepresentation of sum scores in between represented poor information—interpreted as low validity—about the attribute despite large sum-score variance. This tension between reliability and validity became extreme when dichotomous items had equal difficulty and maximum inter-item correlation.

A different way to understand the phenomenon is the following. One may think of items consistent with the Guttman model. Given their marginal distributions, Guttman items have maximum product-moment correlation. We show this as follows. Let π_j denote the proportion of people having a 1 score on item j, and assume arbitrarily that $\pi_j > \pi_k$. Further, let π_{jk} denote the proportion of people having a 1 score on both items j and k and $\pi_{\bar{j}k}$ the proportion having a 0 score on the easier item j indicated by a bar across subscript j, and a 1 score on the more difficult item k; see Table 4.1. For binary variables, the covariance between items equals $\sigma_{jk} = \pi_{jk} - \pi_j \pi_k$ [Equation (3.24)]. In Table 4.1, one may check that $\pi_{jk} + \pi_{\bar{j}k} = \pi_k$, and for the Guttman model, one may recall that it is impossible to have a 0 score on an easy item if one has a 1 score on a more difficult item; hence, $\pi_{\bar{j}k} = 0$, so that $\pi_{jk} = \pi_k$, which is the maximum value. It follows that

$$\sigma_{jk} = \pi_k - \pi_j \pi_k = \pi_k \left(1 - \pi_j\right), \tag{4.50}$$

which is the maximum inter-item covariance, $\sigma_{jk}^{\max}$. It follows from this result that the product-moment correlation also is maximal. Next, if one assumes that $\delta_j = \delta$ for all j, then $\pi_j = \pi$ and $\sigma_j = \sigma$, all j, and one can derive that $\sigma_{jk}^{\max} = \sigma_j \sigma_k = \pi(1 - \pi)$, so that the maximum product-moment correlation equals $\rho_{jk}^{\max} = \dfrac{\sigma_{jk}^{\max}}{\sigma_j \sigma_k} = 1$.

TABLE 4.1

Cross Table for Items j and k

		X_k		
		1	0	
X_j	1	π_{jk}	$\pi_{j\bar{k}}$	π_j
	0	$\pi_{\bar{j}k}$	$\pi_{\bar{j}\bar{k}}$	$1-\pi_j$
		π_k	$1-\pi_k$	1

Note: π_j=proportion correct on item j; π_{jk}=proportion correct on item j and item k; $\pi_{\bar{j}k}$=proportion incorrect on item j and correct on item k; and so on.

We have shown that the Guttman model implies maximum inter-item correlations, and together with equal item difficulties, they represent the situation in which the attenuation paradox is extreme. This becomes obvious as soon as one realizes that in this situation, people can have only one of two sum scores, $X=0$ or $X=J$. For example, for two items ($J=2$) of the Guttman type, and $\pi_j=.5$ ($j=1,2$), let us consider the distribution of the sum score. First, the mean of the sum-score distribution equals $\mu_X = J \times \pi_j = 2 \times .5 = 1$. Second, the variance equals $\sigma_X^2 = \sigma_1^2 + \sigma_2^2 + 2\sigma_{12}$. The item variances and the inter-item covariances all equal $\pi_j(1-\pi_j)=.25$; hence, $\sigma_X^2 = 1$. If the mean squared deviation—the variance—equals 1 and the scale has integer scores 0, 1, 2, then half of the people scored 0 and the other half 2; hence, intermediate scores are absent and the sum-score distribution is perfectly bimodal. Consequently, in this case, the Guttman model implies the attenuation paradox in its extreme form. One may check that if one lets the number of items, J, free, no intermediate sum scores occur and half of the people have sum score $X=0$, whereas the other half have $X=J$. For $\pi_j=\pi$, for all j, one may check that a proportion of people equal to π have $X=0$, whereas proportion $1-\pi$ have $X=J$. For π_j varying across items, intermediate sum scores will show positive frequencies.

The point is the following. A test allowing only two sum scores provides a poor representation of the attribute, even though item reliability is maximal. To have a better attribute representation, lower item reliability is desirable. In IRT parlance, the Rasch model, having less than perfect item discrimination, implies a better representation of sum scores, hence, a better representation of the attribute. The attenuation paradox can be formulated so that it is another way of looking at a probabilistic measurement model breaking down as one considers a limiting situation, and this is what we just did.

Epilogue

The Rasch model or one-parameter logistic model is a restrictive model for the data. Without the sufficiency assumption, the assumptions of unidimensionality, local independence, and monotonicity, together defining the model of monotone homogeneity (Chapter 3), are already restrictive in the sense that there is no *a priori* reason to expect that nature has organized people's responses to even a well-chosen set of items, such that the resulting data **X** are perfectly consistent with these restrictions. Adding the fourth sufficiency assumption thus yielding the Rasch model restricts the model greatly as a tool for representing data. Models never coincide with reality (Box, 1976), but if formulated well, they represent reality's key features while ignoring minor features. To be practically useful, models must not deviate excessively from reality or the data representing reality, or predictions about school success, job performance, therapy suitability, or health-care intervention will fail regularly. As we argued at great length, we prefer to let the theory of the attribute be the guide for meaningful measurement. This perspective does not alter the fact that our experience with the Rasch model has shown that, because it is simple and has well-developed goodness-of-fit methods, a Rasch data analysis can be extremely informative about one's items and may help one to decide which steps one might take next. Examples of such steps are removal of items (e.g., violation of monotonicity), splitting of the item set in different subsets (e.g., violation of unidimensionality), removal of badly fitting person data records (e.g., due to person misfit; e.g., see Meijer & Sijtsma, 2001), splitting of the group in two or more groups (e.g., violation of local independence), or choosing another measurement model to scale the items.

Two and Three-Parameter Logistic Models

We continue with the last suggestion, choosing another, most likely more liberal IRT model than the Rasch model. The two-parameter logistic model and the three-parameter logistic model are particularly popular (Birnbaum, 1968). The two-parameter logistic model is defined as (with $\alpha_j = \alpha_j^*$)

$$P_j(\theta) = \frac{\exp\left[\alpha_j\left(\theta-\delta_j\right)\right]}{1+\exp\left[\alpha_j\left(\theta-\delta_j\right)\right]}. \tag{4.8}$$

Compared to the one-parameter logistic model, the two-parameter logistic model allows the slope parameters to vary across items and this provides a more flexible model. The three-parameter logistic model is an extension of the two-parameter logistic model, because it has a parameter, γ_j, for the lower asymptote of the logistic function, so that

$$P_j(\theta) = \gamma_j + \frac{\left(1-\gamma_j\right)\exp\left[\alpha_j\left(\theta-\delta_j\right)\right]}{1+\exp\left[\alpha_j\left(\theta-\delta_j\right)\right]}. \tag{4.51}$$

For example, $\gamma_j > 0$ is likely to happen when the test consists of multiple-choice items. For the three-parameter normal-ogive model [Equation (4.5)], we argued that the normal-ogive counterpart parameter, c_j, to parameter γ_j in Equation (4.51), may be expected to vary across the items and has to be estimated from the data. The reason is that one cannot expect low-θ persons to guess blindly for the correct answer out of m options, so that one might simply fix $\gamma_j = m^{-1}$ for all items. Items rather contain cues that divert people to the correct option or away from it, which causes $\gamma_j \neq m^{-1}$, and the precise values have to be estimated. Figure 4.10a shows two IRFs for the two-parameter logistic model, and Figure 4.10b shows two IRFs for the three-parameter logistic model.

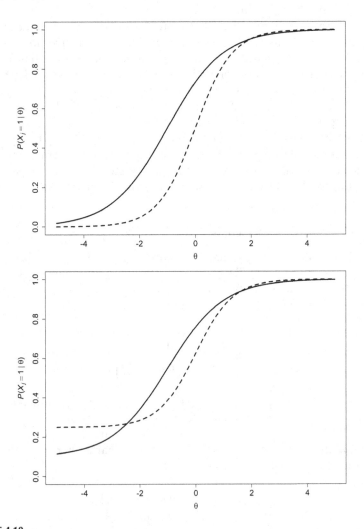

FIGURE 4.10
(a) Two IRFs for the two-parameter logistic model, with $\alpha_j = 1$ and $\delta_j = -1$ (solid curve), and $\alpha_k = 1.5$ and $\delta_k = 0$ (dashed curve). (b) Two IRFs for the three-parameter logistic model, with $\alpha_j = 1$, $\delta_j = -1$, and $\gamma_j = .10$ (solid curve), and $\alpha_k = 2$, $\delta_k = 0$, and $\gamma_k = .25$ (dashed curve).

BOX 4.2   Estimation of the Two-Parameter Logistic Model

Estimation of parameters from the two models is not feasible by means of conditional maximum likelihood. For example, for the two-parameter logistic model, Equation (4.20) has to be modified, so that

$$C_{N,J}(\theta,\delta,\alpha) = \left\{ \prod_{i=1}^{N}\prod_{j=1}^{J}\left\{1+\exp\left[\alpha_j\left(\theta_i-\delta_j\right)\right]\right\}\right\}^{-1}, \qquad (4.52)$$

which, like Equation (4.20), is independent of the data. Furthermore, one has person statistic

$$x_i^* = \sum_{j=1}^{J}\alpha_j x_{ij}, \qquad (4.53)$$

which is sufficient for θ provided the slopes parameters are known, and item statistic

$$s_j^* = s_j = \sum_{i=1}^{N}x_{ij}, \qquad (4.54)$$

which is sufficient for the product $\delta_j^* = \alpha_j\delta_j$ (Fischer, 1974, pp. 203–208). These results are problematic. First, the slope parameters in Equation (4.53) are unknown, whereas sufficient statistics must only depend on the data. Second, statistic s_j^* [Equation (4.54)] estimates a product of two item parameters but does not retrieve information about the slope parameters that one might use to estimate θ_i. The solution comes from maximizing a likelihood that first averages across the distribution of the latent variable θ and then estimates the item parameters. This method is known as marginal maximum likelihood estimation. We assume a distribution $f(\theta|\omega)$, where ω contains the parameters of the distribution, such as the mean and the variance when f represents a normal distribution. The likelihood is integrated over this distribution. For the two-parameter logistic model, we have

$$L(X|\delta,\alpha,\omega) = \int_{-\theta}^{\theta}\prod_{i=1}^{N}\prod_{j=1}^{J}\frac{\exp\left[\alpha_j\left(\theta_i-\delta_j\right)\right]^{x_{ij}}}{1+\exp\left[\alpha_j\left(\theta_i-\delta_j\right)\right]}f(\theta|\omega)d\theta. \qquad (4.55)$$

The resulting marginal likelihood does not contain θ and is maximized with respect to the item parameters δ and α and also with respect to the parameters of $f(\theta|\omega)$, that is, ω. If $f(\theta|\omega)$ is correctly specified, the estimates are consistent. Versions of marginal maximum likelihood estimation in which $f(\theta)$ is estimated from the data and then is used to obtain Equation (4.55) are also available. Next, assuming that the item parameters estimates are the true values, they are inserted in the model's likelihood and estimates of latent variable θ are obtained (Baker, 1992, Chap. 6; Warm, 1989).

Verhelst and Glas (1995) proposed a hybrid model in which the slope parameter is replaced with a slope index, denoted A_j, that the researcher has to specify based on prior knowledge of the functioning of the items in the population of interest. The so-called one-parameter logistic model equals

$$P_j(\theta) = \frac{\exp\left[A_j\left(\theta - \delta_j\right)\right]}{1 + \exp\left[A_j\left(\theta - \delta_j\right)\right]}. \tag{4.56}$$

Specifying the slopes rather than estimating them relieves the researcher of one set of item parameters while maintaining the flexibility of the two-parameter logistic model. Because one only estimates the location parameters, one can use conditional maximum likelihood and proceed as if one uses the Rasch model. In addition, the goodness-of-fit methods of the Rasch model are available. In case of misfit, one can adapt one or more slope indices, re-estimate the model, and assess its fit, until one finds a fitting model.

The information function for two-parameter logistic items equals

$$I_j(\theta) = \frac{\left[P_j'(\theta)\right]^2}{P_j(\theta)\left[1 - P_j(\theta)\right]} = \alpha_j^2\left\{P_j(\theta)\left[1 - P_j(\theta)\right]\right\}. \tag{4.57}$$

Expanding the right-hand side yields

$$I_j(\theta) = \frac{\alpha_j^2 \exp\left(\theta - \delta_j\right)}{\left[1 + \exp\left(\theta - \delta_j\right)\right]^2}. \tag{4.58}$$

The maximum value of Equation (4.58) equals

$$I_j(\theta) = .25\alpha_j^2. \tag{4.59}$$

The test information function again is the sum of the item information functions, and the standard error for $\hat{\theta}$ equals $\sigma_{\hat{\theta}|\theta} = I(\theta)^{-1/2}$.

For the three-parameter logistic model, C. A. W. Glas (personal communication) drew our attention to a version of the item information function that one does not find in the literature but that shows the influence of lower asymptote parameter γ_j more lucidly than other equations. One first defines

$$\psi_j(\theta) = \frac{P_j(\theta) - \gamma_j}{1 - \gamma_j},\qquad(4.60)$$

next notices that

$$1 - P_j(\theta) = (1 - \gamma_j)\left[1 - \psi_j(\theta)\right],\qquad(4.61)$$

and then derives, taking a result in Baker [1992, p. 76, Equation (3.24)] as the starting point,

$$I_j(\theta) = \frac{(1-\gamma_j)^2\,\alpha_j^2\psi_j^2(\theta)\left[1-\psi_j(\theta)\right]^2}{P_j(\theta)\left[1-P_j(\theta)\right]}.\qquad(4.62)$$

If $\gamma_j = 0$ and $\alpha_j = 1$, Equation (4.62) reduces to the item information function for the Rasch model in Equation (4.28). Further, if γ_j increases, then $(1-\gamma_j)^2$ decreases and item information decreases; see the examples in Figure 4.11. In addition, here, the test information function is the sum of the item

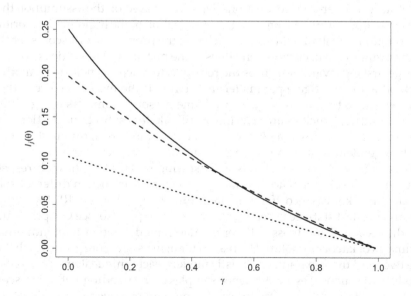

FIGURE 4.11
Information function $I_j(\theta)$ (vertical axis) as a function of lower asymptote γ_j (horizontal axis). For all three curves, slope parameter $\alpha_j = 1$ and location parameter $\delta_j = 0$, and for the solid curve $\theta = 0$, for the dashed curve $\theta = 1$, and for the dotted curve $\theta = 2$.

information functions, and the standard error is obtained as before. Unlike the other models discussed so far, the maximum information for a three-parameter logistic item does not occur at $\theta = \delta_j$, but is obtained at (Birnbaum, 1968, p. 464)

$$\theta_0 = \delta_j + \alpha_j^{-1} \ln\left(\frac{1 + \sqrt{1 + 8\gamma_j}}{2} \right). \tag{4.63}$$

Inserting θ_0 in Equation (4.62) provides the maximum item information. For example, let $\delta_j = 0$, $\alpha_j = 1.5$, and let γ_j be any value greater than .25; then, we find $\theta_0 \approx 0.208$. Inserting the three item parameters and θ_0 in Equation (4.62) yields the maximum item information, $I_j(\theta) \approx 0.348$. Figure 4.12a shows conditional standard errors for two items consistent with the two-parameter logistic model and for the two-item test based on these two items. Figure 4.12b shows similar curves for two items consistent with the three-parameter logistic model.

Goodness-of-fit methods for the two- and three-parameter logistic models have not been developed as well as for the Rasch model. Van der Linden and Hambleton (1997; also, see Hambleton & Swaminathan, 1985, Chap. 8; Hambleton, Swaminathan, & Rogers, 1991, Chap. 4; Yen, 1981) mention several possibilities encountered in the literature on these models. One method is to investigate unidimensionality by means of factor analysis of the matrix of tetrachoric inter-item correlations, which are based on the assumption that the item scores resulted from the dichotomization of the hypothetical normal distributions of latent item scores, such that the proportion of 1 scores for the items results. One uses the eigenvalues of the matrix, and if one does not find a large first eigenvalue, one uses the residuals to reach a conclusion. Another method is to investigate parameter invariance. If the model is correct, then one expects to find approximately the same person parameters using different item subsets, such as an easy and a difficult item subset. In addition, one expects to find approximately the same item parameters in different groups, such as gender groups, educational groups, and score-level groups.

Glas (2016) discussed methods for assessing goodness of fit with respect to the IRFs, local independence, and parameter invariance in different subgroups, also known as differential item functioning. For the IRF, the author discussed a test statistic based on sum-score groups (also, see Chapter 3 and the discussion of goodness of fit for the Rasch model) rather than individual estimates of latent variable θ. The reason for using sum-score groups is that in the two- and three-parameter logistic models, each unique item-score vector produces a unique θ estimate; hence samples corresponding with item-score vectors are too small or even empty to compare observed proportions of 1 scores with model-expected proportions. Because for longer tests, say, $J > 20$, the growing number of sum-score groups may produce information with too much detail, adjacent groups may be combined. However, this summary

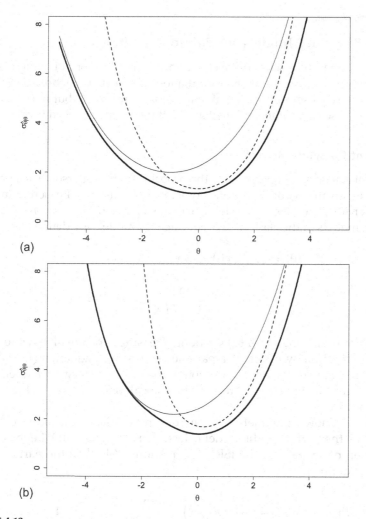

(a)

(b)

FIGURE 4.12
(a) Conditional standard error $\left(\sigma_{\hat{\theta}|\theta}\right)$ for two IRFs for the two-parameter logistic model, with $\alpha_j=1$ and $\delta_j=-1$ (solid curve), and $\alpha_k=1.5$ and $\delta_k=0$ (dashed curve), and for the total test (solid bold curve). (b) Conditional standard error $\left(\sigma_{\hat{\theta}|\theta}\right)$ for two IRFs for the three-parameter logistic model, with $\alpha_j=1$, $\delta_j=-1$, and $\gamma_j=.10$ (solid curve), and $\alpha_k=2$, $\delta_k=0$, and $\gamma_k=.25$ (dashed curve), and for the total test (solid bold curve).

of the data impairs the approximate chi-squared distribution of the statistic Orlando and Thissen (2000) proposed in the marginal maximum likelihood framework to compare observed and expected IRFs. Glas (2016) proposed Lagrange multiplier tests for investigating IRFs, item-parameter invariance, and the distribution of the latent variable as it is used in marginal maximum likelihood estimation. One can also use the Lagrange multiplier tests with polytomous IRT models and multidimensional IRT models.

Some Basic IRT Models for Polytomous Items

In this section, we discuss three classes of IRT models for polytomous items. We also provide nonparametric formulations for each of the three classes. For each class, we discuss the type of items most appropriate but notice that the literature is somewhat inconsistent and that the topic may need further study.

Adjacent Category Models

Adjacent category models provide the probability that a person has a particular score x on an item or a score one point lower, that is, $x-1$. Let acronym ACM stand for adjacent category model. The item-step response functions (ISRFs) of the IRT models in the class of adjacent category models are defined as

$$f_{jx}^{ACM}(\theta) = P\left(X_j = x \mid \theta; X_j = x-1 \vee X_j = x\right)$$

$$= \frac{P\left(X_j = x \mid \theta\right)}{P\left(X_j = x-1 \mid \theta\right) + P\left(X_j = x \mid \theta\right)}, \quad x=1,\ldots,M, \tag{4.64}$$

By assumption, Equation (4.64) is a nondecreasing function of θ. Without further specification by means of a parametric function, Equation (4.64) can be considered the nonparametric version of the adjacent category model, given that θ is unidimensional and that we have local independence at the level of the item scores.

The best-known parametric IRT model in the class of adjacent category models is the partial credit model (Masters, 1982, 1988, 2016). Let δ_{jx} denote a location parameter for the ISRF in Equation (4.64), then the partial credit model is defined as

$$P\left(X_j = x \mid \theta; X_j = x-1 \vee X_j = x\right) = \frac{\exp\left(\theta - \delta_{jx}\right)}{1 + \exp\left(\theta - \delta_{jx}\right)}, \quad x=1,\ldots,M. \tag{4.65}$$

Thus, Equation (4.65) models the ISRF of the partial credit model as a Rasch model [Equation (4.11)], and one needs M Rasch models for M pairs of adjacent item scores, enumerated $(0,1)$, $(1,2)$, ..., $(M-1, M)$. One may notice that, trivially, because for dichotomous items only one pair of item scores is involved, the Rasch model may be written in the form of Equation (4.65) as

$$P\left(X_j = 1 \mid \theta; X_j = 0 \vee X_j = 1\right) = \frac{\exp\left(\theta - \delta_{j1}\right)}{1 + \exp\left(\theta - \delta_{j1}\right)} = \frac{\exp\left(\theta - \delta_j\right)}{1 + \exp\left(\theta - \delta_j\right)}, \tag{4.66}$$

with $\delta_{j1} = \delta_j$. Back to Equation (4.65), the partial credit model conditions on two possibilities, item scores equal to $x-1$ or x, and assumes that the Rasch model governs the probability that the highest of the two scores is obtained. This means that the response process for two adjacent response categories is isolated from the other $M - 1$ answer categories. The consequence is that the M location parameters of item j do not have a fixed order and that their order may vary over items.

Several authors (e.g., Hemker et al., 2001; Van Engelenburg, 1997) have discussed item types eliciting cognitive processes supposedly consistent with the formal structure of the partial credit model. For example, it has been argued that the free ordering of item location parameters suggests an item type consisting of M subtasks that one may approach in an arbitrary order (Van Engelenburg, 1997, pp. 28–30). The M subtasks can be items nested in a testlet, where the testlet is a poem and the subtasks are questions about different aspects of the poem that can be answered without knowing the answer to the other subtasks within the testlet. Van Engelenburg (1997, pp. 32–35) claimed that what we have called the nonparametric adjacent category model is consistent with this item type, but a special case of the partial credit model called the rating scale model (Andrich, 1978a) that we discuss later, was based on the claim that rating scale items can also be modeled with adjacent category models. Tutz (1990) argued that the partial credit model is unfortunate as a cognitive model. His line of reasoning was that the model probability of completing x subtasks given that the person has completed either $x - 1$ or x subtasks is implausible, because then the model assumes that consecutive steps will not be performed, something one cannot know in advance. In addition, Van Engelenburg (1997) argued that it is difficult if not impossible to define task features implying an adjacent category model, and in addition, he claimed that location parameters do not refer to particular features of an item. That is, one cannot know from a model estimated from one's data to which subtasks the item parameters refer, because of the absence of a fixed order in which each person taking the complete item tried them. Thus, considering the partial credit model or other adjacent category models as cognitive models is questionable or at least ambiguous.

The category characteristic curve of the partial credit model refers to the response probability, $P(X_j = x | \theta)$, which combines the information from the M ISRFs in Equation (4.65) as follows.

BOX 4.3   Combining M ISRFs to a Category Characteristic Curve

We show how M ISRFs can be combined to obtain a category characteristic curve for $M=2$, hence, $x=0,1,2$. We use shorthand, $\pi_{x\theta} = \pi_x = P(X_j = x | \theta)$, and, based on Equation (4.64), notice that

$$P\left(X_j = x | \theta; X_j = x-1 \vee X_j = x\right) = \frac{\pi_x}{\pi_{x-1} + \pi_x}. \tag{4.67}$$

Because $M=2$, we consider the cases for $x=1$ and $x=2$. Then, for $x=1$, we have from Equation (4.65) that, writing shorthand $\exp(x) = \exp\left(\theta - \delta_{jx}\right)$, such that exp(1) denotes $\exp(\theta - \delta_{j1})$ but not the numerical value 2.718 and exp(2) denotes $\exp(\theta - \delta_{j2})$ but not the numerical value 7.389, we have

$$\frac{\pi_1}{\pi_0 + \pi_1} = \frac{\exp(\theta - \delta_{j1})}{1 + \exp(\theta - \delta_{j1})} = \frac{\exp(1)}{1 + \exp(1)}. \tag{4.68}$$

Cross-multiplication and solving for π_1 yields

$$\pi_1 = \pi_0 \exp(1). \tag{4.69}$$

Similarly, we develop

$$\frac{\pi_2}{\pi_1 + \pi_2} = \frac{\exp(2)}{1 + \exp(2)}, \tag{4.70}$$

and solve for π_2, which yields

$$\pi_2 = \pi_1 \exp(2). \tag{4.71}$$

Combining Equations (4.69) and (4.71) yields

$$\pi_2 = \pi_0 \exp(1)\exp(2). \tag{4.72}$$

Furthermore, $\pi_0 + \pi_1 + \pi_2 = 1$, and inserting Equations (4.69) and (4.72) and solving for π_0 yields

$$\pi_0 = \left[1 + \exp(1) + \exp(1)\exp(2)\right]^{-1} = \frac{1}{\sum_{z=0}^{2} \exp\left[\sum_{y=1}^{z}\left(\theta - \delta_{jy}\right)\right]}, \tag{4.73}$$

with $\sum_{y=1}^{0}\left(\theta - \delta_{jy}\right) \equiv 0$. Hence, it follows from Equation (4.69) that

$$\pi_1 = \exp(1)\left[1 + \exp(1) + \exp(1)\exp(2)\right]^{-1} = \frac{\exp\left(\theta - \delta_{j1}\right)}{\sum_{z=0}^{2} \exp\left[\sum_{y=1}^{z}\left(\theta - \delta_{jy}\right)\right]}, \tag{4.74}$$

and from Equation (4.72) that

$$\pi_2 = \exp(1)\exp(2)\left[1 + \exp(1) + \exp(1)\exp(2)\right]^{-1}$$

$$= \frac{\exp\left[\sum_{y=1}^{2}\left(\theta - \delta_{j2}\right)\right]}{\sum_{z=0}^{2} \exp\left[\sum_{y=1}^{z}\left(\theta - \delta_{jy}\right)\right]}. \tag{4.75}$$

If one expands for $M > 2$, one will see that in Equation (4.75) changing the summation in the numerator and the first summation in the denominator to run to x rather than 2 yields the desired response probability,

$$P\left(X_j = x \mid \theta\right) = \frac{\exp\left[\sum_{y=1}^{x}\left(\theta - \delta_{jy}\right)\right]}{\sum_{z=0}^{M} \exp\left[\sum_{y=1}^{z}\left(\theta - \delta_{jy}\right)\right]}. \tag{4.76}$$

Figure 4.13 shows the category characteristic curves based on Equation (4.76). The curve for $x=0$ is monotone decreasing; as the latent variable θ grows, it is more unlikely that one will obtain the smallest score possible on the item. The curve for $x=1$ is bell-shaped; as θ grows, the probability that one obtains a score of 1 rather than a score of 0 increases, but beyond a particular point, it is more likely that one obtains a score of 2. The curve for $x=2$ is monotone increasing. The intersection point for the first and the second curves is at $\theta = \delta_{j1}$ and for the second and third curves at $\theta = \delta_{j2}$. Sometimes, the item location parameters are interpreted as thresholds, because they mark the location on the scale where one item score takes over dominance in probability from the other. In Figure 4.13, the maximum for the bell-shaped curve lies above the other two curves, meaning that the location parameters divide the scale into three areas in which a particular item score is

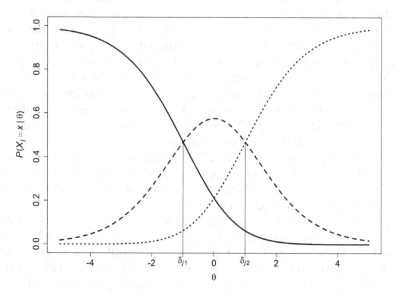

FIGURE 4.13
Category characteristic curves of a three-category item for the partial credit model with $\delta_{j1} = -1$ and $\delta_{j2} = 1$. The solid curve is $P(X_j = 0 \mid \theta)$, the dashed curve is $P(X_j = 1 \mid \theta)$, and the dotted curve is $P(X_j = 2 \mid \theta)$.

dominant in probability. This need not be the case, however, and one may find that $\delta_{j2} < \delta_{j1}$, so that $P(X_j = 1|\theta)$ lies below the other two curves all along θ. Sometimes, this is taken as a sign that the item does not function well, but we do not consider this as a compelling argument unless $P(X_j = 1|\theta)$ has a small value for all θ.

An interesting feature of the location parameters is that they not only locate the category characteristic curves on the latent variable scale but also influence the slope of the response functions and thus the discrimination power (Muraki, 1992). Based on experience, we contend this is easier to see for the conditional expected item score $\mathcal{E}(X|\theta)$ than for the category characteristic curves $P(X_j = x|\theta)$. For example, consider category characteristic curves for $M = 3$ and location parameters $\delta_{j1} = -2$, $\delta_{j2} = 0$, and $\delta_{j3} = 2$ (Figure 4.14a). Moving δ_{j1} closer to the other location parameters results in location parameters $\delta_{j1} = -0.5$, $\delta_{j2} = 0$, and $\delta_{j3} = 2$ (Figure 4.14b). Comparing the two panels does not show clearly that the second item discriminates better than the first item. However, Figure 4.14c shows that replacing the two sets of category characteristic curves $P(X_j = 1|\theta)$ with two conditional expected item scores $\mathcal{E}(X|\theta)$ reveals that the second item discriminates better (Figure 4.14c, dashed curve) than the first item (Figure 4.14c, solid curve).

The partial credit model has M item location parameters for each item, and JM item location parameters for a J-item test. Masters (1982) showed that a conditional maximum likelihood procedure allows for the estimation of the item location parameters independent of the latent-variable distribution, a result similar to that which we outlined in some detail for the Rasch model. That is, one can condition on the sufficient statistics for the item location parameters, which are the numbers of people obtaining score x on item j, which we previously denoted S_j, and the item location parameters drop out of the likelihood. In principle, latent variable values can be estimated independent of the item parameters. By conditioning the likelihood of the model on the sum scores, $X = \sum_j X_j$, on the set of polytomous items, the person parameters θ do not appear in the conditional likelihood, and the sum score thus is a sufficient statistic for estimating θ. Glas and Verhelst (1989; also, see Masters, 2016) discussed a marginal maximum likelihood estimation procedure for the partial credit model that integrates the likelihood across the distribution of the person parameters so that item parameters can be estimated consistently.

For a larger class of ten different Rasch models for polytomous items of which the partial credit model is one model, Glas and Verhelst (1995b) noticed that the properties of parameter separation and conditional maximum likelihood estimation are the defining characteristics of these Rasch models. These authors discussed several goodness-of-fit methods that are generalizations of goodness-of-fit methods for the Rasch model and that can be used for testing the fit of several polytomous Rasch model including the partial credit model to data. Masters (2016)

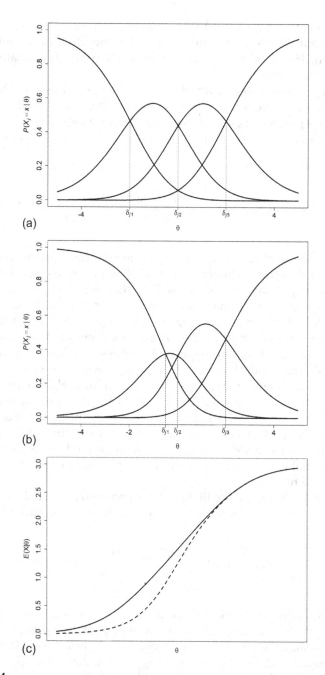

FIGURE 4.14

(a) Category characteristic curves of a four-category item for the partial credit model with $\delta_{j1} = -2$, $\delta_{j2} = 0$, and $\delta_{j3} = 2$. (b) Category characteristic curves of a four-category item for the partial credit model with $\delta_{j1} = -0.5$, $\delta_{j2} = 0$, and $\delta_{j3} = 2$. (c) Item characteristic curves corresponding to panel (a) (solid curve) and panel (b) (dashed curve).

discussed item and person-fit statistics which have unknown sampling properties but which are useful for a first impression of model fit. Van Rijn, Eggen, Hemker, and Sanders (2002) discussed information functions for items and tests.

The rating scale model (Andrich, 1978a) is a special case of the partial credit model. Each item has a location parameter, δ_j, defined as the mean of the M item parameters δ_{jx}, and each item-score combination $(x-1,x)$ has a threshold parameter, τ_x, $x=1,\ldots,M$, so that $\delta_{jx} = \delta_j + \tau_x$. The model equation equals

$$P\left(X_j = x \mid \theta\right) = \frac{\exp\left[\sum_{y=1}^{x}\left(\theta - \delta_j - \tau_y\right)\right]}{\sum_{z=0}^{M}\exp\left[\sum_{y=1}^{z}\left(\theta - \delta_j - \tau_y\right)\right]}. \tag{4.77}$$

This brings the total number of item parameters for a J-item test to $J+M$ rather than JM in the partial credit model. The idea of using a set of identical threshold parameters that is fixed for different items is that different rating scale items usually have the same answer categories, such as "disagree", "mildly disagree", "mildly agree", and "agree", and thus it seems reasonable to hypothesize that relative category locations are fixed across items (Masters & Wright, 1984). This hypothesis implies that the statement preceding the fixed rating scale does not affect the distances between the categories or even that people are capable of using such a set invariably so that it is fixed and the assumption is reasonable. Given that the data are consistent with all other assumptions, the rating scale model tests the assumption of fixed category threshold parameters, $\tau_1,\ldots,\tau_M$. Hence, the rating scale model is a more parsimonious model than the partial credit model, and compared to the latter model its credibility in real-data applications depends on the correctness of the hypothesis.

Compared to the partial credit model, the generalized partial credit model adds a discrimination parameter that depends on the item, such that

$$P\left(X_j = x \mid \theta\right) = \frac{\exp\left[\sum_{y=1}^{x}\alpha_j\left(\theta - \delta_{jy}\right)\right]}{\sum_{z=0}^{M}\exp\left[\sum_{y=1}^{z}\alpha_j\left(\theta - \delta_{jy}\right)\right]} \tag{4.78}$$

(Muraki, 1992). As the discrimination parameter increases, category characteristic curves are steeper and discriminate better between different θ values. Figure 4.15a shows for $x=1, 2$ rather steep bell-shaped curves that suggest that the answer categories corresponding to $x=1, 2$ are strongly related to the latent variable that the item assesses. Figure 4.15b uses the same location parameters but reduces the discrimination parameter from $\alpha_j=1.5$ to $\alpha_j=0.7$, which has the effect of flattening each curve, thus showing the joint effect of location and slope parameters on discrimination.

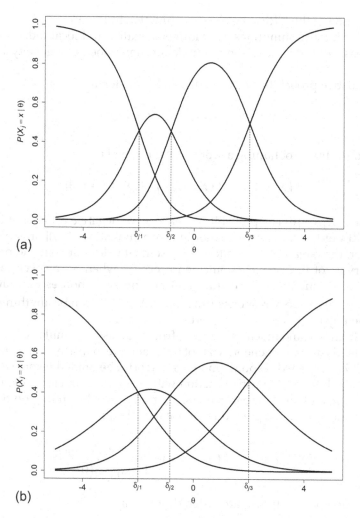

FIGURE 4.15

(a) Category characteristic curves of a four-category item for the generalized partial credit model with $\alpha_j = 1.5$, $\delta_{j1} = -2$, $\delta_{j2} = -0.85$, and $\delta_{j3} = 2$. (b) Category characteristic curves of a four-category item for the generalized partial credit model with $\alpha_j = 0.7$, $\delta_{j1} = -2$, $\delta_{j2} = -0.85$, and $\delta_{j3} = 2$.

Cumulative Probability Models

Let acronym CPM stand for cumulative probability model. In this class of polytomous IRT models nondecreasing ISRFs are defined as

$$f_{jx}^{\text{CPM}}(\theta) = P(X_j \geq x \mid \theta), \quad x = 1, \ldots, M. \tag{4.79}$$

Because for $x = 0$, by definition we have $P(X_j \geq 0 \mid \theta) = 1$, this response probability is uninformative about item responses. Without further parametric specification of the ISRFs, the response function in Equation (4.79), in combination with the assumptions of unidimensionality and local independence, is the basis of the nonparametric model of monotone homogeneity we discussed in Chapter 3.

Cumulative probability $P(X_j \geq x | \theta)$ can be written as

$$P\left(X_j \geq x | \theta\right) = \sum_{y=x}^{M} P(X_j = y | \theta). \tag{4.80}$$

The probability of obtaining a score x then is equal to

$$P\left(X_j = x | \theta\right) = P\left(X_j \geq x | \theta\right) - P(X_j \geq x + 1 | \theta). \tag{4.81}$$

Because $P(X_j = x | \theta)$ is a probability, hence a nonnegative number, it is clear that ISRFs of the same item cannot intersect [Equation (4.81)]. It is possible, however, that adjacent curves touch for some θ values or coincide partially in intervals of θ. Complete coincidence occurs when an answer category is completely ineffective, meaning that no one ever chooses this category. The shape of category response function $P\left(X_j = x | \theta\right)$ can be anything but is restricted by the ISRFs for item scores $x - 1$ and x.

A well-known and much used model from the class of cumulative probability models is the homogeneous case of the graded response model (Samejima, 1969, 1997), or graded response model, for short. The graded response model defines the ISRF as a logistic function with a slope parameter α_j depending on the item, and location parameters, λ_{jx}, $x = 1, \ldots, M$, depending on the item and the item score. The graded response model equals

$$P\left(X_j \geq x | \theta\right) = \sum_{y=x}^{M} P\left(X_j = y | \theta\right) = \frac{\exp\left[\alpha_j\left(\theta - \lambda_{jx}\right)\right]}{1 + \exp\left[\alpha_j\left(\theta - \lambda_{jx}\right)\right]}. \tag{4.82}$$

The location parameters λ_{jx} are ordered, such that

$$\lambda_{j1} \leq \ldots \leq \lambda_{jM}. \tag{4.83}$$

Because the ISRFs of an item must not intersect, an item must have one fixed slope parameter, but different items can have and usually will have different slope parameters. This means that the ISRFs of items j and k, which are $P\left(X_j \geq x | \theta\right)$ and $P\left(X_k \geq y | \theta\right)$, must intersect at

$$\theta = \frac{\alpha_j \lambda_{jx} - \alpha_k \lambda_{ky}}{\alpha_j - \alpha_k}. \tag{4.84}$$

If $\alpha_j = \alpha_k$ and $\alpha_j \lambda_{jx} = \alpha_k \lambda_{ky}$, together implying that $\lambda_{jx} = \lambda_{ky}$, then Equation (4.84) is undefined; hence there is no point of intersection, because the two ISRFs coincide.

We consider an item consistent with the graded response model having three ordered scores, so that $M=2$, and illustrate the category characteristic curves as functions of the distance between the location parameters. Then, writing shorthand $A_{jx} = \left\{ \dfrac{\exp\left[\alpha_j\left(\theta - \lambda_{jx}\right)\right]}{1 + \exp\left[\alpha_j\left(\theta - \lambda_{jx}\right)\right]} \right\}$, one can derive that

$$P\left(X_j = 0 \mid \theta\right) = 1 - P\left(X_j \geq 1 \mid \theta\right) = 1 - A_{j1}, \tag{4.85}$$

$$P\left(X_j = 1 \mid \theta\right) = P\left(X_j \geq 1 \mid \theta\right) - P\left(X_j \geq 2 \mid \theta\right) = A_{j1} - A_{j2}, \tag{4.86}$$

$$P\left(X_j = 2 \mid \theta\right) = P\left(X_j \geq 2 \mid \theta\right) - 0 = A_{j2}. \tag{4.87}$$

One can check that $\sum_{x=0}^{2} P\left(X_j = x \mid \theta\right) = 1$. Figure 4.16a shows the three category characteristic curves for parameter values $\alpha_j=1$, $\lambda_{j1}=-2$, and $\lambda_{j2}=2$. The locations correspond to the intersection points of the response probabilities. When the difference between the locations is smaller, the category characteristic curve for $x=1$ is flatter (Figure 4.16b), until the locations coincide and $P\left(X_j=1\mid\theta\right)=0$ for all θ and the item has become dichotomous. Masters (1982) noticed that for more than three ordered scores the relationship between item location parameters and response curves is more complex.

Van Engelenburg (1997, pp. 30–32) hypothesized that what we call cumulative probability models are consistent with item scores that result from a global assessment task. An example is the rating of a response on a Likert-type item measuring an attitude or a personality trait. One assumes that the tested person forms a general impression of her position on the scale relative to the item. For example, the person determines the degree to which the main character in a text expressed a hostile attitude toward the other characters and then provides a rating on a Likert scale. The steps one takes to arrive at the item score often are of the kind "none", "mild", "considerable", and "strong", which show a natural order. One considers the rating scale options simultaneously to determine one's position. Item score x reflects the position implying that the person also took the previous steps and failed the next steps. However, one may notice that the cognitive actions that this kind of item elicits remain unknown. For example, whether a person actually takes consecutive steps is speculative, and the cognitive processes involved may be subjected to further scrutiny.

Item parameters can be estimated using marginal maximum likelihood in combination with an EM-algorithm (Samejima, 2016); also see Baker (1992, Chap. 8) who discussed both estimation of item parameters and person parameters. Baker (1992, pp. 239–243) discussed the information functions for the graded response model. Although the interest resides with the test information function, the focus is on the item information functions, which are the building blocks of the test information function. Item information

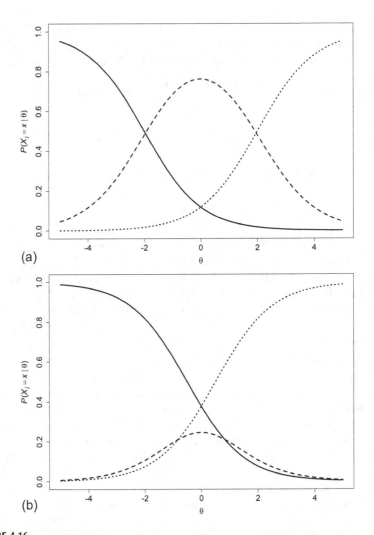

FIGURE 4.16

(a) Category characteristic curves of a three-category item for the graded response model with $\alpha_j=1$, $\lambda_{j1}=-2$, and $\lambda_{j2}=2$. (b) Category characteristic curves of a three-category item for the graded response model with $\alpha_j=1$, $\lambda_{j1}=-0.5$, and $\lambda_{j2}=0.5$. The solid curves are $P(X_j=0|\theta)$, the dashed curves are $P(X_j=1|\theta)$, and the dotted curves are $P(X_j=2|\theta)$.

functions are sums of $M+1$ "information shares" related to the various answer categories in an item (Samejima, 1969). An important distinction is between the information scores in an answer category that contribute to the estimation of latent variable θ, and the share this information contributes to the total item information. The sum of the category information functions does not equal the item information, but the sum of the relative shares does equal the item information. The sum of the item information functions equals the test information function. Samejima (2016) briefly summarized goodness-of-fit methods for the graded response model, and Glas (2010) discussed tests based on the Lagrange multiplier framework.

Continuation Ratio Models

Let acronym CRM stand for continuation ratio model (e.g., Hemker et al., 2001). The third class of polytomous IRT models defines nondecreasing ISRFs as

$$f_{jx}^{CRM}(\theta) = P\left(X_j \geq x \mid \theta; X_j \geq x-1\right) = \frac{P(X_j \geq x \mid \theta)}{P(X_j \geq x-1 \mid \theta)}, \quad x = 1,\ldots,M. \quad (4.88)$$

Without a parametric choice for the ISRF and given unidimensionality and local independence, Equation (4.88) can be considered the nonparametric continuation model. This class of models quantifies the probability of having at least a score of x when one has already come as far as $x-1$.

Tutz (1990) discussed how adjacent category models and cumulative probability models are different from continuation ratio models, but he did not use these names and discussed specific models rather than general classes of models. Specifically, he pointed to the use of boundaries and boundary parameters that define regions separated by boundaries in the first two classes of models. For example, in the Rasch model for dichotomous items, assume a latent continuous random variable Z_{ij} for the combination of person i and item j, and assume that ε_{ij} reflects random measurement error, then

$$Z_{ij} = \theta_i + \varepsilon_{ij}. \quad (4.89)$$

The response mechanism is

$$X_{ij} = 1 \text{ if and only if } Z_{ij} \geq \delta_j. \quad (4.90)$$

By assuming a particular logistic distribution for random measurement error ε_{ij} (Tutz, 1990), the Rasch model follows immediately. One may notice that this conceptualization of the response process was also used in Equations (4.3) and (4.4) for the normal-ogive model. As an aside, one may relate the conceptualization in Equations (4.3) and (4.4) to the Rasch paradox, in the sense that the addition of a random error to the response process saves the Rasch model from being deterministic. It may also be noticed that the difference, $Z_{ij} - \delta_j$, determines the magnitude of the influence of random error on the item score.

For the graded response model [Equation (4.82)], a latent response process can be imagined as

$$X_{ij} = x_{ij} \text{ if and only if } \delta_{j,x_{ij}-1} \leq Z_{ij} \leq \delta_{jx_{ij}}. \quad (4.91)$$

Let us assume a particular logistic distribution for random error ε_{ij} that contains a slope parameter depending on the item; then in combination with Equations (4.89) and (4.91) this distribution for ε_{ij} produces Samejima's graded response model. For the partial credit model, the same mechanism applies as in the Rasch model, because the partial credit model isolates adjacent pairs of item scores and models the response process as if it were a Rasch process. Tutz (1990) objected that this conceptualization, which refers to $P(X_j = x \mid \theta; X_j = x-1 \vee X_j = x)$, assumes that the item score is either $x-1$ or x but not another score. Thus, if one has evaluated $x=0$ and $x=1$, higher scores seem to be out of the question. However, this is true only if one insists that the different steps in solving or answering an item are executed in a fixed order. With adjacent category models, this is not the case, and one can consider these models as describing item responses that represent performance on a set of tasks that are nested within a common context, such as a text about a Shakespearian play, each of which can be solved or answered without having knowledge of the other subtasks.

For so-called sequential response processes, Tutz (1990) defined the general sequential model as follows. We assume that

$$Z_{ijx} = \theta_i + \varepsilon_{ijx}, \tag{4.92}$$

and that

$$X_{ij} \geq x_{ij} \text{ if and only if } Z_{ijx} \geq \delta_{jx}, \text{ given that } X_{ij} \geq x_{ij} - 1. \tag{4.93}$$

A choice of a monotone distribution function for the random error produces the general sequential model. Let us assume that $f_{jx}^{\mathrm{CRM}}(\theta)=1$ if $x<1$, and $f_{jx}^{\mathrm{CRM}}(\theta)=0$ if $x>M$. Given these definitions, Samejima (1972, Chap. 4) showed that for continuation ratio models, the category characteristic curve can be written as

$$P(X_j = x \mid \theta) = \prod_{y=0}^{x} f_{jy}^{\mathrm{CRM}}(\theta)\left[1 - f_{j,x+1}^{\mathrm{CRM}}(\theta)\right]. \tag{4.94}$$

Equation (4.94) shows that the probability of having a score of x is the product of x ISRFs for the first x subtasks on which the person succeeded, and one probability of failing the $(x+1)$st subtask.

Typically, here, an item may be conceived of as a fixed sequence of M subtasks. Failure on the $(x+1)$st subtask implies that the final item score equals x. Thus, the subtasks of the item have to be executed in a fixed order, and failure on one subtask implies failure on the subsequent subtasks. For example, in a text comprehension item, the test administrator may first check whether the tested person has understood the topic of the text (if not, $x=0$). Next, the test administrator checks whether the person has understood a particular fact about an event explicitly described in the text (if not, $x=1$). Finally, the test administrator checks whether the person has understood the implicitly mentioned intention of the main character portrayed (if not, $x=2$; otherwise, $x=3$). As far as we know, research scrutinizing the cognitive processes involved and the fit of these processes with the formal structure of continuation ratio models has not been done.

Let β_{jx} be a location parameter; then the sequential model or stepwise Rasch model is

$$P\left(X_j \geq x \mid \theta; X_j \geq x-1\right) = \frac{\exp\left(\theta - \beta_{jx}\right)}{1 + \exp\left(\theta - \beta_{jx}\right)} \qquad (4.95)$$

(also, see Verhelst, Glas, & De Vries, 1997). Item parameter β_{jx} gives the difficulty of going from step $x-1$ to step x, that is, solving the subtask associated with this transition, and given that the previous subtasks were all solved or answered positively.

Glas (1988b) showed that in the context of the sequential model conditional maximum likelihood estimation is unfeasible. Tutz (2016) discussed joint maximum likelihood and marginal maximum likelihood for estimating the parameters of the complete model. Glas (2010) discussed goodness-of-fit methods based on the Lagrange multiplier framework and the Bayesian framework for the shape of the response functions, the invariance of item parameters, and local independence. Kim (2002, p. 18) discussed the information functions for the continuation ratio model.

Finally, two interesting issues remain. The first issue is that it is not clear whether a particular class of polytomous models can claim to be uniquely suited for modeling a particular response type consistent with particular item types or that classes overlap in this respect. The second issue is whether the three classes of polytomous IRT models yield clearly discernable results when applied to real data. Verhelst et al. (1997) concluded that their steps model and the partial credit model could not be distinguished when both were used to analyze the data from a geography test. More research is needed with respect to this issue.

Filling in the Taxonomy

Hemker et al. (1997; Hemker et al., 2001; also, see Hemker, 1996; Hemker et al., 1996; Van der Ark, Hemker, & Sijtsma, 2002) presented a hierarchical taxonomy in which all IRT models for polytomous items that one can characterize as an adjacent category model, a cumulative probability model, or a continuation ratio model find a place. The proofs underlying the hierarchy are many, and sometimes the proofs are quite involved. The proofs rely on ordering properties such as monotone likelihood ratio of the sum score by the latent variable and of the latent variable by the sum score, and implications of monotone likelihood ratio, which are stochastic ordering of the sum score by the latent variable, and of the latent variable by the sum score [SOL; Equation (3.17)]. Hemker and his colleagues studied these ordering properties at the levels of scores of individual items and sum scores, and examples and counterexamples served to exclude particular IRT models from possessing particular ordering properties. We refer the reader interested in the analytical underpinnings of the hierarchical taxonomy to the original sources but refrain here from a full-fledged treatment of the proofs and limit ourselves to a discussion of the taxonomy and its meaning.

First, the taxonomy refers to IRT models for ordered item scores ($M+1 \geq 3$), also called polytomous item scores in this monograph, and thus excludes IRT models for dichotomous item scores. We will later see why dichotomous-item models are excluded and come back to this feature after we have defined three nonparametric IRT models for polytomous items. Using the nomenclature adopted by Hemker and colleagues, the definitions are the following.

Nonparametric Partial Credit Model (Hemker et al., 1996). This model is based on three assumptions: (1) The latent variable θ is unidimensional; (2) the item scores are locally independent; and (3) the ISRF,

$$f_{jx}^{ACM}(\theta) = P\left(X_j = x \mid \theta; X_j = x-1 \vee X_j = x\right)$$

$$= \frac{P\left(X_j = x \mid \theta\right)}{P\left(X_j = x-1 \mid \theta\right) + P\left(X_j = x \mid \theta\right)}, \quad x = 1, \ldots, M, \tag{4.64}$$

is nondecreasing in latent variable θ.

Nonparametric Graded Response Model (Hemker et al., 1996). The assumptions of this model are: (1) The latent variable θ is unidimensional; (2) the item scores are locally independent; and (3) the ISRF,

$$f_{jx}^{CPM}(\theta) = P\left(X_j \geq x \mid \theta\right) = \sum_{y=x}^{M} P(X_j = y \mid \theta), \quad x = 1, \ldots, M, \tag{4.79, 4.80}$$

is nondecreasing in θ.

Nonparametric Sequential Model (Hemker, 1996, Chap. 6; Van der Ark et al., 2002). The assumptions of this model are: (1) The latent variable θ is unidimensional; (2) the item scores are locally independent; and (3) the ISRF,

$$f_{jx}^{CRM}(\theta) = P\left(X_j \geq x \mid \theta; X_j \geq x-1\right) = \frac{P(X_j \geq x \mid \theta)}{P(X_j \geq x-1 \mid \theta)}, \quad x = 1, \ldots, M, \tag{4.88}$$

is nondecreasing in θ.

Hemker (1996, Chap. 6; Van der Ark et al., 2002) proved that the nonpara-metric partial credit model (np-PCM) is a special case of the nonparamet-ric sequential model (np-SM). In addition, Hemker et al. (1997) proved that the nonparametric sequential model is a special case of the nonparametric graded response model (np-GRM). Formally, the relations are such that

$$np\text{-}PCM \Rightarrow np\text{-}SM \Rightarrow np\text{-}GRM. \tag{4.96}$$

The Venn diagram in Figure 4.17 shows that the set corresponding to the np-GRM encompasses the others sets.

Obviously, the parametric versions of the np-GRM, the np-SM, and the np-PCM are special cases of their nonparametric "mother" versions, but the next results provide more insights:

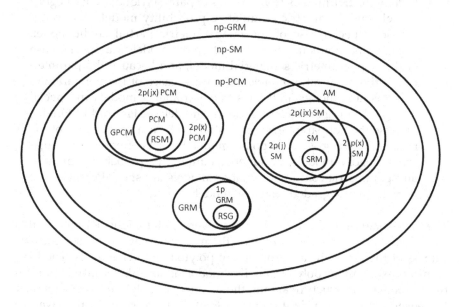

FIGURE 4.17
Venn diagram showing the hierarchical relationships among various nonparametric and para-metric IRT models for polytomous items. np-GRM = nonparametric graded response model; np-SM = nonparametric sequential model; np-PCM = nonparametric partial credit model; 2p(jx) PCM = 2p(jx) partial credit model; GPCM = generalized partial credit model (2p(j) PCM); 2p(x) PCM = 2p(x) partial credit model; PCM = partial credit model (1p PCM); RSM = rating scale model; AM = acceleration model; 2p(jx) SM = 2p(jx) sequential model; 2p(j) SM = 2p(j) sequen-tial model; 2p(x) SM = 2p(x) sequential model; SM = sequential model (1p SM); SRM = sequen-tial Rasch model; GRM = graded response model (2p(j) GRM); 1p GRM = 1p graded response model; RSG = rating-scale graded response model. Note. Based on Figure 6 in Hemker et al. (2001).

1. The three sets of parametric IRT models for polytomous items that subsume under the three nonparametric "mother" IRT models do not overlap. That is, mathematically, there is no way in which a parametric IRT model, say, from the set of parametric adjacent category models implies any of the parametric IRT models from the sets of parametric cumulative probability models and parametric continuation ratio models. Each parametric model from one set excludes each parametric model from the other two sets. Hence, the clusters of sets in Figure 4.17 that represent the three kinds of parametric models do not overlap; their intersections are empty.

2. Each of the IRT models from the set of parametric continuation ratio models is a mathematical special case of the nonparametric sequential model; hence, in the Venn diagram, the cluster of these models falls within the set representing the nonparametric sequential model.

3. Each of the IRT models from the sets of parametric adjacent category models and parametric cumulative probability models is a mathematical special case of the nonparametric partial credit model. Because the nonparametric partial credit model is a special case of the nonparametric sequential model, and because the parametric continuation ratio models are special cases of the nonparametric sequential model, all three sets of parametric models are special cases of the nonparametric sequential model; see Figure 4.17. Finally, because the nonparametric sequential model is a special case of the nonparametric graded response model, the sets of parametric adjacent category models, parametric cumulative probability models, and parametric continuation ration models are special cases of the nonparametric graded response model.

The Venn diagram in Figure 4.17 shows several polytomous IRT models that we did not discuss. We briefly explain them after we have discussed the usefulness of the hierarchical taxonomy of polytomous IRT models. The hierarchical taxonomy of polytomous IRT models is useful, because it clarifies the mathematical structure among the diversity of polytomous IRT models. Moreover, it provides insight into the use of the sum score X which we discussed in Chapters 2 and 3 and which proved to be a sufficient statistic for the latent variable θ in the Rasch model, the partial credit model, and the special cases of the partial credit model, such as the rating scale model.

In Chapter 3, we discussed the ordering property of SOL [Equation (3.17)] and noticed that this is a useful ordering property when one has the sum score as a point of departure and needs to make inferences about the unknown latent variable. With the exception of the partial credit model and its special cases in which sum score X is a sufficient statistic for latent variable θ, for other polytomous IRT models this ordering property does not hold.

However, we saw that a property called weak stochastic ordering of latent variable θ by means of sum score X [weak SOL; Equation (3.20)] holds for the model of monotone homogeneity, here called the nonparametric graded response model. Hence, given Figure 4.17, because all polytomous IRT models are special cases of the nonparametric graded response model, weak SOL is an ordering property of all models in the Venn diagram.

In Chapter 3, we did not consider the mirror image of SOL, which is stochastic ordering of sum score X by latent variable θ, abbreviated SOM (Hemker et al., 1996). Let $\theta_v < \theta_w$, then SOM entails

$$P(X \geq x \mid \theta_v) \leq P(X \geq x \mid \theta_w). \tag{4.97}$$

Thus, given that one knows the latent variable values for a group of persons (this is the variable on which one conditions), then, except for random measurement error, their ordering with respect to θ also reflects their ordering with respect to the sum score X. As an ordering property, SOM it is not as valuable as SOL, which implies the ordering on the latent variable given the sum score (Chapter 3). However, SOM has some theoretical value because SOM holds for the most general model, which is the nonparametric graded response model; hence, SOM holds for all polytomous IRT models in the Venn diagram in Figure 4.17.

The models in Figure 4.17 that we did not discuss are the following. In the set of adjacent category models, in order of decreasing generality, we already identified and discussed the generalized partial credit model [Equation (4.78)], the partial credit model [Equation (4.76)], and the rating scale model [Equation (4.77)]. More general than the generalized partial credit model is a model that has a slope parameter α_{jx} for each combination of j and x (Hemker et al., 1996). This is the 2p(jx)-partial credit model (2p refers to two-parameter, (jx) indicates that the slope parameter varies across j and x). If one assumes nominal response categories, this model is identical to the nominal categories model (Bock, 1972, 1997; Glas, 1999; Thissen & Cai, 2016). Thissen and Steinberg (1984, 1997) discussed a version of this nominal model that allows for guessing in case of multiple-choice items. Thissen and Cai (2016) discussed several applications of the model to real-data problems. Because it deviates from the other models discussed, we do not include the nominal categories model in the taxonomy, but because the model has received considerable attention, we discuss it separately in Box 4.4.

BOX 4.4   The Nominal Categories Model

We follow the discussion Mellenbergh (1995) provided. Let an item have $M+1$ nominal categories and item scores $x = 0,\dots,M$, and let the probability that person i chooses or is assigned to category x of item j be

denoted $P\left(X_{ijx}=1|\theta\right)=P_{ijx}$, for short. The model Bock (1972) proposed
equals

$$P_{ijx}=\frac{\exp\left(\delta_{jx}+\alpha_{jx}\theta_i\right)}{\sum_{x=0}^{M}\exp\left(\delta_{jx}+\alpha_{jx}\theta_i\right)}. \tag{4.98}$$

Mellenbergh (1995) proposed to restrict the item parameters by fixing
the parameters of the first category of each item to 0, that is, $\alpha_{j0}=\delta_{j0}=0$,
for $j=1,\dots,J$. To construct a nominal response variable, one has to formulate the model such that no order, let alone metric information,
is retrievable. This is done by redefining response variable X_j as M
dichotomous response variables. One category can be chosen to be the
benchmark or the reference category, and because we are dealing with
nominal responses the choice is arbitrary; one may think of dummy
coding in regression analysis. Thus, one takes the first category as reference and compares each of the other M categories independent of one
another with category 1 by means of the odds of being in category x
($x=1,\dots,M$) relative to being in category 1; that is, P_{ijx}/P_{ij0}. Then, recall-
ing that $\alpha_{j0}=\delta_{j0}=0$, for all j, the log odds of Bock's model equals

$$\ln\left(\frac{P_{ijx}}{P_{ij0}}\right)=\delta_{jx}+\alpha_{jx}\theta_i-\left(\delta_{j0}+\alpha_{j0}\theta_i\right)=\delta_{jx}+\alpha_{jx}\theta_i,\ x=1,\dots,M. \tag{4.99}$$

So for each item the model has M equations in which δ_{jx} is the intercept or location parameter representing the value of the log odds when
$\theta_i=0$, and α_{jx} is the slope or discrimination parameter indicating the
rate of increase of the log odds as θ_i increases by one unit. Hence, α_{jx}
is a measure of the discrimination between categories x and 1. The log
odds intersect the θ axis at $\theta_0=-\delta_{jx}/\alpha_{jx}$; for $\theta>\theta_0$, category x is more
probable than reference category 1, and for $\theta<\theta_0$, category 1 is the most
probable.

Finally, when one restricts the slope parameters to have values that are
dependent on the item score, $\alpha_{jx}=\alpha_x$, the 2p(x)-partial credit model results.
Noticing that the generalized partial credit model has restriction $\alpha_{jx}=\alpha_j$,
and thus might be called the 2p(j)-partial credit model, we have a partial
order (ties within braces) by increasing generality:

rating scale model $\Rightarrow$ partial credit model

$\Rightarrow\{2\mathrm{p}(j)-\text{partial credit model},2\mathrm{p}(x)-\text{partial credit model}\}$ (4.100)

$\Rightarrow 2\mathrm{p}(jx)-\text{partial credit model}$

Figure 4.17 shows this partial order in a set-theoretical representation.

In the set of cumulative probability models, two special cases are added to the graded response model. We notice that the 2p(jx)-graded response model in which α_{jx} is defined for all combinations of item and item score, and the 2p(x)-graded response model in which $\alpha_{jx} = \alpha_x$, do not exist. The reason is that ISRFs of the same item would cross for varying slope parameters, and Equation (4.81) shows that this is impossible, because the difference between any two ISRFs must be non-negative. One could restrict slope parameters to be equal across items, so that $\alpha_j = \alpha$, and the one-parameter (i.e., 1p) graded response model results (Hemker et al., 1996). A further restriction is to decompose the location parameter λ_{jx} in an item-dependent location parameter ζ_j and a score-dependent parameter κ_x, with $\sum_x \kappa_x = 0$, so that $\lambda_{jx} = \zeta_j + \kappa_x$. This is the rating-scale graded response model (Hemker et al., 2001). The three models are ordered by increasing generality as:

rating-scale graded response model $\Rightarrow$ 1p-graded response model

$$\Rightarrow \text{graded response model} \tag{4.101}$$

See Figure 4.17 for the set-theoretical representation of Equation (4.101).

Finally, in the set of continuation ratio models, we discuss six models in order of decreasing generality. The acceleration model (Samejima, 1995) is the most general parametric model. Let α_{jx} be a slope parameter associated with score x of item j, and let $\xi_j \geq 0$ be the acceleration parameter. Constant $D = 1.7$ scales the response function to the normal ogive [see Equation (4.10)]. Then, the acceleration model equals

$$P\left(X_j \geq x \mid \theta; X_j \geq x-1\right) = \left\{ \frac{\exp\left[D\alpha_{jx}\left(\theta - \beta_{jx}\right)\right]}{1 + \exp\left[D\alpha_{jx}\left(\theta - \beta_{jx}\right)\right]} \right\}^{\xi_j}. \tag{4.102}$$

When $\xi_j \neq 1$, the response function in Equation (4.102) is not a logistic function. For $\xi_j < 1$, the entire curve is "pushed down" and for $\xi_j > 1$, the entire curve is "lifted up". Both effects add to the effect that the slope parameter has on the steepness of the curve in the inflection point and render the curve non-symmetric. Hemker et al. (2001) discussed that for $\xi_j = 1$, the acceleration model becomes the two-parameter sequential model, abbreviated 2p(jx)-sequential model, with parameters α_{jx} and β_{jx}, both dependent on combinations of item and item-score. Restricting $\alpha_{jx} = \alpha_j$, implying slopes depend only on items, produces the 2p(j)-sequential model, and restricting $\alpha_{jx} = \alpha_x$, implying slopes depend only on item scores, produces the 2p(x)-sequential model. The 2p(j)-sequential model and the 2p(x)-sequential model cannot be ordered hierarchically. Restricting $\alpha_{jx} = 1$ produces the sequential Rasch model [Equation (4.95)] or 1p-sequential model. Tutz (1990) defined a further specialization of the sequential model, which decomposes location parameter β_{jx} into an item-dependent location parameter υ_j and a step location parameter ς_x, so that $\beta_{jx} = \upsilon_j + \varsigma_x$, and $\sum_x \varsigma_x = 0$, comparable to the rating scale model [Equation (4.77)] and called sequential rating scale model. The partial order (ties within braces) by increasing generality of the sequential models is:

$$\text{sequential rating scale model} \Rightarrow \text{sequential Rasch model}$$

$$\Rightarrow \{2p(j)\text{-sequential model}, 2p(x)\text{-sequential model}\} \qquad (4.103)$$

$$\Rightarrow 2p(jx)\text{-sequential model} \Rightarrow \text{acceleration model}$$

One finds the set-theoretical representation of Equation (4.103) in Figure 4.17.

Finally, we notice that dichotomous IRT models do not have a place in the taxonomy, because the three different classes of nonparametric models coincide when item scores are binary. Specifically, for adjacent category models, if $x = 0,1$, then

$$P\left(X_j = x \mid \theta; X_j = x-1 \vee X_j = x\right) = \frac{P\left(X_j = x \mid \theta\right)}{P\left(X_j = x-1 \mid \theta\right) + P\left(X_j = x \mid \theta\right)}$$

$$= \frac{P\left(X_j = 1 \mid \theta\right)}{P\left(X_j = 0 \mid \theta\right) + P\left(X_j = 1 \mid \theta\right)} \qquad (4.104)$$

$$= P\left(X_j = 1 \mid \theta\right).$$

For cumulative probability models,

$$P\left(X_j \geq x \mid \theta\right) = \sum_{y=x}^{M} P(X_j = y \mid \theta) = \sum_{y=1}^{1} P\left(X_j = 1 \mid \theta\right) = P\left(X_j = 1 \mid \theta\right). \qquad (4.105)$$

For continuation ratio models,

$$P\left(X_j \geq x \mid \theta; X_j \geq x-1\right) = \frac{P(X_j \geq x \mid \theta)}{P(X_j \geq x-1 \mid \theta)} = \frac{P(X_j = 1 \mid \theta)}{P(X_j \geq 0 \mid \theta)} \qquad (4.106)$$

$$= P(X_j = 1 \mid \theta).$$

If we assume that there is no practical difference between normal-ogive and logistic counterpart models, and we focus on the latter, we distinguish and combine two hierarchies of dichotomous IRT models in one Venn diagram. The models we consider are the monotone homogeneity model (MHM; Chapter 3), the double monotonicity model (DMM; Chapter 3), the one-parameter logistic model (1PLM) or Rasch model, the two-parameter logistic model (2PLM), and the three-parameter logistic model (3PLM). One order by increasing generality is:

$$1\text{PLM} \Rightarrow 2\text{PLM} \Rightarrow 3\text{PLM} \Rightarrow \text{MHM} \qquad (4.107)$$

and another order by increasing generality is:

$$1\text{PLM} \Rightarrow \text{DMM} \Rightarrow \text{MHM} \qquad (4.108)$$

Clearly, the MHM is the most general model and the 1PLM the most restrictive model. The DMM does not allow IRFs to intersect, and the 3PLM and the 2PLM do. The 3PLM has nonintersecting IRFs when the lower asymptotes are ordered such that for any item pair, $\gamma_j > \gamma_k$ corresponds with the opposite ordering of the item locations, $\delta_j < \delta_k$, while the slope parameters are equal, $\alpha_j = \alpha_k$. One may check that curves for which $\alpha_j \neq \alpha_k$ always intersect. Thus, this special case of the 3PLM, for which $\gamma_j > \gamma_k$, $\delta_j < \delta_k$, and $\alpha_j = \alpha_k$, forms the intersection of the sets of DMMs and 3PLMs. For the 2PLM to have non-intersecting IRFs, slope parameters must be equal; that is, $\alpha_j = \alpha_k$. The intersection of DMMs and 2PLMs contains these models. One may notice that by rescaling the distribution of the latent variable, one can force $\alpha_j = \alpha_k = 1$, which constitutes the 1PLM. The Venn diagram in Figure 4.18 shows the hierarchical relations of the dichotomous IRT models.

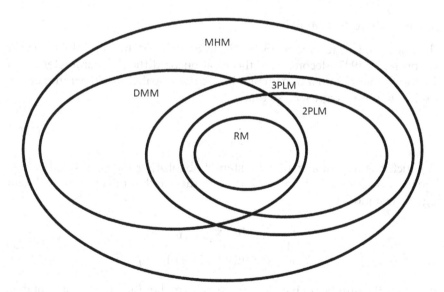

FIGURE 4.18
Venn diagram showing the hierarchical relationships among various nonparametric and parametric IRT models for dichotomous items. MHM=monotone homogeneity model; DMM=double monotonicity model; 3PLM=three-parameter logistic model; 2PLM=two-parameter logistic model; RM=Rasch model.

IRT Models for Special Purposes

We briefly discuss IRT models that serve special purposes. The literature on parametric IRT is simply too overwhelming to provide a full discussion in this chapter. We sacrifice completeness for conceptual clarity, with an eye toward practical use in psychological measurement, and trust that colleagues will recognize the necessity of making choices from a horn of plenty. Having said this, we notice that a substantial part of IRT has been devoted to the blending of measurement and cognitive processes. This is done mainly by expressing the latent variable, θ, as a linear combination of covariates, or the location or difficulty parameter, δ_j, as a linear combination of parameters quantifying the cognitive components and how they cooperate to produce the item response. We discuss examples of both approaches. Another way of digging deeper into the psychology of measurement is by addressing the measurement's dimensionality. We also consider this topic. In Chapter 5, we address the topic of cognitive modeling more extensively when we discuss cognitive diagnostic modeling, and in Chapter 6, we discuss models for response times. Both approaches add to the richness of latent variable modeling of cognitive processes. De Boeck and Wilson (2004) provided an overview of several explanatory IRT models following a generalized linear and nonlinear approach.

Linear Logistic Test Model

The linear logistic test model (Scheiblechner, 1972; Fischer, 1973, 1997; Fischer & Formann, 1982) decomposes the location or difficulty parameter, δ_j, of the Rasch model into a linear combination of so-called basic parameters, η_f, $f = 1, \ldots, F$, with weights w_{jf} so that

$$\delta_j = \sum_{f=1}^{F} w_{jf} \eta_f + c, \tag{4.109}$$

in which c is a normalization constant. Substituting the right-hand side of Equation (4.109) for δ_j in Equation (4.11) of the Rasch model yields the linear logistic test model,

$$P_j(\theta) = \frac{\exp\left(\theta - \sum_{f=1}^{F} w_{jf} \eta_f - c\right)}{1 + \exp\left(\theta - \sum_{f=1}^{F} w_{jf} \eta_f - c\right)}. \tag{4.110}$$

Typically, the number of basic parameters is smaller than the number of item difficulty parameters, so that $F < J$. The model explains the difficulty of the items from the difficulty contributions of a smaller number of basic operations, which can have different meanings across contexts. One context is that in which the solution of an arithmetic problem may require three logical operations, and the operations each have varying difficulty, represented by η_1, η_2, and η_3. Some items may have a simpler structure, requiring only subsets of operations, in this case, only one or two, and other items may require an operation to be repeated multiple times to solve the item. The researcher has to specify the weights for each item-operation combination in advance, and it is here in particular that one should better start from a theory about the attribute of interest than follow a trial-and-error strategy. One defines matrix $\mathbf{W}$ that has order $J \times F$ and has elements w_{if} representing the weights that hypothesize which operations are needed ($w_{if}=1$ or $w_{if}>1$ when the operations have to be used multiple times) or which operations are not needed ($w_{if}=0$) to solve item j. Matrix $\mathbf{W}$ thus represents the hypothesis about the attribute to be tested, provided of course that the theory allows a formalization as in Equation (4.110).

Other contexts in which the linear logistic model is applied are feasible (e.g., see Janssen, 2016, for an overview of empirical-data studies). Fischer (1997) mentions basic parameters representing therapeutic treatments given to persons between testing occasions, educational treatments given to students, and effects of experimental conditions. The linear logistic test model can be generalized to the partial credit model and the rating scale model, that is, to polytomous items (e.g., Fischer & Ponocny, 1994). Another interesting application is the measurement of change effects that are equal for all persons. Here, the situation is a test that is administered at T different time points with corresponding virtual item locations, δ_{jt}^*, $t=1,\ldots,T$. Changes in these virtual item locations indicate change in the persons on latent variable, θ. For J items and T measurement occasions, one can write the $J \times T$ virtual item location parameters as

$$\delta_{j1}^* = \delta_j,$$

$$\delta_{j2}^* = \delta_j + \eta_1,$$

$$\ldots$$

$$\delta_{jt}^* = \delta_j + \eta_1 + \eta_2 + \cdots + \eta_{t-1},$$

$$\ldots$$

Thus, $J \times T$ virtual item location parameters are replaced with $J-1$ item location parameters and $T-1$ basic parameters (Fischer, 1995b, 1997).

Generalized Rasch Model with Manifest Predictors

Zwinderman (1991) discussed a Rasch model in which the latent variable was replaced with a linear combination of manifest covariates and later (Zwinderman, 1997) generalized the approach to the unidimensional polytomous Rasch model (Andersen, 1973b), a model comparable to the partial credit model. The manifest covariates are random variables that we denote Y_h and enumerate $h = 1,...,H$. We assume the following linear model,

$$\theta_i = \sum_{h=1}^{H} Y_{ih} + \epsilon_i, \tag{4.111}$$

in which ϵ_i is the residual. This linear combination of manifest variables is substituted for the latent variable in Equation (4.11), yielding

$$P_j(\theta) = \frac{\exp\left(\sum_{h=1}^{H} Y_{ih} + \epsilon_i - \delta_j\right)}{1 + \exp\left(\sum_{h=1}^{H} Y_{ih} + \epsilon_i - \delta_j\right)}. \tag{4.112}$$

One may first use the Rasch model to estimate the item location parameters and then use the estimates as if they were parameter values. Given that the Rasch model fits the data adequately, one may next estimate the logistic regression model in Equation (4.112) so that the relationship of the set of covariates or explanatory variables with the response probability is estimated. The reason to do this using Equation (4.112), thus circumventing having to estimate latent variable θ, is that estimates of θ tend to be biased for extreme values and short tests (Lord, 1983). In addition, Croon (2002) showed that using estimated latent variable values in a structural model as if they were observed values results in biased estimates of the structural parameters.

Zwinderman (1991) discussed a medical example in which 11 (i.e., $J = 11$) artificial patients suffer in different degrees from bronchitis, and 434 general practitioners (i.e., $N = 434$) vary with respect to their inclination to prescribe antibiotics. The author found that the Rasch model fitted the data and estimated the variance σ_θ^2. He considered the estimated item locations, $\hat{\delta}_1,...,\hat{\delta}_{11}$, as given and used them in Equation (4.112) as fixed, together with six covariates. The covariates were age, sex, experience as a general practitioner, kind of practice in which the general practitioner worked, the university where she was trained, and whether she ran a pharmacy. The fourth and fifth covariates required dummy coding. Regression weights for the four covariates without dummy coding and the dummy variables were estimated, and each was tested for significance. The residual variance, σ_ϵ^2, was estimated, and the multiple correlation between the latent variable, θ, and the linear combination of covariates, $\tilde{\theta} = \sum_{h=1}^{H} Y_{ih}$, was estimated from

$$R_{\theta\tilde{\theta}} = 1 - \frac{\hat{\sigma}_\epsilon^2}{\hat{\sigma}_\theta^2}. \tag{4.113}$$

Once more, we emphasize that this correlation was obtained by avoiding the estimation of the biased latent variable that would have attenuated the required correlation. Zwinderman found that male practitioners and practitioners running a pharmacy were more inclined to prescribe antibiotics and that quite some differences existed between universities where one received one's training. The multiple correlation between the latent variable and the weighted sum of the covariates equaled .19.

Multidimensional IRT Models

Psychological, social science, and health measurements are never purely unidimensional. The items used to assess particular attributes elicit reactions that come from different, usually unknown sources, and the result is multidimensional measurement. Often the multidimensionality is unwanted and undesirable, because it disturbs the assessment of the attribute of interest. Sometimes, multidimensionality is typical of the intended measurement, for example, when arithmetic items are presented in the context of practical problems, such as shopping, redecorating a room, and reading a map. Then, arithmetic ability is not the only influence on test performance, and in addition, spatial-orientation ability and language skills play a role. Reckase (1997, 2016) discussed a multidimensional IRT model for dichotomous items. Let $\theta = (\theta_1, \ldots, \theta_Q)$ denote a vector with Q latent variables, $q = 1, \ldots, Q$, and let $\alpha_j = (\alpha_{j1}, \ldots, \alpha_{jQ})$ be a vector containing slope parameters for the latent variables. Then, the multidimensional IRT model is equal to

$$P\left(X_j = 1 \mid \theta, \alpha_j, \delta_j, \gamma_j\right) = \gamma_j + \frac{\left(1 - \gamma_j\right) \exp\left(\alpha_j' \theta + \delta_j\right)}{1 + \exp\left(\alpha_j' \theta + \delta_j\right)}. \tag{4.114}$$

Figure 4.19 shows an example of a three-dimensional item response surface based on two latent variables. Equation (4.114) represents a compensatory model, because high or low values on one latent variable can be balanced by low or high values on other latent variables (Reckase, 2016) to obtain a high probability of giving the positive response. Adams, Wilson, and Wu (2016) discussed a model that defines a vector $\delta_j = (\delta_1, \ldots, \delta_Q)$, the elements of which are weighted by vector $\mathbf{w}_j = (w_1, \ldots, w_Q)$, so that upon substitution in Equation (4.114) the exponent equals $\alpha_j' \theta + \mathbf{w}_j' \delta_j$, whereas the lower asymptote is ignored, that is, $\gamma_j = 0$. The model is estimable and has sufficient statistics when the researcher fixes rather than estimates the weights in α_j and $\mathbf{w}_j$ so as to hypothesize the latent variables and item properties needed for item j. Weights are often 0s and 1s.

Reckase (1997) recommended overestimating the number of latent variables to obtain the best possible interpretation rather than emphasizing dimensionality reduction using a small number of latent variables, possibly too few. This approach is exploratory rather than confirmatory, but one might as well use the model, as any model, in a confirmatory way. Comparable to factors in factor analysis, latent variables in Equation (4.114) may or may not be orthogonal. Slope parameter α_{jq} is related to the slope of the item

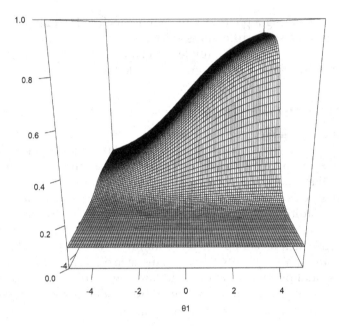

FIGURE 4.19
Item response surface of the two-dimensional three-parameter logistic model with $\alpha_{j1} = 0,5$, $\alpha_{j2} = 2$, $\delta_j = 0$, and $\gamma_j = .10$, Plotted in 3D perspective.

response surface in the direction of the θ_q axis. Depending on the attribute in which one is interested, item discrimination can vary within the same item, and especially if the item only measures the attribute of interest weakly does it contribute little to the measurement of that attribute, even if it measures other attributes more reliably. For example, a context item intended to measure arithmetic ability may do so only weakly, whereas it much better assesses language skills. The item location parameter δ_j has an interpretation somewhat different from that in simple unidimensional IRT models and is related to the distance from the origin in Figure 4.19 to the point on the item response surface where the surface has its steepest slope. The lower asymptote, γ_j, has the same interpretation as in unidimensional IRT models.

Embretson (1997, 2016; also, Whitely, 1980) discussed several non-compensatory IRT models in a confirmatory context. The multicomponent latent trait model requires for successful performance on an item the conjunction of successful performances on several subtasks, each of which follows a separate unidimensional IRT model (e.g., the Rasch model). The model is based on the idea that a person solves an item by solving a sequence of subtasks explicitly, each modeled by an IRT model. Thus, instead of solving "$\sqrt{12^2 - 63} = \ldots$" one solves three sub items in a fixed order, without having the opportunity to return to a previous sub item to withdraw a previous answer: "$12^2 = \ldots$", "$144 - 63 = \ldots$", and "$\sqrt{81} = \ldots$". Each of the sub items assumes a unique attribute to solve the sub item, collected as latent variables in vector $\boldsymbol{\theta}$, and each sub item has a particular difficulty and difficulty parameters are collected in $\boldsymbol{\delta}_j$. Further, let sub items be indexed $r = 1,\ldots,R$, let g denote the probability of solving the item by guessing, and let s denote the probability of attempting to solve the whole task other than by guessing. Then, the probability that person i has item j correct according to the multicomponent latent trait model equals

$$P\left(X_{ij} = 1 \mid \boldsymbol{\theta}, \boldsymbol{\delta}_j\right) = \left(s - g\right)\prod_{r=1}^{R}P\left(X_{ijr} = 1 \mid \theta_{ir}, \delta_{jr}\right) + g. \tag{4.115}$$

One person parameter and one sub-item parameter characterize each sub item, whereas across sub items all person and sub-item parameters combine in a non-compensatory way to define the probability that the item is correct. The response probabilities for the sub items, $P\left(X_{ijr} = 1 \mid \theta_{ir}, \delta_{jr}\right)$, are Rasch modeled. The model is non-compensatory, because the response probabilities for the sub items are all numbers between 0 and 1, so that their product cannot exceed the smallest of the probabilities. Hence, if one probability is small, the probability of solving the item correctly is even smaller, and there is no way to make up for the weakest link in the chain. For example, if two probabilities are .9 and the third only .1, their product is smaller than .1, that is, $.9 \times .9 \times .1 = .081$. Different variations on the model have been proposed. For example, the general component latent trait model expresses the sub-item parameters δ_{jr} as linear combinations of sub-item features, comparable to the linear logistic test model.

Hooker, Finkelman, and Schwartzman (2009) pointed out a paradoxical phenomenon for multidimensional IRT models. Let two persons have the same scores on a set of J_0 multidimensional items. This necessarily produces the same set of scores on the dimensions that the items assess. Next, let the two persons attempt the same $(J_0 + 1)$st item, and assume that the first person provides the correct answer and the second person provides an incorrect answer; then, it can happen that the first person obtains the lower overall test result. This is the paradoxical result: By answering the next item correct, one can lower one's overall test result, and, likewise, an incorrect answer can increase one's overall test result. The authors showed that all makes sense mathematically and is not even a rare result. They provided the next intuitive explanation for a two-dimensional IRT model. Suppose the first J_0 items require both arithmetic and language abilities and both persons have some items correct and some incorrect but such that their item-score patterns are equal. Let the $(J_0 + 1)$st item require a high level of arithmetic ability and almost no language ability. Then, if John solved the $(J_0 + 1)$st item correctly, this suggests that with the first J_0 items he barely relied on his language ability, which, on second thoughts, must have been low, also producing incorrect answers when items required language ability for their solution. On the other hand, Mary demonstrated lower arithmetic ability but better language ability than John did. The net result can be that the same dimension scores obtained after J_0 items change to approximately the same arithmetic scores for John and Mary but a markedly lower language score for John after J_0+1 items. If John's and Mary's results are used for selection purposes, John may fail whereas Mary passes, even though John had the same J_0 items correct plus one extra, a result that is hard to communicate to John.

Hooker et al.'s (2009) paradox raised quite some discussion (e.g., Jordan & Spiess, 2012, 2018; Van der Linden, 2012). Van Rijn and Rijmen (2012) argued that the paradox is only of interest if one considers a test a contest in which a person tries to solve as many items as she can. However, not all testing is a contest. One alternative way to look at measurement is as a means to produce formally correct conclusions about people's performance and to select the measurement model that accomplishes this, and another way to look at measurement is as a means to collect information about the cognitive diagnosis of individuals. In these alternative views, the search is for the best measurement values irrespective of what people might expect, such as simple counts of positive answers, or for information not necessarily in terms of number-correct counts that is useful for a diagnosis of abilities and skills, respectively. Without challenging the usefulness of these alternative measurement endeavors, we notice that the main use of tests and questionnaires is to measure individual performance or behavior and that it helps when communication with the tested person is in terms of understandable language, simultaneously transmitting correct conclusions. Number-correct and number-of-endorsements (or its credit version) is understandable to people but from the perspective of measurement not always correct. We already noticed that the sum score does not stochastically order the latent variable in polytomous unidimensional IRT models other than the partial credit model and its special cases, and the paradox Hooker et al. identified is somewhat similar although mathematically different, in that particular items can distort a latent-variable ordering expected based on the number-correct logic. Van der Ark (2005) showed for the stochastic ordering property that for most cases, scale distortions are modest, and for the multidimensional IRT paradox, robustness research might produce similar conclusions. The best advice is probably to include only items in scales that mitigate or circumvent these problems.

Data Example: Transitive Reasoning

The large number of parametric IRT models makes it virtually impossible to use each of them to analyze one data set to discuss their differences and commonalities. We expect differences to appear but often to be difficult to interpret and unrelated to meaningful item or group features. Such differences are not only caused by small differences between models but also by the messiness of typical test data (Chapter 1), even if they have been collected carefully using a well-designed set of items. Hence, using many different models to analyze one's data may create confusion rather than provide clarity. In addition, we notice that in real-data analyses usually one model or a small number of nested models are used. We follow this strategy, using the Rasch model and the nonparametric model of double monotonicity (Chapter 3), in which the Rasch model is nested, for some of the preliminary analyses before we use the flow chart in Figure 4.7 to analyze the data by means of the Rasch model.

Bouwmeester and Sijtsma (2004) developed a transitive reasoning scale to investigate the plausibility of three theories about transitive reasoning; see Chapter 1 for more details. The authors explained transitive reasoning as follows (ibid., p. 123):

> Suppose an experimenter shows a child two sticks, A and B, which differ in length, Y, such that $Y_A > Y_B$. Next, stick B is compared with another stick C, which differs in length from B, such that $Y_B > Y_C$. In this example the length relations $Y_A > Y_B$ and $Y_B > Y_C$ are the premises. When the child is asked, without being given the opportunity to visually compare this pair of sticks, which is longer, stick A or stick C, (s)he may or may not be able to give the correct answer. When a child is able to infer the unknown relation ($Y_A > Y_C$) using the information of the premises ($Y_A > Y_B$ and $Y_B > Y_C$), (s)he is capable of transitive reasoning.

The transitive reasoning scale consists of 16 dichotomous items constructed based on a $2 \times 2 \times 4$ facets design. The first facet is *type of content*, which is either physical (the child can see the differences) or verbal (the child is told about the differences). The second facet is *presentation form*, which is simultaneous (all premise information is presented at once) or successive (the information is presented in consecutive steps). The third facet is *task format*, which is either three strictly ordered objects ($Y_A > Y_B > Y_C$), four equal objects ($Y_A = Y_B = Y_C = Y_D$), five strictly ordered objects ($Y_A > Y_B > Y_C > Y_D > Y_E$), or four objects with a mixed relation ($Y_A = Y_B > Y_C = Y_D$). Figure 1.1 shows two example items for transitive reasoning. Figure 1.1 (upper panel) shows an item characterized by physical content, successive presentation, and task format with three strictly ordered objects (this is item 6 from the transitive-reasoning scale). The left-hand side shows the two stimuli that are presented in successive order, and the right-hand side shows the picture that is presented while the supervisor asks "which of the two sticks is longer?" Similarly, Figure 1.1 (lower panel) shows an item characterized by verbal content, successive presentation, and task format with three strictly ordered objects (this item looks like item 12 in the transitive-reasoning scale but uses different animals). For item 12, the supervisor first describes the ordinal age relation between the rabbit and the fox and then the ordinal age relation between the fox and the deer. Finally, the supervisor asks which animal is older, the rabbit or the deer. The data we analyzed consist of the item scores that 606 children from the age of 6 to 13 from six elementary schools in the Netherlands produced for the 16 items (for details, see Bouwmeester & Sijtsma, 2004). Table 4.2 shows the items and proportion of 1 scores for each item, which are the estimated π values.

The initial question we addressed was whether the 16 items constitute a scale that is consistent with the Rasch model. We used software package MIRT (Glas, 2010) to obtain global fit statistic R_{1c} [Equation (4.33)] for assessing monotonicity and sufficiency ($R_{1c} = 218.68$, $df = 60$, $p < .0001$), global fit statistic R_{2c} [Equation (4.40)] for assessing unidimensionality and local independence ($R_{2c} = 497.46$, $df = 60$, $p < .001$), and global likelihood ratio test LR [Equation (4.47)] for assessing invariance of item parameters in separate groups ($LR = 204.80$, $df = 60$, $p < .0001$). Based on these goodness-of-fit results, we must reject the null hypothesis that the Rasch model is consistent with the data for the 16-item transitive-reasoning set. In the next data-analysis step, we aimed at selecting items that constitute one or more scales consistent with the Rasch model.

TABLE 4.2

Items for Transitive Reasoning, Item Facets, Estimated π Values, and Estimated Item Location Parameters (and Standard Errors, SE) for Seven Items Constituting a Scale Consistent with the Rasch Model

Item	Facet			$\hat{\pi}_j$	$\hat{\delta}_j$	SE
	Type of content	Presentation form	Task format			
1	Verbal	Simultaneous	$Y_A > Y_B > Y_C$	.72	−1.11	(0.09)
2[a]	Physical	Successive	$Y_A = Y_B > Y_C = Y_D$	.48		
3[a]	Verbal	Successive	$Y_A = Y_B = Y_C = Y_D$	.87		
4	Physical	Simultaneous	$Y_A > Y_B > Y_C > Y_D > Y_E$	.55	−0.22	(0.08)
5	Verbal	Simultaneous	$Y_A = Y_B > Y_C = Y_D$	.24	1.38	(0.10)
6[d]	Physical	Successive	$Y_A > Y_B > Y_C$	.42		
7[a]	Verbal	Simultaneous	$Y_A = Y_B = Y_C = Y_D$	.87		
8	Physical	Successive	$Y_A > Y_B > Y_C > Y_D > Y_E$	.52	−0.08	(0.08)
9[a]	Verbal	Successive	$Y_A = Y_B = Y_C = Y_D$	.77		
10[c]	Physical	Simultaneous	$Y_A > Y_B > Y_C > Y_D > Y_E$	.51		
11[d]	Verbal	Simultaneous	$Y_A = Y_B > Y_C = Y_D$	.38		
12	Physical	Successive	$Y_A > Y_B > Y_C$	.62	−0.55	(0.09)
13[d]	Verbal	Simultaneous	$Y_A > Y_B > Y_C$	.70		
14	Physical	Successive	$Y_A = Y_B > Y_C = Y_D$	.40	0.48	(0.08)
15	Verbal	Successive	$Y_A > Y_B > Y_C > Y_D > Y_E$	.49	0.09	(0.08)
16[b]	Physical	Simultaneous	$Y_A = Y_B = Y_C = Y_D$	.94		

Note: $\hat{\pi}_j$ = estimated proportion 1 scores, $\hat{\delta}_j$ = estimated location parameter, a = item removed due to negative scalability coefficients, b = item removed due to violation of manifest monotonicity, c = item removed due to violation of manifest invariant item ordering, d = item removed in Rasch analysis.

Because the Rasch model is a special case of the nonparametric double monotonicity model (Chapter 3; Figure 4.18), items that are inconsistent with the double monotonicity model are also inconsistent with the Rasch model. We used Mokken scale analysis (Chapter 3) to remove the items that were inconsistent with the double monotonicity model. First, because items 2, 3, 7, and 9 had negative H_{jk} values [Equation (3.25)] with at least one of the remaining items, we removed these items from the initial 16-item set. Second, because item 16 showed significant violations of manifest monotonicity [Equation (3.58)], it was removed next, leaving 11 items for further analysis. Third, item 10 showed six violations of manifest invariant item ordering [Equation (3.89)]; hence, it was the sixth item removed from the set. After we removed six items in total, the remaining ten-item set did not show evidence of violations of double monotonicity.

We used the flow diagram in Figure 4.7 to select the Rasch items from the remaining ten-item set. In Step 1, aimed at assessing monotonicity and sufficiency simultaneously for a set of items, we used global statistic R_{1c} for evaluating monotonicity and sufficiency of the whole ten-item set. In Step 2, we used local statistic U_j [Equation (4.38)] for identifying items from the item set that were inconsistent with monotonicity and sufficiency, and we removed the item with the highest absolute U_j value; then, we returned to Step 1. Once we identified an item set that was consistent with monotonicity and sufficiency, in Step 3, we used global statistic R_{2c} for assessing whether the item set was consistent with unidimensionality and local independence. In Step 4, using the splitter-item methodology, we removed the item producing the highest LR statistic, based on the statistical test with the nominal Type I error rate set at $\alpha = .05$, without controlling for multiple testing. Thus, we sooner concluded that an item deviated from the others in the provisional scale than when using a corrected critical value that by definition is smaller than the critical value we used. Our strategy was based on the belief that, in general, a researcher should be prepared to identify misfitting items, study them, and possibly remove them from the scale rather than ignore misfitting items and include them in the scale uncritically. Controlling the familywise Type I error rate reduces the power to detect item misfit and should be avoided because it tends to include misfitting items too easily. This is not to say that we strive to reject as many items as possible, but we recommend to study items for the way in which they are different from other items and finally to make a decision based on theoretical considerations including item content. We conducted all analyses by means of the software package MIRT (Glas, 2010).

Table 4.3 shows the results of following the flow chart for an item analysis by means of the Rasch model (Figure 4.7). The global test statistic R_{1c} assessing monotonicity and sufficiency was significant, hence rejecting the null hypothesis that the Rasch model holds for the ten-item set (Table 4.3, first row). Next, inspecting local statistic U_j for each item in the ten-item set, we found that item 6 has the largest U_j value (i.e., $U_6 = 2.17$; Table 4.3, second

TABLE 4.3

Step-by-Step Rasch Analysis Following Flow Chart in
Figure 4.7

Step	Item removed	R_{1c} (df)	U_j	R_{2c} (df)	LR (df)
1		30.14* (18)			
2	6		2.17*		
1		34.32 (32)			
3				66.51* (32)	
4	13				28.14* (7)
1		31.51 (28)			
3				39.33* (24)	
4	11				13.10* (6)
1		20.18 (18)			
3				27.67 (18)	

Note: Asterisk * means significant at level $\alpha = .05$ without correction
for number of significance tests.

row), suggesting that the discrimination power of item 6 was too low compared with the other items. Based on this result, we rejected item 6 from the
ten-item set. For the resulting nine-item set, statistic R_{1c} was not significant
(Table 4.3, third row). In the next step, we investigated unidimensionality
and local independence for the nine-item set.

Global test statistic R_{2c} was significant, suggesting the nine-item set as
a whole was inconsistent with the assumptions of unidimensionality and
local independence (Table 4.3, fourth row). For each of the items in the nine-
item set, we used the splitter-item method in which each item consecutively
served as the criterion for defining two subgroups for identifying items or
item subsets that are inconsistent with unidimensionality and local independence. For splitter item 13, we found the highest significant LR value
(Table 4.3, fifth row), suggesting a violation of unidimensionality and local
independence. We removed item 13 and continued with the resulting eight-
item set. Statistic R_{1c} of the eight-item set was not significant (Table 4.3, sixth
row), but statistic R_{2c} was significant (Table 4.3, seventh row). Item 11 had the
highest significant LR value (Table 4.3, eighth row) and was removed from
the item set. Table 4.3 (ninth and tenth rows) shows nonsignificant R_{1c} and
R_{2c} values, suggesting that the Rasch model is consistent with the seven-item
subset.

Table 4.2 (last two columns) shows the location parameters estimated by
means of condition maximum likelihood and their standard errors for the
remaining items: 1, 4, 5, 8, 12, 14, and 15. Both types of task content, both
types of presentation form, and three of the four item types are represented.
For the Rasch analysis, these item characteristics do not seem to be relevant.

We conclude by noticing that the Rasch scaling procedure (Figure 4.7) is
not compelling in the sense that it provides the only possibility for an item

analysis. Different procedures may be devised, and during our procedure, different decisions may be taken resulting in different sets of items. For example, omitting the Mokken scale analysis, and starting the scaling procedure in Figure 4.7 with all 16 items, results in a different seven-item scale consisting of items 1, 4, 11, 12, 14, 15, and 16. This dependence on methodology suggests that one needs to add judgment based on knowledge about the target attribute to the statistical results to reach a conclusion about the final scale.

Discussion

The number or articles, chapters and books on parametric IRT models defies imagination, and so does the number of different IRT models. For example, only recently several books appeared discussing parametric IRT extensively (e.g., Bartolucci, Bacci, & Gnaldi, 2016; Mislevy, 2018; Nering & Ostini, 2011; Raykov & Marcoulides, 2011). We have only discussed a modest number of models and limited ourselves to the best known. Compared to nonparametric IRT, parametric IRT is much more diverse, but the point remains that the various models are all variations on one theme, which is a monotone, locally independent basis model. In Chapter 3, the diversity among the different nonparametric IRT models resided with weakening the basic assumptions until particular ordering properties of the scale break down (Chapter 3), and allowing or not allowing IRFs to intersect. Parametric IRT models vary with respect to the parametric response functions and the number of item parameters, possibly defining the item parameters as functions of basic parameters, and assuming a multidimensional rather than a unidimensional latent variable that may or may not depend on functions of other variables.

The attraction of parametric IRT models for scale construction is that they effectively summarize item characteristics and facilitate the use of the information function to estimate a standard error for the latent variable that is a function of the latent variable. Describing items using only one or two parameters (but sometimes more parameters) frees one of the sometimes-tedious highly detailed data analysis the results of which are difficult to summarize or interpret, which is typical of nonparametric IRT. A scale-dependent standard error is a wonderful tool for ascertaining measurement precision that reveals that some scales have greater precision for some people than for others. The logit scale of latent variable θ is convenient for equating purposes in educational measurement but less so for interpretation and communication of measurement results to tested people and other interested parties. For unidimensional, locally independent, monotone IRT models, sum score X stochastically orders latent variable θ [Equation (3.17)], so that the sum score can easily replace the latent variable score. For polytomous IRT models, weak stochastic ordering [Equation (3.20)] must replace this ordering property. However, the use of the simple sum score seems to do little harm to the correctness of the measurement results, because latent variable θ orders the sum score stochastically [Equation (4.97)] in all polytomous IRT models.

In the absence of well-founded attribute theories, it is difficult to see how different IRT models can accommodate different types of attributes. Multidimensionality rather than unidimensionality, and reliance of item parameters on a smaller set of foundational parameters describing features of cognitive processes leading to item responses, are probably the features that discriminate between theories for different attributes the best. On the other hand, it is difficult to imagine why the operationalization of an attribute would restrict the slope parameters of a set of items to be equal. Different models have different features often compatible with technical details of testing. For example, the three-parameter logistics models have a lower-asymptote parameter consistent with items allowing a positive response probability even for very low latent variable values, such as multiple-choice items. Most striking, of course, are the different models for dichotomously and polytomously scored items. The three different classes of polytomous IRT models seem to quantify different response processes, but one has to know whether people actually use such processes to pick the appropriate model to analyze the data. Verhelst et al. (1997) argued that models from the three classes, even though formulated differently, resemble one another so much that they are likely indistinguishable in real-data analysis.

5

Latent Class Models and Cognitive Diagnostic Models

Nomina si nescis, perit et cognitio rerum.

—**Carolus Linnaeus**

Introduction

The previous chapters discussed the idea of cumulative scaling. This is the idea that, for dichotomous items, a larger number of correct answers to cognitive problems or positive responses to trait and attitude statements implies a higher scale value. For polytomous items, the corresponding principle is that the more the response approaches the correct response, or the more the response indicates stronger possession of a trait or an attitude, the higher the scale value. Classical test theory uses the sum score and item response theory the latent-variable score to express an individual's position on a cumulative scale. Test constructors and test users often use normed scores derived from sum scores whenever they think the normed scores facilitate a more meaningful interpretation of test performance. Because of the difficulty of their interpretation for non-specialists and the communication problems this may cause, one regularly transforms latent-variable scales to sum scores or normed scales derived from sum scores in an effort to facilitate interpretation and improve communication with the tested person. The main difference between CTT and IRT is that CTT simply assumes, without formal justification, that the sum score is a cumulative measure, whereas IRT justifies a cumulative latent-variable scale by means of a testable set of assumptions that imply this property. In defense of the more intuitive CTT approach, one could refer to the sum score's capacity to predict future school, job, and therapy success in numerous successful applications, even in the absence of a formal justification for the sum score's use and merely based on test constructors' craftsmanship and the use of simple correlational methods. Thus, one could use a practical rather than a formal argument to justify use of the sum score or a transformation thereof: It works, so why worry? On the other hand, IRT provides the more powerful tools for deepening, investigating, and justifying CTT's intuition and then expanding on it, facilitating techniques such as equating and (computerized) adaptive testing. These techniques are then theoretically justifiable rather than based on experience in the form of empirical data from particular applications that, in the absence of a theoretical foundation, might be incidentally valid for the specific application rather than generally valid.

In this chapter, we abandon the idea of cumulative scaling and focus on two related models for classifying people based on typical patterns of behaviors rather than counts of correct solutions, endorsed statements, and numbers of credit points expressing the quality of solutions or the degree of endorsement. Classification as a measurement approach, in particular, a nominal measurement model, through the years has found little resonance in psychology. The main reason probably is psychology's orientation towards the exact sciences, in older days, physics, and nowadays, biology and medicine, and with this orientation a desire to conceive of psychological attributes as quantities rather than anything else. We have no appetite for refueling the numerous debates about attributes as quantities and measurement on interval and ratio scales. One may consult Michell (1999) for a critical review of the history of measurement in psychology. We rather wish to remind the reader of the underdeveloped state of many psychological attribute theories and the messy character of many psychological data sets, justifying, if anything, an open attitude to other options. One of these options is the latent class model, another option is the cognitive diagnostic model, and both facilitate nominal scales for classification.

Classification as a strategy to create order out of chaos and thus promote knowledge and insight of course is not foreign to psychology. The Diagnostic and Statistical Manual of Mental Disorders (DSM-V; American Psychiatric Association, 2013), which is the standard for diagnosis in psychiatry, is amply used in clinical psychology and illustrates our point. However, as a measurement strategy in psychology, classification is rather atypical. For example, for measuring the mental disorders identified by the DSM-V, both clinical psychological and psychiatric researchers alike predominantly use cumulative scales and classify people, when needed, using cut-scores dividing the scale in two or more mutually exclusive and exhaustive classes. For example, the Type D Scale-14 (DS14; Denollet, 2005) that we analyzed in Chapters 2 and 3 is a typical example of a continuous scale, which is nevertheless used for classifying people as Type D personality when they score at least ten points on both the negative affectivity and social inhibition subscales. Rather exceptionally, Meehl (Ruscio, Haslam, & Ruscio, 2006; Waller & Meehl, 1998) introduced taxometrics in psychology as a statistical method to divide people into a taxon, exhibiting a clinical syndrome, and a non-taxon, which represents the complementary group, showing absence of the syndrome. For several reasons, not discussed here, taxometrics never conquered a strong foothold in psychology. In the psychometric context, De Boeck, Wilson, and Acton

(2005; also, see von Davier, Naemi, & Roberts, 2012) extensively discussed the issue of continuity versus discreteness or, similarly, dimension versus class, with respect to the measurement of psychological attributes.

We discuss nominal measurement as the classification of people based on observable patterns of behaviors, usually scores on J items, and assume that these typical patterns each are expressions of attributes. Because the attributes are unobservable and inferred indirectly from the pattern of item scores, the classes representing attributes or particular attribute states are latent and only indirectly derivable from the score patterns on the J items. Unrestricted latent class models (LCMs) classify people based on their vector of J *observed item scores*. Classifying children in groups identifiable by their use of a particular solution strategy for a cognitive problem, reflected by a unique pattern of J item scores, is a typical example of a developmental, nominal scale constructed using the unrestricted LCM (e.g., Bouwmeester & Sijtsma, 2007; Formann, 2003; Jansen & Van der Maas, 1997, 2002). We discuss the case of proportional reasoning in more detail in the first half of this chapter on LCMs and as an example, re-analyze data from Van Maanen, Been, and Sijtsma (1989). Rost and Langeheine (1997) and Hagenaars and McCutcheon (2002) provided anthologies of LCM applications to several social-science research areas. Unrestricted LCMs can be specialized or restricted in numerous ways that may be tailored to test a particular hypothesis about the attribute. Cognitive diagnostic models (CDMs) classify people based on a pattern of *latent variable scores* that reflect whether a person does or does not possess the attributes necessary for responding positively to the items. These models are usually applied in cognition and education but also in clinical psychology identifying patients' disorder profiles based on anxiety, somatoform disorder, thought disorder, and major depression (de la Torre, Van der Ark, & Rossi, 2018). We discuss this example in more detail in the second half of this chapter, which is devoted to CDMs, and re-analyze the data.

The number of specialized LCMs and CDMs is impressive and far too large to discuss in this monograph. Therefore, as in the previous chapters, we focus on the principles of a set of models and mention only a few of these models, and we discuss how the models can produce scales. As with parametric IRT models, because of its technical nature and because so much has already been published about it, we give little attention to estimating parameters. However, investigating the fit of a model to the data is of paramount importance for constructing scales based on theories, and this topic receives ample attention. Moreover, we discuss briefly how LCMs are related to nonparametric and parametric IRT and how CDMs are both LCMs but also define a set of parametric IRT models while having relations to nonparametric IRT as well. Clearly, it is all too easy to get lost in the myriad of psychometric models, showing both the wealth of modern-day psychometrics and the urgency of developing psychological attribute theory that helps the researcher to elbow her way toward the model most appropriate to produce the right scale for the particular target attribute.

Latent Class Model

An Example: Proportional Reasoning

Because psychologists do not use the LCM as frequently as they use CTT or IRT, we begin this section with the measurement of proportional reasoning by means of developmental stages through which children move, progressively using qualitatively different rules for solving so-called balance problems (Hofman, Visser, Jansen, & Van der Maas, 2015; Jansen & Van der Maas, 1997; Siegler, 1976, 1981). A balance problem shows a classical balance for measuring weight, with two arms that rest on a fulcrum that is fixed, and weights on both sides that may be equal or different, at distances to the fulcrum that may be equal or different. The tested person has to say whether the scale is in equilibrium or whether it tips to the left or to the right when the fulcrum is released. To solve the problem, it suffices to know that if the products of weight and distance on either side of the fulcrum are equal, the scale is in equilibrium; otherwise, it tips to the side with the greatest product, also called the torque. In Figure 5.1, the product on the left equals $3 \times 4 = 12$, and on the right it equals $4 \times 2 = 8$; hence, the scale tips to the left. However, this is how an expert and a computer program solve these problems, but children use strategies or rules that are incomplete and contain errors, and only when their cognitive abilities develop do the rules become more advanced, although in individual cases perfection may never be attained. An example of a rule, say, Rule 1, is considering only the weights—in the figure, the number of blocks on the pegs—and ignoring the distances, and predicting that the side with the largest weight tips. Another rule (Rule 2) considers distances only when weights are equal and predicts that in that case the side with the largest distance tips; otherwise, the side with the largest weight tips

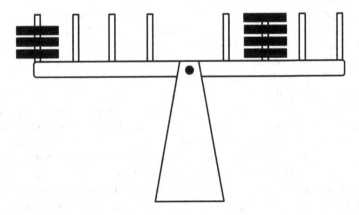

FIGURE 5.1
Example of balance scale problem.

no matter the distances. A final rule (Rule 3) adds weights and distance units and predicts that the side with the largest sum tips; otherwise, the scale is in equilibrium.

The combination of an item and a rule allows the derivation of the probability that a correct answer is given. Ignoring random measurement error, for the example item Figure 5.1, the three rules produce the correct answer as follows. For Rule 1, if one considers weights only, then one predicts that the scale tips to the right; this answer is incorrect, and assuming the response process is deterministic, the probability of a correct answer equals 0. For Rule 2, if one considers distances only when weights are equal, which is not the case in Figure 5.1, and otherwise the side with the largest weight tips, one predicts that the scale tips to the right, with the same conclusion as for Rule 1. For Rule 3, if one adds weight and distance units and predicts the scale is in equilibrium when both sums are equal, then for Figure 5.1, in which the first sum equals $3 + 4 = 7$ and the second sum equals $4 + 2 = 6$, one predicts that the scale tips to the left. Here, an incorrect strategy produces a correct answer and, assuming the response process is deterministic, a probability of a correct answer equal to 1. A test consisting of a well-chosen set of balance problems allows one to make the distinction between the different rules by their unique patterns of item-response probabilities. The LCM can be used to test whether a data set collected from a sample of children who tried to solve the set of balance problems contains the evidence of the hypothesized rule use and if so, to allocate each child to the most probable rule, in technical LCM parlance called a latent class.

Hofman et al. (2015) analyzed the data collected from 805 children who solved 26 balance problems (Jansen & Van der Maas, 2002). The authors used two approaches. First, they used an exploratory approach by means of an unrestricted LCM, allowing them to choose the solution with the number of classes that best fitted the observed data, following a multinomial distribution. Using this strategy, researchers can let the data suggest how many rules children actually used to solve the items, and allow typical characteristics of the sample to influence their conclusions. Such characteristics may refer to covariates such as schools' educational methods and parental socioeconomic status. Second, the authors tested the hypothesis that response probabilities of solving problems correctly for different items of the same type were equal across these items for children using the same rule. This is an example of a restricted LCM aimed at testing a hypothesis about response probabilities, comparable to fixing factor loadings in the confirmatory use of factor analysis. For this purpose, the authors used a logit formulation of the LCM (Bouwmeester, Sijtsma, & Vermunt, 2004).

After we have discussed the LCM, the real-data example will use the balance-scale data collected and analyzed by Van Maanen et al. (1989) using cluster analysis that Jansen and Van der Maas (1997) re-analyzed by means of the LCM. Vermunt and Magidson (2002) discussed the similarities and the differences between cluster analysis and the LCM.

Introduction

Like IRT models, LCMs define response probabilities for each item, conditional upon a latent variable. Two differences with IRT exist (Heinen, 1993, pp. 28–29; also, see Heinen, 1996). First, whereas IRT models define a continuous latent variable, the LCM assumes a discrete latent variable but refrains from specifying its dimensionality, thus defining unordered measurement values that identify latent classes without ordering them, let alone quantify distances between the classes. One can use the LCM to identify the latent classes based on the data, where one unique pattern of response probabilities on J observable variables—here, variables are items—characterizes a particular latent class. The number of latent classes identified defines the number of discrete values of the latent variable that one distinguishes. Second, like IRT, the LCM assumes item-response probabilities conditional on the latent variable, but unlike IRT, the value on which one conditions defines a latent class, where latent classes are unordered. Hence, LCMs have IRFs that relate the probability of obtaining a particular score on a particular item to a discrete latent class rather than a continuous latent variable. One typically searches for the set of response probabilities and the number of latent classes and their class size or class probability that produce the best fit of the LCM to the data. This approach renders most applications of the LCM exploratory and unrestricted LCM analysis a data-reduction method. Our viewpoint is that in a measurement context, unless one knows absolutely nothing about an attribute, one should strive to use the LCM to test hypotheses about the attribute structure, leading to scales matching that structure or indications on how to improve the structure (Chapter 1).

The latter, confirmatory viewpoint entails identifying a priori the number of latent classes one expects and possibly imposing restrictions on conditional probabilities if these follow from theoretical considerations. Such restricted models can be tested empirically (McCutcheon, 2002; also, Goodman, 2002), thus providing a tool for theory development. In this sense, one can compare the LCM to confirmatory factor analysis (Chapter 2), latent classes replacing factors and conditional item-score probabilities replacing loadings of items on factors. One can impose restrictions on the number of latent classes and the conditional probability structure similar to the number of factors and the loading structure in factor analysis. We do not reject the exploratory use of the LCM or factor analysis but would like to encourage their confirmatory, theory-driven possibilities as a means to stimulate theory development and instrument construction based on attribute theory.

Heinen (1993) studied the relations between the LCM and IRT in depth. He used the log-linear model version of both to clarify, for example, that versions of the LCM restricting the relation between the response probability and the latent variable produced discrete versions of the nominal categories model (Bock, 1972; Thissen & Cai, 2016) and the partial credit model

(Masters, 1982, 2016). Heinen demonstrated that these restricted LCMs use functions to relate response probabilities to the latent variables, thus mitigating the second of the two alleged differences between the models. He also argued that some parametric IRT models only estimate a limited number of latent variable values. Examples are the one-parameter logistic model (Rasch, 1960) in which the sum score is a sufficient statistic, allowing the estimation of $J+1$ different latent variable values, and the partial credit model (Masters, 1982) that, for the same reason, allows the estimation of $M(J+1)$ different latent-variable values. In other IRT models that estimate latent-variable values based in the various J-item score vectors, theoretically the number of different latent-variable values can be immense, but in practice, the finite sample size limits the number of latent-variable values that can be estimated. Thus, because test length, item scoring, and sample size are finite, the continuous latent variable in the IRT model is discrete in practice. Together the two counterarguments take away the sharp boundary between the LCM and IRT.

We acknowledge the value of studying similarities and differences between different measurement models. For example, revealing particular similarities may open the possibility of using estimation methods anticipated previously to be of value only for one approach with the other approach (Heinen, 1993). However, we also notice that such exercises, no matter how informative in certain respects, may unintendedly conceal the practical value each approach can have separately. For example, in a world in which we would only have IRT models with continuous scales and no attribute theories to guide the choice of measurement models other than continuous ones, it would be extremely difficult to imagine that some attributes are discrete and require discrete, nominal scales. This is precisely what the LCM suggests, however, and the fact that in several research areas one thinks of attributes as discrete rather than continuous tremendously helps researchers in other areas to consider this possibility. In addition, the availability of attribute theories or the lack thereof in a research area is another determinant for using a particular measurement model. Such attribute theories represent a different level of knowledge than the level formal measurement models provide. We argue that the value of any formal measurement model is realized only when the choice and the use of a model are guided by arguments suggested at a different knowledge level, such as a theory that predicts different cognitive strategies typical of a small number of developmental stages (Jansen & Van der Maas, 1997, 2002). When strategies do not differentiate response probabilities across developmental stages, an LCM with ordered latent classes (Croon, 1990; Vermunt, 2001) may be used to study a developmental order. LCMs are special cases of mixture models (e.g., Von Davier & Rost, 2016), which describe continuous response models for nominal classes, and the methodology of mixture models may be used to model variation between subgroups with respect to the latent variable. Provided it has a theoretical foundation, this kind of reasoning gives the LCM its place in measurement next to models for continuous scales and in addition lets it rise above mere data fitting.

The Unrestricted Model

We start by discussing the unrestricted LCM. Latent class modeling is aimed at identifying a limited number of nominal classes, and our starting point that latent variable θ has W discrete values denoted $w = 1, \ldots, W$ reflects this intention. Then, local independence is defined as in Equation (3.2) but with the difference that in Equation (3.2), local independence was defined for continuous θ, whereas here θ is discrete, that is,

$$P(\mathbf{X} = \mathbf{x} \mid \theta = w) = \prod_{j=1}^{J} P(X_j = x_j \mid \theta = w). \tag{5.1}$$

Latent class $\theta = w$ is characterized by a set of J response probabilities. Response probabilities may vary and usually will vary across items. If the number of latent classes is small, say, no more than five for dichotomous items (ignoring polytomous items for the moment), this means that for each item in the target population only five response probabilities are distinguished. This is different from IRT, where the number of response probabilities for an item varies with continuous θ and thus is infinitely large. Only assuming a discrete latent variable and local independence would provide too little structure to restrict the structure of one's data (Suppes & Zanotti, 1981), and having a limited number of latent classes, W, further restricts the LCM. First, we notice that latent classes are mutually exclusive and exhaustive; that is, a person is always a member of one latent class. The probability that a randomly chosen person is in class $\theta = w$ is denoted by $P(\theta = w)$, and because latent classes are mutually exclusive and exhaustive,

$$\sum_{w=1}^{W} P(\theta = w) = 1. \tag{5.2}$$

Second, we write the joint probability of having a particular pattern of item scores, denoted $\mathbf{X} = (X_1, \ldots, X_J)$ with realization $\mathbf{x} = (x_1, \ldots, x_J)$, and being in class $\theta = w$, as $P(\mathbf{X} = \mathbf{x} \wedge \theta = w)$. Third, using the well-known rule for the probability of events A and B, that $P(A \wedge B) = P(B)P(A \mid B)$, we write this joint probability as the product of the probability of being in class $\theta = w$, $P(\theta = w)$, and the probability of obtaining score pattern $\mathbf{X} = \mathbf{x}$ conditional on class membership, $P(\mathbf{X} = \mathbf{x} \mid \theta = w)$, that is,

$$P(\mathbf{X} = \mathbf{x} \wedge \theta = w) = P(\theta = w)P(\mathbf{X} = \mathbf{x} \mid \theta = w). \tag{5.3}$$

Fourth, applying local independence [Equation (5.1)] to conditional probability $P(\mathbf{X} = \mathbf{x} \mid \theta = w)$ in Equation (5.3), and summing across discrete classes yields the foundational equation of the LCM,

$$P(\mathbf{X} = \mathbf{x}) = \sum_{w=1}^{W} P(\theta = w) \prod_{j=1}^{J} P(X_j = x_j \mid \theta = w). \tag{5.4}$$

The W class-specific response probabilities for each of the J items, $P(X_j = x | \theta = w)$, appear on the right-hand side in Equation (5.4). They are unordered and do not constitute a function relating response probabilities to latent variable θ, comparable to the continuous IRF in IRT.

One may wonder how to determine the number of classes W in the absence of an attribute theory that hypothesizes W. In practice, because the classes are unknown, hence latent, researchers typically use the LCM [Equation (5.4)] in an exploratory fashion, and the quest usually is for the number W that explains the data structure best using the model in Equation (5.4). For the number W that explains the data best, one estimates the class probabilities, $P(\theta = w)$, for each latent class w, and the item-response probabilities, $P(X_j = x | \theta = w)$, for each item j and each latent class w.

The substantive interpretation of the W latent classes is derived from the typical pattern of response probabilities in a latent class, denoted as

$$\left(P(X_1 = 1 | \theta = w), \ldots, P(X_J = 1 | \theta = w) \right), \quad w = 1, \ldots, W. \tag{5.5}$$

These patterns vary across latent classes. Interpretation after having seen these probabilities carries the risk of reading something in the probability pattern that is not there, a danger typical of exploration but difficult to avoid when theoretical expectations have not been expressed before taking notice of the probabilities. At the end of the discussion about the theory of LCMs, we discuss an example of how the LCM was used to test a theory about proportional reasoning in the subsection on real-data analysis.

How are individuals assigned to latent classes? For this purpose, we need theoretical class probabilities, $P(\theta = w)$, which we previously also called class sizes and are also known as class weights and mixing proportions. We will use the terminology of class probability throughout. Let us assume that the theoretical class probabilities, $P(\theta = w)$, and the item-response probabilities, $P(X_j = x | \theta = w)$, are known, then in conjunction with Bayes' theorem,

$$P(B | A) = \frac{P(A | B) P(B)}{P(A)},$$

one can determine the probability that person i belongs to class w as

$$P(\theta = w | \mathbf{X}_i = \mathbf{x}_i) = \frac{P(\mathbf{X}_i = \mathbf{x}_i | \theta = w) P(\theta = w)}{P(\mathbf{X}_i = \mathbf{x}_i)}. \tag{5.6}$$

For each person, there are W such probabilities, and it appears reasonable to assign person i to the class for which her class-membership probability is greatest.

In practice, assigning individuals to a most-likely class means that, using the shorthand notation $\hat{P}_{w|x_i} = \hat{P}(\theta = w | \mathbf{X}_i = \mathbf{x}_i)$ for estimated class-membership probabilities, for person i one finds the maximum a posteriori (MAP) estimator,

$$\hat{P}^*_{w|\mathbf{x}_i} = \underset{w}{\mathrm{argmax}}\left(\hat{P}_{1|\mathbf{x}_i}, \ldots, \hat{P}_{W|\mathbf{x}_i}\right). \qquad (5.7)$$

Class assignment based on Equation (5.7) is also called modal assignment (Vermunt, 2010). Using modal assignment, one can assign all persons that responded to the J items to their most probable or modal latent class. Thus, people do not receive sum scores or norm scores derived thereof but become members of particular latent classes. Some persons will have two or more relatively large probabilities, $\hat{P}_{w|\mathbf{x}_i}$, so that an assignment based on the MAP estimator is not convincing, and this subgroup may be considered a hybrid group when it comes to psychological diagnosis. Proportional assignment (Vermunt, 2010) recognizes the hybrid nature of latent classes but produces a larger number of classification errors. Random assignment (Vermunt, 2010) classifies people in a latent class by randomly drawing a class number from the multinomial distribution defined by the posterior distribution and is sometimes used for missing data imputation, but this is not a serious candidate for assigning individuals in a diagnostic context.

Similar to reliability in CTT and the information function in IRT, in the LCM, several measures are available to express the precision with which the LCM assigns people to W latent classes. One measure is the proportion misclassified, denoted E, and this measure is defined as follows. If $\hat{P}^*_{w|\mathbf{x}_i}$ [Equation (5.7)] gives the MAP estimator for person i's most probable latent class, then the probability of assigning person i incorrectly equals $1 - \hat{P}^*_{w|\mathbf{x}_i}$. In fact, this is the probability of misclassification for any person with item-score pattern $\mathbf{x}_i$. Indexing a particular item-score pattern by $u = 1, \ldots, U$, for J dichotomous items, one may notice that $U = 2^J$. Weighing the probability of misclassification by the estimated probability of obtaining item-score pattern $\mathbf{x}_u$, denoted $\hat{P}_{\mathbf{x}_u}$, yields

$$E = \sum_{u=1}^{U} \hat{P}_{\mathbf{x}_u}\left(1 - \hat{P}^*_{w|\mathbf{x}_u}\right). \qquad (5.8)$$

One may notice that $(1 - E)$ is the proportion correctly classified, and depending on the application, one might as well report the proportion correctly classified.

An alternative measure of classifying uncertainty is entropy. The measure of entropy Shannon (1948) proposed refers to the amount of "choice" involved in the selection of an event out of W possible events or how uncertain one is of the outcome. If there are W possible events, and $W - 1$ have probability 0, so that one has probability 1, then entropy equals 0, reflecting no choice and maximum certainty about the outcome. If all W events have the same probability, then entropy is maximal, reflecting maximal choice options and thus minimal certainty about the outcome. The more the individual event probabilities tend to be equal, the greater the entropy. Considering assignment of people to latent classes, for person i ($i = 1, ..., N$), we use her W class assignment probabilities determined according to Equation (5.6) and collected in the argument of Equation (5.7), that is, $\hat{P}_{1|x_i}, ..., \hat{P}_{W|x_i}$. For all N persons that are classified, we collect the W class assignment probabilities in the two-dimensional matrix denoted $\mathbf{P}$. Then, entropy denoted $EN(\mathbf{P})$ is defined as

$$EN(\mathbf{P}) = -\sum_{i=1}^{N} \sum_{w=1}^{W} \hat{P}_{w|x_i} \log \hat{P}_{w|x_i} \qquad (5.9)$$

(Cover & Thomas, 2006, p. 14; Shannon, 1948). Because $0 < \hat{P}_{w|x_i} < 1$, it follows that $\log \hat{P}_{w|x_i} < 0$, and $EN(\mathbf{P}) \geq 0$. Entropy $EN(\mathbf{P})$ does not attain values on a fixed interval. An entropy measure defined on a fixed interval is obtained by comparing $EN(\mathbf{P})$ to the maximum entropy when all class assignment probabilities are equal, so that for person i, we have $\mathbf{P}_{i(0)} = (W^{-1}, ..., W^{-1})$, and $EN(\mathbf{P}_{(0)}) = \log W^{-1}$. Moreover, because $EN(\mathbf{P}) \leq EN(\mathbf{P}_{(0)})$, reflecting that the entropy of a distribution with unequal assignment probabilities has a lower entropy level than the uniform distribution, one may define relative reduction of entropy, RE, as

$$RE = 1 - \frac{EN(\mathbf{P})}{N \log W}, \qquad (5.10)$$

which attains values between 0 and 1. Values close to 0 indicate low certainty and values close to 1 indicate high certainty.

Restricted Models

The unrestricted LCM is interesting but allows the researcher only the number of latent classes, W, to choose but not to influence the estimation of the conditional item-response probabilities. Thus, if the theory about an attribute specifies the number of latent classes, as in proportional reasoning (Jansen & Van der Maas, 1997, 2002), the researcher hypothesizes W, and estimates the J conditional item-response probabilities freely within each latent class. However, with proportional reasoning, each strategy implied a particular item-response probability pattern based on the combination of the strategy and the physical item characteristics. Estimating all the conditional item-response probabilities freely and then comparing them to the expected probabilities is one way of testing the attribute theory (e.g., Hofman et al., 2015), and the other is to specify the probabilities *a priori* as a restricted LCM and then test the restricted model as a hypothesis. The LCM can be restricted, thus allowing interesting possibilities for testing assumptions of a theory about an attribute, in two ways. Equality restrictions set conditional item-response probabilities equal to one another but without specifying numerical values, and deterministic restrictions assign particular numerical values to conditional item-response probabilities. The former possibility is relevant for the proportional-reasoning items, as we will see later. Another possibility is to impose order restrictions on the latent classes (Croon, 1990; Hoijtink & Molenaar, 1997), a possibility we encounter later.

One can impose equality restrictions on conditional item-response probabilities in a particular latent class, across different latent classes, or both in and across classes. As a hypothetical example, imagine a test that measures arithmetic ability by means of items, half of which concern addition problems and the other half subtraction items. Answers to items are scored incorrect/correct. We assume we know that in a particular developmental group one can distinguish two latent classes, both mastering addition to the same degree but with considerable variation existing between latent classes with respect to the mastery of subtraction. Given this prior knowledge, to estimate all parameters freely is unduly non-committal, and to equate the item response probabilities of the addition items between the two latent classes is more consistent with prior knowledge. There are several possibilities to do this, but we use the following line of reasoning. We know that addition is mastered to the same degree in both latent classes, but we have no reason to assume that within latent classes different addition items are equally difficult, and we do know that the two latent classes master addition to the same degree, but we do not know precisely to what degree. If the test contains ten addition items, indexed $j = 1, ..., 10$, then we impose the following restrictions, using shorthand $P_{jw} = P(X_j = x | \theta = w)$, as follows,

$$P_{j1} = P_{j2}, \quad j = 1, \dots, 10. \tag{5.11}$$

For the sake of illustration, we also consider two different hypotheses. First, we have additional knowledge leading us to hypothesize that different addition items are equally difficult in the same group, but that the items' difficulty level may vary between the two groups; then, a reasonable restriction on the ten item response probabilities in the two latent classes is that they are equal in each group without restricting them to be equal between the two groups, so that

$$P_{1w} = P_{2w} = \cdots = P_{10w}, \quad w = 1, 2. \tag{5.12}$$

Second, we assume that we have additional knowledge enabling us to hypothesize that mastery of addition is equal for all ten items and in each group, where the groups do differ with respect to subtraction skills that we ignore here. This would imply a restriction like

$$P_{jw} = P, \quad j = 1,\dots,10, \, w = 1,2, \tag{5.13}$$

and P has a particular numerical value. Obviously, equality restrictions on the parameters reduce the number of parameters that one estimates and increase the parsimony of the model as Equations (5.11), (5.12) and (5.13) illustrate, rendering it more interesting than the unrestricted LCM.

Deterministic restrictions equate a parameter to a particular numerical value, and the cognitive strategies in combination with the proportional reasoning items that Jansen and Van der Maas (1997, 2002) studied provided excellent examples. Often, restrictions equate item-response probabilities to 0 or 1, but guessing probabilities are another possibility. If theory about arithmetic ability predicts that (hypothetical) strategies 1 and 3 imply that item 7 will be failed, strategy 2 that one will guess for the right answer with probability .33 given three response probabilities, one correct and the other two incorrect, and strategies 4 and 5 imply that the person will succeed at item 7, then one can hypothesize

$$P_{71} = P_{73} = 0; \quad P_{72} = 0.33; \quad P_{74} = P_{75} = 1. \tag{5.14}$$

One can impose deterministic restrictions like these on all items and then test whether the hypothesized restricted LCM fits the data.

Finally, as another example of a restricted LCM, we discuss an early LCM (Proctor, 1970), which provides a possible probabilistic explanation for failure of the Guttman model [Chapter 4; Equations (4.2a) and (4.2b)] to explain particular model-inconsistent observed item-score vectors. We assume that the J dichotomous items in the test are ordered from difficult to easy, so that in an item-score vector $\mathbf{x}$, 1s never precede 0s, because this would mean that after having succeeded at an item the person failed an easier item; that is, in the Guttman model, one expects item-score vectors in which score pairs (00), (01), and (11) occur but not (10). Given J items, the Guttman model allows $J+1$ item-score vectors, and from the Guttman model's perspective, the other item-score vectors one finds in a data set represent errors. Proctor (1970) simply assumed that the probability of an error happening at the item level equals a, irrespective of the item's difficulty and the person's attribute level. Obviously, the Proctor model is a simple model that will not fit real data, but its simplicity renders the model a fine didactic example of a restricted LCM, which is the reason why we discuss it.

Let $J = 4$, and notice that the $J + 1 = 5$ Guttman item-score vectors are

$$\mathbf{y}_1 = (0000), \quad \mathbf{y}_2 = (0001), \quad \mathbf{y}_3 = (0011), \quad \mathbf{y}_4 = (0111), \quad \mathbf{y}_5 = (1111). \tag{5.15}$$

We assume that these five vectors represent five latent classes, with class probabilities $P_w = P(\theta = w)$, $w = 1, \dots, 5$. Consider an observed vector $\mathbf{x}_u = (0101)$, where subscript u enumerates the different item-score vectors and attains values $u=1, \ldots, 2^J$, and compute its probability given each of the five latent classes. From the perspective of class 1, the two 1 scores in $\mathbf{x}_u$ represent errors, both with probability a, whereas the two 0 scores represent permissible scores, both with probability $1-a$. Thus, for class 1, we find

$$P(\mathbf{x}_u|\theta=1)=(1-a)a(1-a)a=(1-a)^2 a^2.$$

Following the same line of reasoning, for class 2, we find

$$P(\mathbf{x}_u|\theta=2)=(1-a)^3 a,$$

and similarly for the other three classes. The complete restricted LCM formulation is

$$P(\mathbf{x}_u)=P_1(1-a)^2 a^2 + P_2(1-a)^3 a + P_3(1-a)^2 a^2$$
$$+P_4(1-a)^3 a + P_5(1-a)^2 a^2. \tag{5.16}$$

In each product in Equation (5.16), the exponent of the error probability, a, indicates the number of errors present in the observed item-score vector compared to the Guttman item-score vector typical of a particular latent class. One obtains a general formulation of the Proctor model using indicator variables as follows. Let $d_{wuj}=0$ if $y_{wj}=x_{uj}$, and $d_{wuj}=1$ if $y_{wj}\neq x_{uj}$; then, $d_{wu}=\sum_{j=1}^J d_{wuj}$ counts the number of errors present in item-score vector $\mathbf{x}_u$ relative to item-score vector $\mathbf{y}_w$. Using this notation, the Proctor LCM equals

$$P(\mathbf{x}_u)=\sum_{w=1}^W P_w(1-a)^{J-d_{wu}} a^{d_{wu}}. \tag{5.17}$$

For the Proctor model, Equation (5.17) provides the probability of a particular item-score vector in cell u of a multinomial distribution. Proctor (1970) used the probabilities estimated from the data to compute the cell frequencies one expects based on the restricted LCM [Equation (5.17)] and compared them with the observed frequencies by means of a chi-squared test. To obtain a more flexible model, Proctor (1970) suggested introducing error rates that vary across items, and other restricted LCMs.

Estimation

The parameters of the LCM, which are the class probabilities, $P_w=P(\theta=w)$, $w=1, \ldots, W$, and the conditional item-score probabilities, $P_{jw}=P(X_j=x|\theta=w)$, $j=1, \ldots, J, x=0, \ldots, M$, and $w=1, \ldots, W$, can be estimated by means of a maximum likelihood procedure. Let the frequency of people having item-score vector be denoted $n_{x_1 x_2 \ldots x_J}=n_x$, and let the number of different item-score vectors be indexed $u = 1, \ldots, M^J$, so that frequencies are enumerated $n_{\mathbf{x}_u}$. The likelihood can be defined across all persons and, taking equal item-score vectors together, across the cells of the J-dimensional cross tables. Assuming observations are independent and identically distributed, the likelihood then equals

$$L(\mathbf{X}) = \prod_{i=1}^{N} P(\mathbf{x}_i) = \prod_{u=1}^{M^J} P(\mathbf{x}_u)^{n_{\mathbf{x}_u}}. \tag{5.18}$$

Let an item score in $\mathbf{x}_u$ be denoted x_{ju}, and let $P_{juw} = P(X_{ju} = x_{ju} | \theta = w)$. Inserting the LCM [Equation (5.4)] for $P(\mathbf{x}_u)$ yields

$$L(\mathbf{X}) = \prod_{u=1}^{M^J} P(\mathbf{x}_u)^{n_{\mathbf{x}_u}} = \prod_{u=1}^{M^J} \left(\sum_{w=1}^{W} P_w \prod_{j=1}^{J} P_{juw} \right)^{n_{\mathbf{x}_u}}. \tag{5.19}$$

Computationally, it is more convenient to work with the logarithm of the likelihood, but because they have been explained elsewhere (e.g., Eliason, 1993) extensively, we refrain from discussing further computational steps. The likelihood usually is maximized using the Expectation-Maximization (EM) algorithm, but other possibilities such as Newton–Raphson (NR) exist. EM is stable and relatively slow, and NR is relatively unstable but fast. The two procedures are often combined, so that EM is used to find a stable pre-result and next NR is used to swiftly find the final solution. See Dempster, Laird, and Rubin (1977) for the seminal article on EM, and Wasserman (2004, pp. 142–146) for a short explanation of EM and NR.

Equations (5.18) and (5.19) show that for $n_{\mathbf{x}_u} = 0$, the corresponding term equals 1, hence does not affect the likelihood, showing that zero-cells in the cross table do not negatively affect the estimation of LCM parameters. Following Heinen (1993) and Vermunt and Magidson (2004), a few remarks are in order. First, models may be unidentified, and this may be true even if the number of degrees of freedom is zero or positive. Vermunt and Magidson (2004) provided a few cases where one knows for sure that the model is identifiable, but a characteristic of the LCM is that general rules for deciding whether a model is identifiable do not exist. Unidentified models do not have a set of unique parameter values, meaning that multiple sets of parameters produce the same maximum of the likelihood. In practice, one resolves this problem by adapting the model in such a way that the new model is identifiable. Adapting the model may be problematic if it is not supported by theory about the attribute but may be acceptable in an exploratory context. Second, one may think of likelihood functions as foggy, hilly landscapes, often in many dimensions, with several peaks but only one global maximum. For simplicity, let us imagine three dimensions. One wishes to find the global maximum but, depending on where one starts climbing, that is, which starting values one chooses, one may land on a local maximum but cannot know this, because one does not see the other peaks. Several sets of starting values should predominantly lead one to the same solution, which then is hoped to be the global maximum. Third, boundary values, which are probabilities equal to 0 or 1, may produce computational problems, such as leading the algorithm to a local maximum, but also render confidence intervals and statistical tests for the parameters meaningless. Galindo Garre and Vermunt (2006) suggested using a Bayesian approach to estimation and showed that their approach avoids the problems better than other approaches.

Goodness-of-Fit Methods

Goodness-of-fit methods typically compare frequencies $n_{x_1x_2...x_J} = n_x$ of people having a particular score pattern on the J items in the test with the corresponding expected frequencies $e_{x_1x_2...x_J} = e_x$ predicted by the estimated LCM; see the Proctor (1970) model as an example. Greater inconsistency between n_x and e_x frequencies implies greater misfit. For the assessment of model fit with respect to LCMs, one predominantly uses methods assessing the fit of a model to another model; this is relative fit. Models having fewer parameters than other models are more parsimonious and fit worse to the data or relative to other models that have more parameters. An important distinction is that between nested and non-nested models. Several authors (Heinen, 1993; Magidson & Vermunt, 2004; McCutcheon, 2002) discussed these and other methods. Given the omnipresence of such discussions, we only give a brief outline.

Likelihood Statistic G². The first type of fit statistic we discuss assesses the fit of the cell frequencies expected under the unrestricted LCM to the observed cell frequencies one assumes are sampled from a multinomial distribution. To obtain expected cell frequencies, one first computes the probability of being in the cell corresponding to item-score pattern $x = (x_1, x_2, ..., x_J)$, and inserts estimates of the class and item-response probabilities,

$$\hat{P}(\mathbf{X} = \mathbf{x}) = \sum_{w=1}^{W} \hat{P}(\theta = w) \prod_{j=1}^{J} \hat{P}(X_j = x_j \mid \theta = w).$$ (5.20)

Next, multiplying cell estimate $\hat{P}(\mathbf{X} = \mathbf{x})$ with sample size N provides one with the expected cell frequency,

$$e_x = N\hat{P}(\mathbf{X} = \mathbf{x}).$$ (5.21)

The family of fit statistics known as power divergence statistics have a limiting chi-squared distribution if certain regularity conditions are met (Cressie & Read, 1984). The most well-known members of this family are the likelihood statistic (G^2) and Pearson's chi-square statistic (X^2). Both statistics assess absolute fit; this is the fit of the model to the data. For the purpose of illustration, we discuss the likelihood statistic G^2. As before, let u index item-score vectors x. Let

$$\ell_{\text{LCM(W)}} = \sum_u n_{x_u} \ln\left(\frac{e_{x_u}}{N}\right) \tag{5.22}$$

be the log-likelihood of the unrestricted latent class model with W latent classes, and let

$$\ell_{\text{Data}} = \sum_u n_{x_u} \ln\left(\frac{n_{x_u}}{N}\right) \tag{5.23}$$

be the log-likelihood of the unrestricted multinomial distribution that produced the data; then, statistic G^2 equals -2 times the difference between the log-likelihood of the model that is most restrictive, which is the LCM, and the model that is the most general, which are the data. Thus statistic G^2 is defined as

$$G^2 = -2\left(\ell_{\text{LCM(W)}} - \ell_{\text{Data}}\right) = -2\sum_u n_{x_u} \ln\left(\frac{e_{x_u}}{n_{x_u}}\right). \tag{5.24}$$

The number of degrees of freedom, denoted df, equals the number of free parameters of the model that is the most general minus the number of free parameters of the model that is the most restrictive. The number of free parameters of the multinomial distribution equals the number of cells minus 1, which equals $[M+1]^J - 1$. The number of free parameters for the unrestricted LCM equals $[W-1] + WJM$ and is determined as follows. Constraints necessary to identify the parameters limit the number of model parameters that can be estimated freely. The constraints are

$$\sum_{w=1}^{W} P(\theta = w) = 1, \tag{5.25}$$

and

$$\sum_{x=0}^{M} P(X_j = x \mid \theta = w) = 1, \quad \text{for } w = 1,\dots,W. \tag{5.26}$$

Equation (5.25) implies that $(W-1)$ class probabilities can be estimated freely. Given $M+1$ scores for each item, Equation (5.26) implies that M conditional response probabilities in each latent class (i.e., WJM conditional response probabilities in total) can be estimated freely. As a result, the number of degrees of freedom equals

$$df = \left[(M+1)^J - 1 \right] - \left[WJM + (W-1) \right]. \tag{5.27}$$

If the asymptotic chi-square test of the model renders a probability of exceedance or *p*-value smaller than significance level α, the unrestricted LCM does not fit the data.

For larger numbers of items, the number of degrees of freedom in Equation (5.27) can be enormous. For example, 20 items each having five response categories correspond with $M^J = 5^{20} \approx 9.54 \times 10^{13}$ cells to accommodate each item-score pattern $\mathbf{x}$, and an almost equally large number of degrees of freedom. Even if samples are huge, say, $N = 1,000,000$, only a fraction of the cells in the example are non-empty, rendering the use of statistic G^2 for assessing the fit of the model to the data problematic (Magidson & Vermunt, 2004). This problem is often referred to as the *curse of dimensionality*. As a result, one can no longer trust statistic G^2 [Equation (5.24)] to follow an asymptotic chi-square distribution. To obtain trustworthy probabilities of exceedance, Vermunt and Magidson (2002) recommended investigating the bivariate residuals. This topic returns when we discuss CDMs.

Statistic G^2 can also be used to compare two nested models. Let ℓ_{LCM} and ℓ_{LCM^*} denote the log-likelihood of a relatively general LCM and a relatively restrictive LCM*, respectively. For example, the general LCM could have W latent classes and the restrictive LCM* could have W^* ($W^* < W$) latent classes. Alternatively, the general LCM could have W latent classes and the restrictive LCM* also could have W latent classes but with equality constraints on some of the parameters. For example, one may constrain the response probabilities of item j to be equal across the W latent classes, that is

$$P\left(X_j = x \mid \theta = 1\right) = P\left(X_j = x \mid \theta = 2\right) = \cdots = P\left(X_j = x \mid \theta = W\right). \tag{5.28}$$

Let $e_{\mathbf{x}_u}$ and $e^*_{\mathbf{x}_u}$ denote the expected cell frequencies of the general LCM and the restrictive LCM*, respectively. Similar to Equation (5.24)

$$G^2 = -2\left(\ell_{\text{LCM}^*} - \ell_{\text{LCM}}\right) = -2\sum_u n_{\mathbf{x}_u} \ln\left(\frac{e^*_{\mathbf{x}_u}}{e_{\mathbf{x}_u}}\right). \tag{5.29}$$

Similar to statistic G^2 in Equation (5.24), the number of degrees of freedom for G^2 in Equation (5.29) equals the difference of the number of free parameters between the general and restricted models. For example, comparing models with W and W^* latent classes ($W^* < W$),

$$df = \left[WJM + (W-1) \right] - \left[W^*JM + (W^*-1) \right] = (W - W^*)(MJ + 1). \tag{5.30}$$

The number of degrees of freedom is relatively small, so that G^2 does not suffer from the curse of dimensionality. For example, comparing a six-class and a five-class LCM using data obtained from 20 five-category items produces $df = 1(5 \times 20 + 1) = 101$. However, two words of caution are relevant here. First, the regularity conditions do not hold when statistic G^2 [Equation (5.29)] is used to compare two unrestricted LCMs with different numbers of latent classes and, consequently, statistic G^2 [Equation (5.29)] does not follow a chi-squared distribution (e.g., Holt & Macready, 1989). McLachlan (1987; also, see Langeheine, Pannekoek, & Van de Pol, 1996; Tekle, Gudicha, & Vermunt, 2016) suggested using a parametric bootstrap procedure to approximate the distribution of G^2 under the restrictive LCM. Second, the results of model comparison can only be trusted when the model that is the most general is the true model. Hence, if the more general LCM has poor fit to the data, the model should not be used as a baseline model for more restrictive LCMs. For practical data analysis, given that the research context often is exploratory more than confirmatory and that models by definition represent simplifications of a hypothesized truth, taking the unrestricted LCM as a baseline model and then working progressively towards including restrictions one by one to learn where the consistencies and the inconsistencies with respect to the hypothesized structure are seems to be a useful strategy.

Assessing Individual Items. Psychological measurement procedures, such as CTT and nonparametric and parametric IRT, all address the issue of item fit, for example, because different items tend to contribute differently to the reliability or the standard error of measurement of the sum score or the standard error of the latent-variable value. At the instruments' construction phase, the researcher may wish to delete items contributing little or replace such items with other items. In the LCM, items contributing little to people's reliable class membership allocation need to be identified and possibly removed from the test or replaced with items contributing more. The search thus is for items that contribute little to the classification of people in the W latent classes. An approach is to test whether the distribution of item score X_j is the same across latent classes (Magidson & Vermunt, 2004); different distributions suggest that item j does not contribute to the reliable allocation of people to latent classes. Specifically, similar to Equation (5.11), one tests whether

$$P\left(X_j = x \mid \theta = 1\right) = \cdots = P\left(X_j = x \mid \theta = W\right), \quad x = 0, \ldots, M. \qquad (5.28)$$

To test the null hypothesis in Equation (5.28), one uses the log-linear transformation of the LCM (e.g., Haberman, 1978; Heinen, 1993, p. 61) in which the log-linear model includes a latent variable, θ. From now on, we let u represent an effect and no longer an index for item-score vectors. Because of local independence, all higher-order effects are 0—that is, all effects involving two or more items—and the model only contains parameters referring to effect of item-score variables, $u_{x_j}^j$, the latent variable, u_w^θ, and the association between individual items and the latent variable, which in this case represents class membership, $u_{x_j w}^{j\theta}$; hence, the model is

$$\ln P\left(\mathbf{X} = \mathbf{x} \wedge \theta = w\right) = u + \sum_{j=1}^{J} u_{x_j}^{j} + u_{w}^{\theta} + \sum_{j=1}^{J} u_{x_j w}^{j\theta},$$ (5.31)

where u is an intercept parameter. In Box 5.1, we transform Equation (5.31) into a conditional probability of the kind in Equation (5.28), which is $P(X_j = x | \theta = w)$, and this transformation proves helpful in testing the null hypothesis of equal item-score distributions across latent classes.

BOX 5.1   Transforming the Log-Linear Model
into a Conditional Probability

Following Heinen (1993, p. 62), first, we use the rule for conditional probability, $P(A|B) = P(A \wedge B)/P(B)$, to write

$$P\left(X_j = x | \theta = w\right) = \frac{P\left(X_j = x \wedge \theta = w\right)}{P(\theta = w)}.$$ (5.32)

Second, we rewrite the numerator and the denominator in Equation (5.32), and we do this using the general result that $P(A) = \sum_b P(A \wedge B)$. We notice that for J items denoted j, $l_1, \ldots, l_{J-1}$, summing across the items $l_1, \ldots, l_{J-1}$ for the numerator in Equation (5.32), we may write

$$P\left(X_j = x_j \wedge \theta = w\right)$$

$$= \sum_{x_{l_1}=0}^{M} \cdots \sum_{x_{l_{J-1}}=0}^{M} P\left(X_j = x_j \wedge X_{l_1} = x_{l_1} \wedge \cdots \wedge X_{l_{J-1}} = x_{l_{J-1}} \wedge \theta = w\right),$$ (5.33)

and for the denominator, letting the item index run across $j = 1, \ldots, J$, and summing across the J items, we may write

$$P(\theta = w) = \sum_{x_1=0}^{M} \cdots \sum_{x_J=0}^{M} P\left(X_1 = x_1 \wedge \cdots \wedge X_J = x_J \wedge \theta = w\right).$$ (5.34)

Third, by taking the exponential of both sides of an equality as in Equation (5.34), and using the result that $\exp(\ln P) = P$, we write the log-linear model in Equation (5.31) as

$$P(\mathbf{X} = \mathbf{x} \wedge \theta = w) = \exp\left(u + \sum_{j=1}^{J} u_{x_j}^{j} + u_{w}^{\theta} + \sum_{j=1}^{J} u_{x_j w}^{j\theta}\right).$$ (5.35)

Using the general result that for any numbers z, a, and b, we have that $z^{a+b} = z^a z^b$, hence, also $\exp(a+b) = \exp(a)\exp(b)$, and we write

$$P(\mathbf{X} = \mathbf{x} \wedge \theta = w) = \exp\left(u_{x_j}^j + u_{x_j w}^{j\theta}\right)\exp\left(u + \sum_{l:l\neq j} u_{x_l}^l + u_w^\theta + \sum_{l:l\neq j} u_{x_l w}^{l\theta}\right). \quad (5.36)$$

Using this result, we write the numerator in Equation (5.32) which was expanded in Equation (5.33), as

$$P\left(X_j = x_j \wedge \theta = w\right)$$

$$= \exp\left(u_{x_j}^j + u_{x_j w}^{j\theta}\right) \sum_{x_1=0}^{M} \cdots \sum_{x_{I_{J-1}}=0}^{M} \exp\left(u + \sum_{l:l\neq j} u_{x_l}^l + u_w^\theta + \sum_{l:l\neq j} u_{x_l w}^{l\theta}\right). \quad (5.37)$$

Likewise, we rewrite the denominator in Equation (5.32) which was expanded in Equation (5.34), as

$$P(\theta = w)$$

$$= \sum_{x_j=0}^{M} \exp\left(u_{x_j}^j + u_{x_j w}^{j\theta}\right) \sum_{x_1=0}^{M} \cdots \sum_{x_{I_{J-1}}=0}^{M} \exp\left(u + \sum_{l:l\neq j} u_{x_l}^l + u_w^\theta + \sum_{l:l\neq j} u_{x_l w}^{l\theta}\right). \quad (5.38)$$

One may notice that the parts of Equations (5.37) and (5.38) that do not refer to item j are precisely equal and cancel upon division. The result is an expression for $P(X_j = x_j | \theta = w)$ that proves important enough to include it in the main text, which is where we proceed; see Equation (5.39).

Dividing $P(X_j = x_j \wedge \theta = w)$ [Equation (5.37)] by $P(\theta = w)$ [Equation (5.38)] yields the result

$$P\left(X_j = x_j | \theta = w\right) = \frac{P\left(X_j = x_j \wedge \theta = w\right)}{P(\theta = w)} = \frac{\exp\left(u_{x_j}^j + u_{x_j w}^{j\theta}\right)}{\sum_{x_j=0}^{M} \exp\left(u_{x_j}^j + u_{x_j w}^{j\theta}\right)}. \quad (5.39)$$

Equation (5.39) is a logistic formulation of the class-specific response probability, comparable to many logistic IRT models (Chapter 4) but with a limited number of W discrete latent-variable values. To identify the model, the log-linear parameters are estimated using restrictions, such as sums of parameters equal to 0; see Heinen (1993, pp. 34–37).

We rephrase the null hypothesis formulated in Equation (5.28) using the association parameters in Equation (5.39) as

$$u^{j\theta}_{x_j1} = \cdots = u^{j\theta}_{x_jW}, \quad x_j = 0,\ldots,M. \tag{5.40}$$

The model including the restrictions from Equation (5.40) is denoted $LCM^*_{(j)}$. Because the models are nested, the unrestricted model denoted LCM can be compared with the restricted model by means of the difference $G^2 = G^2_{LCM} - G^2_{LCM^*_{(j)}}$; also, see Equation (5.29). The logic for testing the difference between two nested models was discussed in the previous subsection and is now applied to item assessment. Thus, the difference between the statistics follows an asymptotic chi-square distribution with degrees of freedom equal to the difference between the degrees of freedom for the unrestricted model and the restricted model. If one suspects item j of malfunctioning, one can test a targeted null hypothesis for item j, but an exploratory strategy is to test Equation (5.40) for each item separately.

Information Fit Measures. The best-known information fit measures are the Akaike information index (AIC; Akaike, 1974) and the Bayesian information index (BIC; Schwartz, 1978); for a comparison, see Vrieze (2012). These measures can be used for comparing unrestricted LCMs with different numbers of latent classes, thus trying to determine the correct value for W. The measures are also suited for comparing restricted with unrestricted models. The AIC and BIC indices are based on information theory (Shannon, 1948). AIC is a measure of the quality of statistical models for a given set of data that assesses each model relative to competing models for the same data. Let K be the number of parameters that one estimates for a model, and let L_M denote the likelihood for model M that was maximized to obtain the K parameter estimates. Then, AIC weighs the maximized likelihood with the complexity of the model, based on the principle that scientific theories must be as sparse as possible while maintaining their explanatory power, which is expressed by

$$AIC = G^2_M - 2 \times df. \tag{5.41}$$

In Equation (5.41), G^2_M expresses model fit and $2 \times df$ expresses model complexity. A model that explains the data relatively well has a relatively low G^2 value, expressing little misfit, hence, good fit, and AIC corrects the fit by adding a penalty for the model's complexity, expressed as $2 \times df$. One can improve the fit—that is, further reduce misfit G^2—somewhat by adding more parameters to the model, but this tends to increase the penalty more than it reduces the misfit; hence, then AIC is increased. Thus, improving fit at all costs means unnecessarily complicating the model to the detriment of parsimony, producing greater AIC. On the other hand, trying to simplify the model by leaving out essential parameters produces a model that is too simple to explain the data well, and the small penalty due to few parameters estimated is counterbalanced by worse fit, hence higher AIC. The researcher needs to find the right balance between misfit and complexity. Based on the same data set and a collection of models, such as different LCMs that may or may not be nested, the best-fitting model has the smallest AIC value and thus finds the best balance between misfit or bias on the one hand, and complexity or variance on the other hand. Most importantly, one does not know whether the best-fitting model actually represents reality. One only knows that of the models considered, the best-fitting model is closest to reality but not how close. In this sense, AIC provides information about relative fit.

Sample size introduces uncertainty, and the BIC uses this information to define an index closely resembling AIC (albeit based on a different theoretical background), as

$$\text{BIC} = G_M^2 - \ln(N) \times df. \tag{5.42}$$

BIC is based on the idea that one model is true, and that if one assumes known prior probabilities, one can use Bayes' theorem to determine the posterior probability of each of the competing models given the data. Given several assumptions that are beyond the scope of our discussion and that we do not discuss here, one may argue that the model that has the largest posterior probability also has the smallest BIC value when $\ln(N)$ is used instead of factor 2 as in AIC (Dziak, Coffman, Lanza, & Li, 2017; Wasserman, 2004, pp. 218–223). We emphasize that the use of factor $\ln(N)$ only is evident when a particular Bayesian line of reasoning including several assumptions is accepted, but we refrain from delving into that topic. The result is a penalty term that depends both on sample size and model complexity in terms of number of estimated parameters. Clearly, as sample size grows, the penalty term grows, because one has more certainty about the model's complexity, and a larger $\ln(N)$ term diminishes the probability that one will choose a model too complex. Given the same misfit and number of parameters, as N grows, one will eventually reject the model in favor of models having a better balance given this particular sample. Taking the natural logarithm of N reflects that the decreasing influence of sample size levels off as N grows. Similar to model selection using AIC, one chooses the model that has the smallest BIC. Other model selection methods are available, and we briefly mention a few in the section on cognitive diagnostic models.

Ordered LCM and Testing Monotonicity in Nonparametric IRT

LCMs have been extended with explanatory structures, such as regression models (Bouwmeester, Sijtsma, & Vermunt, 2004), multilevel models (Henry & Muthén, 2010; Vermunt, 2003), and factor models (Magidson & Vermunt, 2001), but also IRT models such as the partial credit model (e.g., Bouwmeester et al., 2007; see Chapter 1), and led to numerous applications in a variety of research areas. In an effort to tie the LCM to the monotone homogeneity model (Chapter 3), several authors (Croon 1990, 1991; Hoijtink & Molenaar, 1997; Ligtvoet & Vermunt, 2012; Van Onna, 2002; Vermunt, 2001) used ordered LCM analysis to investigate the monotonicity assumption of the monotone homogeneity model. The unconstrained LCM [Equation (5.4)] is typically estimated using an EM algorithm (Dempster et al., 1977) but can also be estimated using a Gibbs sampler (e.g., Wasserman, 2004, pp. 411–419). Both estimation methods yield estimates for the class probabilities, $P(\theta = w)$, and the item response probabilities, $P(X_j = x | \theta = w)$. Ligtvoet and Vermunt (2012) explained how to use the LCM to test the monotonicity assumption of the monotone homogeneity model (Chapter 3), when one rephrases monotonicity as follows. We replace continuous latent variable θ with discrete latent variable $\theta = w$, $w = 1, \ldots, W$, and define expectation

$$\mathcal{E}\left(X_j \mid \theta\right) = \sum_{x=1}^{M} P\left(X_j \geq x \mid \theta = w\right). \tag{5.43}$$

We assume that consecutive values of $\mathcal{E}(X_j | \theta)$ are non-decreasing in θ. The conditional expected item score, $\mathcal{E}\left(X_j | \theta\right)$, summarizes the M item step response functions, $P(X_j \geq x | \theta)$, for each item. This summary goes at the expense of information at the lower aggregation level about non-monotonicity or monotonicity [Equation (3.8)] but simplifies the investigation of monotonicity. Ligtvoet and Vermunt (2012) discussed a Bayesian method for estimating the model parameter estimates. These authors also discuss a statistic for assessing the fit of the constrained LCM relative to competing models, and two different statistics for assessing the fit of individual items. Their model fitting strategy can be used to assess items' monotonicity assumption but also for investigating the assumption of an invariant item ordering, that is, an ordering of conditional expected item scores $\mathcal{E}(X_j | \theta)$ that is invariant across latent classes with the exception of possible ties [Chapter 3, Equation (3.81)].

Data Example: Proportional Reasoning by Means of the Balance Scale

The data consist of the dichotomous scores of 484 children on 25 balance problems (i.e., items) and were collected by Van Maanen et al. (1989). For each item, the three possible responses were "balance tips over to the right", "balance is in equilibrium", and "balance tips over to the left". Correct responses were scored 1, and incorrect responses were scored 0. The item set consisted of five groups of five items each: Weight items, distance items, conflict-balance items, conflict-weight items, and conflict-distance items (Table 5.1). The main research question is whether the data support hypothesized strategies for solving balance problems (Siegler, 1981). Van Maanen et al. (1989) and Jansen and Van der Maas (1997) investigated the research question using these data. The items together measure the attribute of proportional reasoning.

Based on Siegler (1981), Van Maanen et al. (1989) hypothesized five strategies (Strategies R1 to R5; Table 5.2) that children use to solve the balance problems.

TABLE 5.1

Five Subsets of Balance-Problem Items

Item subset	Description	Picture
Weight items	Weights differ but distances are the same. Balance tips over to the side of the largest weight.	
Distance items	Distances differ but weights are the same. Balance tips over to the side of the largest distance.	
Conflict-balance items	Both distances and weights differ. Balance is in equilibrium.	
Conflict-weight items	Both distances and weights differ. Balance tips over to the side of the largest weight.	
Conflict-distance items	Both distances and weights differ. Balance tips over to the side of the largest distance.	

TABLE 5.2

Strategies for Solving Balance Problems Investigated in Three Publications

Publication	Strategy					
	R0	R1	R2	R3	R4	R5
Van Maanen, Been, & Sijtsma (1989)		X	X	X	X	X
Jansen & Van der Maas (1997)		X	X	X	X	
This chapter	X	X	X	X	X	

Note: RO = no strategy, children provide random answers; R1 = children consider only the weights; R2 = if the weights are unequal, children consider only the weights. If the weights are equal, children consider the distances; R3 = children consider weights and distances but cannot handle conflict items; R4 (torque rule) = children consider weights and distances and compute the torque if necessary; R5 (buggy rule) = children consider weights and distances. For conflict items, they use a non-optimal strategy to compute the torque (see Jansen & Van der Maas, 1997, for details). Siegler (1981) suggested strategies R1, R2, R3, and R4.

Each strategy produces a unique pattern of item scores, and typical item-score patterns suggest the strategy that the child employed. Four strategies are formally incorrect and when used sometimes produce correct answers and in other cases incorrect answers, depending on features of the item. One strategy (i.e., Strategy R4, Torque rule) is formally correct and, apart from measurement error, thus was expected to produce only correct answers. Van Maanen et al. (1989) used the linear logistic model [Scheiblechner, 1972; Chapter 4, Equation (4.110)] for testing the hypotheses that a subgroup consistently uses one strategy and that different subgroups use different strategies. Because the Rasch model encompasses the linear logistic model, Van Maanen et al. (1989) first estimated the Rasch model from the data collected from the complete group of children answering all 25 items. They found that the Rasch model did not fit. Next, the authors investigated the fit of the Rasch model in subgroups characterized by the use of a unique strategy. They identified these subgroups using cluster analysis on a subset of five items, one item per item type. Using the remaining 20 items, the Rasch model fitted in two of the four remaining subgroups. The linear logistic test model formalizing the strategy typical of a particular subgroup did not fit in any of the four strategy subgroups. Because subgroups predominantly produced the same item-score pattern, sum scores showed little variance, which is a condition that mimics local independence, suggesting that an IRT model actually fits the data when in fact the fit is a consequence of small variance. This artifact renders the applicability of IRT models perhaps inappropriate (Wood, 1978), and recognizing this problem, Van Maanen et al. argued that an LCM might have been a feasible alternative to test the presence of the strategies in homogeneous subgroups. We notice that in the late 1980s, the LCM was available only for very small numbers of items and technically not ready to analyze a test consisting of 25 items.

A decade later, better estimation procedures and software for the LCM were available, and Jansen and Van der Maas (1997) used them to re-analyze the data Van Maanen et al. (1989) collected. Because we focus on the LCM, we refrain from discussing the five strategies in detail but provide a brief description in the note of Table 5.2 and trust the reader gained an adequate impression when we discussed the example strategies using Figure 5.1 as an example item. Jansen and Van der Maas (1997) argued that strategies R4 and R5 cannot be distinguished by means of the available data and hypothesized Siegler's original four strategies (Table 5.2). Following the suggestion that Van Maanen et al. made, Jansen and van der Maas used the LCM to analyze the data. Their results partly supported Siegler's strategies but only after they removed the data of 12 students who together produced seven of the 17 observed item-score patterns of the weight items. When including these 12 data records, the LCMs did not fit well, and the support for Siegler's strategies weakened. We redid the analysis of Jansen and van der Maas and investigated whether adding an extra strategy entailing random guessing and denoted RO (Table 5.2) accommodated the unclassifiable data records. Table 5.3 shows the expected probabilities for each item subset and each strategy subgroup.

First, we analyzed the item scores for each five-item subset separately. We estimated the hypothesized latent classes, once with and once without constraining the response probabilities to be equal, and used the freely available program LEM (Vermunt, 1997). Let LCM(W) denote an unconstrained LCM with W latent classes. Table 5.4 shows the hypothesized latent classes derived from Table 5.3. For example, for the weight items, we expected that

TABLE 5.3

Expected Values of $P(X_j = 1 | \theta = w)$ for Each Combination of Item Subset and Strategy Subgroup

	Strategy				
Item subset	R0	R1	R2	R3	R4
Weight items	1/3	1	1	1	1
Distance items	1/3	0	1	1	1
Conflict-balance items	1/3	0	0	1/3	1
Conflict-weight items	1/3	1	1	1/3	1
Conflict-distance items	1/3	0	0	1/3	1

Note: R0 = no strategy, children provide random answers; R1 = children consider only the weights; R2 = if the weights are unequal, children consider only the weights. If the weights are equal, children consider the distances; R3 = children consider weights and distances but cannot handle conflict items; R4 (torque rule) = children consider weights and distances and compute the torque if necessary.

TABLE 5.4

Distribution of the Strategies Over the Expected Latent Classes for Each Subset of Balance-Problem Items

	Expected latent classes		
Item subset	LC 1 (correct)	LC 2 (random)	LC 3 (incorrect)
Weight items	R1, R2, R3, R4	R0	NA
Distance items	R2, R3, R4	R0	R1
Conflict-balance items	R4	R0, R3	R1, R2
Conflict-weight items	R1, R2, R4	R0, R3	NA
Conflict-distance items	R4	R0, R3	R1, R2

Note: LC 1 (correct) = latent class of children using a strategy resulting in the correct response, response probabilities tend to 1; LC 2 (random) = latent class of children using no strategy or a strategy resulting in a random guess, response probabilities approximately equal 1/3; LC 3 (incorrect) = latent class of children using a strategy resulting in the incorrect response, response probabilities tend to 0; NA = not available.

an LCM(2) would fit the data. We expected that the first latent class consisted of data records produced by children who employ a strategy that results in the correct solution (i.e., strategies R1, R2, R3, and R4). Hence, the response probabilities are high. We also expected that the second latent class consisted of data records produced by children who employ a strategy that results in guessing (i.e., strategy R0); hence, the response probabilities are approximately equal to .33.

Table 5.5 shows the fit of the hypothesized LCM for each five-item subset. For the weight items, the unconstrained LCM(2) fitted well (Table 5.5, first row, BIC = 970). Class 1 consisted of almost all cases, except the 12 item-score patterns that Jansen and Van der Maas (1997) deleted from their LCM analysis; these 12 item-score patterns constituted Class 2. The model suggests that for the weight items, 2.8% of the sample used the random strategy, and 97.2% used one of the strategies Siegler proposed, resulting in response probabilities greater than .93. When constraining response probabilities to be equal, the absolute fit of the model deteriorated (Table 5.5, second row), but the relative fit (BIC = 910) improved.

For the distance items, the unconstrained LCM(3) showed good fit (Table 5.5, third row) and was also the best fitting unconstrained model in terms of relative fit (BIC = 1,947). The response probabilities of Class 1 were higher than .96, and those of Class 3 were lower than .61 for all items, both results as expected, whereas the response probabilities of Class 2 ranged from .573 to .753 showing more variation than the expected random

TABLE 5.5

Results from Latent Class Analyses

						LC 1		LC 2		LC 3	
Items	W	M	G^2	df	p	P_w	$P_{j|w}$	P_w	$P_{j|w}$	P_w	$P_{j|w}$
W	2	U	26.4	20	.153	.972	.938–.982	.028	.299–.579		
	2	C	43.8	28	.029	.978	.963	.022	.418		
D	3	U	14.6	14	.403	.571	.961–.990	.205	.573–.753	.224	.005–.061
	3	C	30.3	26	.256	.565	.975	.211	.672	.224	.042
CB	3	U	35.3	14	.001	.412	.821–.917	.332	.289–.552	.247	.000–.029
	3	C	69.4	26	.000	.392	.882	.357	.374	.251	.006
CW	2	U	78.0	20	.000	.403	.897–.973	.597	.139–.692		
	2	C	466	28	.000	.333	.972	.667	.451		
CD	3	U	11.4	14	.652	.055	.515–1.00	.228	.145–.588	.717	.000–.338
	3	C	296	26	.000	.162	.629	.265	.116	.573	.114

Note: W = number of latent classes; M = model: Either unconstrained (U) or constrained (C); G^2 = likelihood ratio statistic; df = degrees of freedom; p = p-value; LC = Latent Class (see note Table 5.4); $P_w = P(\theta = w)$, class probability; $P_{j|w} = P(X_j = 1|\theta = w)$, probability of a correct response. For unconstrained models the range of response probabilities is provided; W = weight items; D = distance items; CB = conflict-balance items; CW = conflict-weight items; CD = conflict-distance items.

probabilities. For the weight items, Class 2 was hypothesized to contain the data records of children who guessed randomly. However, for the weight items, Class 2 contained 3% of the data records, and for the distance items, Class 2 contained 21% of the data records. Hence, Class 2 does not have the same interpretation for weight items as for distance items. Constraining response probabilities to be equal produced a well-fitting LCM (Table 5.5, fourth row; BIC = 1,889) that had approximately the same interpretation as the unconstrained LCM.

For the conflict-balance items, for $W = 1, ..., 5$, the five corresponding LCMs showed poor fit. The unconstrained LCM(3) (Table 5.5, fifth row) had the best relative fit (BIC = 2,483). Constraining the probabilities deteriorated the absolute fit (Table 5.5, sixth row) but produced a lower BIC value (BIC = 2,443). Both models supported the hypothesized strategies and suggested that for the conflict-balance items approximately 40% of the students used strategy R4 (Class 1), 35% strategies R0 or R3 (Class 2), and 25% strategies R1 or R2 (Class 3). However, due to the poor absolute fit of the models, these results should be interpreted with caution.

For the conflict-weight items, an LCM(2) was expected (Table 5.4), but both the unconstrained and the constrained LCM(2) showed poor fit (Table 5.5, seventh and eighth rows). However, among all constrained and unconstrained LCMs with $W = 1, ..., 5$ classes we estimated, the unconstrained LCM(2) had the best relative fit (BIC = 2,585). The model suggests that for the conflict-weight items, 33% of the students used strategies R1, R2, or R4 (Class 1), and 67% used strategies R0 and R3 (Class 2). Because the model did not fit well, the results should be interpreted with caution.

For the conflict-distance items, the unconstrained LCM(3) fitted well (Table 5.5, ninth row; BIC = 2,068) and the constrained LCM(3) showed considerable misfit (Table 5.5, tenth row, BIC = 2,278). The misfit of the constrained LCM(3) was mainly due to the large variance in $\hat{P}_{j|w}$ values across items within a latent class (see Table 5.5, ninth row). Due to this heterogeneity, the classes could not be interpreted in terms of Siegler's strategies. Also, because more than 70% of the students belonged to Class 3 and thus had very low probabilities of having items 1, 2, 3, and 4 correct and a probability of .33 of having item 5 correct, the conflict-distance items were more difficult than the other item types.

To summarize the results, our findings supported the hypothesized structure only for the weight and the distance items. Compared to the weight items, more children responded randomly to the distance items. For the conflict-balance items and the conflict-weight items, the LCMs supported the hypothesized structure but models fitted poorly. For the conflict-distance items, the results failed to support the hypothesized structure. The great variation in item difficulty that these balance problems showed may have complicated the response structure and provides a topic for further investigation. We only found limited support for a problem-solving strategy resulting in random responding.

Discussion

Applications of the LCM are found primarily in the social sciences, in particular, sociology, but also in genetics (Nyholt et al., 2004), health (Larsen, Pedersen, Friis, Glümer, & Lasgaard, 2017), and literature science (Van Rees, Vermunt, & Verboord, 1999). Like principal component analysis and factor analysis in psychology, researchers in other areas use the LCM as an exploratory method to find the best description of their data, balancing a preferably small number of latent classes with the interpretability of each of the classes based on the conditional item-response probabilities typical of each of the latent classes. This strategy of selecting the best fitting model—model selection, for short—is found in many research areas and seems to outnumber the applications in which one actually tests models and, by means of the models, hypotheses about reality. For the construction of measurement instruments, we prefer the latter hypothesis-testing approach, aimed at constructing tests based on theory rather than the fit of one of many models to the data. Instrument construction has seen a few favorable examples that might pave the way for future applications.

As a bridge to CDMs, we notice that the Guttman model's (Guttman, 1944, 1950) assumption that people who succeed at an item cannot fail an easier item bothered many researchers already since the 1940s (for an overview, see Mokken, 1971, Chap. 2), because numerous real-data sets repeatedly showed people failing items while succeeding at items that were more difficult. The Proctor (1970) model we discussed was only one of the many answers to the deterministic Guttman model that tried to incorporate uncertainty into the response process to items. One main direction was IRT, where perhaps Mokken's (Mokken, 1971) nonparametric approach constituted the most obvious attempt; see Chapter 3. Prior to formulating IRT models as probabilistic improvements of the Guttman scalogram, Mokken (1971, Chap. 2) extensively discussed the large collection of reproducibility coefficients that had been proposed thus far to quantify the degree to which a data set diverted from an ideal data set consistent with the Guttman scalogram. His item-pair scalability coefficient H_{jk}, item scalability coefficient H_j, and J-item scalability coefficient H, which we discussed in Chapter 3, were adaptations of a coefficient of reproducibility that Loevinger (1948) previously proposed, in a modernized appearance thought to be useful as quality indices in the context of probabilistic IRT. Parametric IRT less frequently linked its models to its deterministic predecessor, but several authors have noted the model's undeniable heritage for IRT (e.g., Andrich, 1988b, pp. 39–41; Fischer, 1974, pp. 137–144; Hulin, Drasgow, & Parsons, 1983, p. 16; Millsap, 2011).

The other main direction in which researchers tackled the Guttman model's inconsistency with real data was the LCM. Goodman (1975) proposed an intrinsically unscalable latent class. Proctor (1970) discussed an LCM that had one parameter to account for item scores that were unexpected based on the particular latent class. Dayton and Macready (1976) discussed an LCM that allowed two kinds of errors to occur at particular rates. One error type was an incorrect answer when the latent class predicted a correct answer. This error was called an omission, meaning the person forgot the correct answer or the correct answer otherwise had slipped the person's attention. The other error type was a correct answer when the latent class predicted an incorrect answer. This error type was called an intrusion, typically occurring when the person guessed for the right answer and guessed correctly. Omission and intrusion parameters were constant across items, thus defining a model less restrictive than Proctor's, but with just two parameters, it was still highly restrictive. The authors discussed a third model, called the item-error model, which had a parameter equal for intrusion and omission but varying across items, thus constituting a compromise between the models of Proctor and Dayton and Macready. Rindskopf (1983) discussed a unified latent class approach that included the older models as well as models not considered thus far. Haertel (1989) introduced a more flexible set of models, which he called *binary skills models* and which we consider the start of CDMs, even though the breakthrough of interest in CDMs only came in the new millennium. We will get back to Haertel's work in the next subsection. Nichols, Chipman, and Brennan (1995) provided a relevant collection of chapters on psychometric models for cognitive processes studied thus far.

Cognitive Diagnostic Model

An Example: Identifying Patients' Disorder Profiles Using the MCMI-III

We start this section discussing an application of a CDM to clinical data, thus defying the C in CDM and showing that CDMs are also applied in other research areas than the cognitive or the educational domains where they are usually applied. The purpose of the CDM application (de la Torre et al., 2018) that we discuss here is to identify patients' disorder profiles based on their responses to 44 items measuring anxiety, somatoform disorder, thought disorder, and major depression. The items were included in a longer, much-used clinical inventory, which is the Millon Clinical Multiaxial Inventory-III (MCMI-III; Millon, Millon, Davis, & Grossman, 2009). For each item, based on available theory, it was determined a priori which subset of disorders the item identifies, and with four disorders, one can readily check that there are $2^4 = 16$ possible subsets, including absence of all disorders and presence of all disorders. Three examples are the following. Item 1 from the MCMI-III refers to a loss of physical strength, and someone endorsing it is assumed to suffer from the second disorder, which is somatoform, and the fourth disorder, which is major depression. Item 12 (numbered Item 76 in the MCMI-III) refers to suffering from obsessive thoughts, driven by the first disorder, which is anxiety, and the third disorder, which is thought disorder. Finally, Item 37 (numbered Item 150 in the MCMI-III) refers to feeling depressed as a new day begins, which refers only to the fourth disorder, which is major depression. The Q-matrix contains this information and serves as the hypothesis one tests with respect to the underlying theory of the psychological property that the test or questionnaire measures. Elements of the Q-matrix signify whether the symptom (i.e., item; row) refers to the disorder (columns), and for the three example items, the rows are (0101), (1010), and (0001), respectively.

Based on her responses to the 44 items, and incorporating the relations between the symptoms (items) and the disorders, for each patient one estimates her typical syndrome of disorders as a vector with 1s for disorders the patient possesses and 0s for the disorders she does not possess. Avoiding formal notation at this stage of the discussion, one first estimates a model that specifies how the probability of an item endorsement is related to main and possibly interaction effects of the disorders. If the model fits the data well, one can use the estimated model parameters for the main and interaction effects to estimate posterior probabilities that a patient possesses a particular disorder based on her pattern of scores on the 44 items that represent symptoms. The posterior probabilities can be rounded to 0 and 1 to obtain one of the 16 patterns reflecting absence and presence of disorders in a particular patient. Later, we will re-analyze the data that de la Torre et al. (2018) used to illustrate the use of CDMs.

Introduction

CDMs allow the assessment of the possession or the absence of traits to endorse items in an inventory, the mastery or the non-mastery of skills needed to solve cognitive problems, and the presence or the absence of clinical disorders in patients. We will generically use the term attribute to summarize traits, skills, disorders, and other person influences on responses to items. CDMs are applied most frequently to cognitive items in an educational context, but applications are also known in the evaluation and diagnosis of pathological gambling (Templin & Henson, 2006), the understanding and scoring of situational judgment tests (Sorrel et al., 2016), and the assessment of anxiety, somatoform disorder, thought disorder, and major depression (de la Torre et al., 2018). Several models are available that have in common that they assume that the response to an item depends on the availability of a set of latent attributes. For different items, different albeit partly overlapping subsets of latent attributes may be required. The most important difference between the models we discuss next is how they assume that attributes combine in producing a response to an item that requires the availability of a particular subset of latent attributes. Typically, this combination of attributes takes place at an unobservable, hence, latent level and produces a latent item response. This latent item response is the input for the probability of a particular observable item score and, depending on the latent item responses, CDMs model these response probabilities. Each person is characterized by a pattern of latent attributes, and each pattern defines a pattern of J item-score probabilities. These item-score probabilities define the latent classes, and different CDMs differentially restrict the set of latent classes.

The way the latent attributes combine to produce a latent item response can vary quite a lot and needs some careful explanation, because terminologies used can be confusing. Two distinctions are common. They are the distinction between non-compensatory and compensatory models and between conjunctive and disjunctive models. It is important to notice that the distinction between conjunctive and disjunctive models does not coincide with the distinction between non-compensatory and compensatory models and that different authors discuss the distinctions differently. A scan of the literature shows that the use of the terminology for CDMs is not as unequivocal as it is for, for example, multidimensional IRT models (Chapter 4), where a high standing on one latent variable can compensate for a low standing on another latent variable to produce a particular response probability on the item. With CDMs, for most models the binary character of the latent attributes—one possesses the attribute or misses it completely—renders such compensation impossible. We discuss the distinction that Henson, Templin, and Willse (2009) made between non-compensatory and compensatory models and, within the former class, between conjunctive and disjunctive models to give an impression of the kind of CDM models available. We discuss their terminology but notice other discussions are available (e.g., DiBello, Roussos, & Stout, 2007; Roussos, Templin, & Henson, 2007).

In non-compensatory models, the relation between a latent attribute and the item score depends on possession or absence of the other latent attributes. For example, we may assume that to solve a particular item, one needs all the latent attributes, and if present, together they produce a latent response in the form of a latent item score, which equals 1. The latent item score can be considered a stopover that is free of error and which thus is not a real item score, and the CDM then uses the latent item score as input to define the probability that leads to an observable item score of 1. This probability is expected to be high. However, if one attribute is missing, the latent item score equals 0, and at the model level, the success probability drops dramatically. The reason why item-score probabilities are not 1 or 0 is due to the probabilistic process or the error process that the CDMs define. This is an example of a conjunctive response process in which all latent attributes typically combine to produce a latent response equal to 1, and it is important to notice that the definition of the response process takes place at the latent level.

The logical version of the conjunctive response process is the following. Logically, let A and B denote two states that can be either true or false, then the logical conjunction of the two states, denoted $A \wedge B$, is true if and only if both A is true and B is true. In the example, we used this logic at the latent level. An example of a conjunctive process occurs when one solves an

arithmetic problem in a series of non-observable or latent steps that all have
to be correct to arrive at a correct answer. A concrete example is the arithmetic problem, $\sqrt{(72/9)-4} = \cdots$, which necessitates the execution of the series
of steps $72/9 = \cdots$, $8-4 = \cdots$, and $\sqrt{4} = \cdots$. Succeeding on each of the three steps
produces a latent item score equal to 1, whereas failing at least one of the
steps produces a latent item score equal to 0. A latent item score equal to 1
may slip into an observed item score of 0 due to, for example, a lapse of concentration, and a latent item score of 0 may become an observable item score
of 1 when, for example, the person does not trust her answer and guesses.

Another example of a non-compensatory process occurs when the possession of different subsets of attributes may produce a latent item score equal
to 1 but the absence of all rather than one or a subset of the attributes produces a latent item score equal to 0. Let us consider a case with only two
latent attributes and reduce it to its logical structure. Logically, the disjunction or the separation of the two states, denoted $A \vee B$, is true if and only if
A is true, or B is true, or both A is true and B is true; if both A is false and
B is false, the logical disjunction is false. Now, possessing one of the two
attributes is sufficient for solving the problem, possessing both is also sufficient, but possessing none produces failure. To see that the response process
is non-compensatory, hence, the relation between a latent attribute and the
item score depends on possession or absence of the other latent attributes,
one has to recognize that possessing more than one of the required attributes does not change the latent item score, nor does it affect the probability
of a particular item score. Thus, the relation between any required attribute
and the item is absent when the person possesses at least one of the other
required attributes. An example is the availability of different strategies
to solve a particular problem when it does not matter which strategy one
chooses. Let us consider the case where different strategies may be used to
solve the same type of arithmetic problem. For example, to find a common
denominator for two fractions, a/b and c/d, where $b \neq d$, one may use cross
multiplication, $(a \cdot d)/(b \cdot d) = (b \cdot c)/(b \cdot d)$, resulting in common denominators
and numerators that can be compared directly. This strategy, as we will call
it, produces latent item scores equal to 1, but between latency and reality,
many disturbing processes may intervene, such as arithmetic errors, accidentally misplacing b and d in the numerators, and concentration failure,
producing observable errors, hence, 0 item scores. Another, fictitious and less
efficient strategy, is to take multiples of the largest of b and d (e.g., $b > d$), say,
$2b$, $3b$, and so on, and check in each step whether the multiple can be divided
by d. If this is true for tb, so that $(tb/d) = e$, which is an integer, then multiply a
and b by t, and c and d by e, so that $(t \cdot a)/(t \cdot b)$ and $(e \cdot c)/(e \cdot d)$, where $t \cdot b = e \cdot d$.
For example, take 2/3 and 5/7, then $t = 7$ and $e = 3$. Both strategies are logically
correct, hence, interchangeable, leading to latent item scores equal to 1, but
likely produce different observable item scores due to different intervening
processes. Interestingly, the strategies used to solve balance-scale problems

(Hofman et al., 2015; Jansen & Van der Maas, 1997; Siegler, 1981) are inadequate examples here, because several of these strategies are logically incorrect, thus producing different subsets of correct and incorrect answers at the latent level. Hence, the strategies for solving balance-scale problems cannot replace one another in this example.

Henson et al. (2009) argued that a model is compensatory if the conditional relation between any attribute and an item response does not depend on possessing or failing the other attributes relevant to the item. Consider just two attributes relevant for a particular item. In a compensatory model, the second-order interaction of two attributes and an item response is absent. Thus, failure of a particular attribute can be compensated by the presence of another attribute. In a non-compensatory model, one also has to consider the second and possibly higher-order interactions. For example, in the model that requires presence of all relevant attributes for a particular item, the relation between one particular attribute and the item response depends on the presence or the absence of all other attributes. If they are present, the latent item response equals 1, and if they are absent, the latent item response equals 0. The complexity of the terminologies is reflected by the fact that a disjunctive model can also represent the extreme case of a compensatory model. One may think of the example in which we considered alternative, logically correct strategies for finding common denominators for fractions. If one masters one, one does not need to master the other, and item-response probabilities may hardly be affected.

Thus far, we defined models following rather strict logic, but real response processes may have different features. For example, for the moment ignoring CDMs for binary latent attributes and allowing continuous latent attributes in our discussion, compensatory models may assume that interaction between different attributes occurs when the level of one attribute is high enough to compensate for a low level of another attribute or when the presence of one attribute compensates for the absence of another attribute to produce a success with positive probability. Different combinations of (levels of) attributes produce different item-score probabilities. In what follows, we try to characterize the CDMs we discuss but acknowledge different characterizations may be considered, if deemed desirable at all.

Models

Some notation needed for the CDMs is the following. Let $\mathbf{X}_i = (X_{i1}, \ldots, X_{iJ})$ be the vector with the binary (incorrect/correct) scores of person i on the J items in the test. To be consistent with many publications in cognitive diagnostic modeling, we will no longer use index k to enumerate items but in this section rather to enumerate latent attributes. Let $\boldsymbol{\alpha}_i = (\alpha_{i1}, \ldots, \alpha_{iK})$ be the binary latent attribute vector, where $\alpha_{ik} = 1$ means that person i possesses attribute k, and $\alpha_{ik} = 0$ that the person does not possess the attribute. Another building block of CDMs is the two-dimensional Q-matrix, which contains for each item (rows) and attribute (columns) elements q_{jk} indicating whether item j requires attribute k for its solution ($q_{jk}=1$) or not ($q_{jk}=0$). Different CDMs define different ways in which vector $\boldsymbol{\alpha}_i$ and matrix $\mathbf{Q}$ can be combined to produce a latent item-score vector, $\boldsymbol{\xi}_i = (\xi_{i1},\ldots,\xi_{iJ})$, that may be considered ideal and has to be compared with the real-data item-score vector, $\mathbf{X}_i$, to determine whether the model fits the data.

Several authors have defined different CDMs, and the reader may sometimes find it difficult to tell the models apart when they look alike at first sight. The most important differences from a psychological perspective are with the way in which the models assume attributes combine to produce a response, and we try to clarify this for the reader. However, models can also differ in the way they define particular parameters. For example, some models assume that particular parameters are typical of items whereas other models assume that the same parameters are typical of attributes. This requires some leniency of the reader. The diversity of CDMs is comparable with the wealth of models found, for example, in IRT (Chapter 4) and demonstrates the field's flexibility, allowing the researcher some choice when modeling her data.

Deterministic Input, Noisy "AND" Gate Model. The deterministic input, noisy "AND" gate model (DINA model: Junker & Sijtsma, 2001b; also Haertel, 1989; Macready & Dayton, 1977) is a non-compensatory, conjunctive model that defines the binary latent response variable, ξ_{ij}, to indicate whether person i possesses all the attributes needed for solving item j ($\xi_{ij}=1$) or not ($\xi_{ij}=0$), and equals

$$\xi_{ij} = \prod_{k=1}^{K} \alpha_{ik}^{q_{jk}}. \tag{5.44}$$

Equation (5.44) shows that if item j requires an attribute k (i.e., $q_{jk}=1$) that person i lacks (i.e., $\alpha_{ik}=0$), then $\alpha_{ik}^{q_{jk}} = 0^1 = 0$, yielding $\xi_{ij}=0$; otherwise, $\alpha_{ik}^{q_{jk}} = 1$: that is, item j requires attribute k which person i possesses ($1^1=1$), and item j does not require attribute k irrespective of whether person i possesses attribute k ($0^0=1$ and $1^0=1$). Thus, only if all K terms $\alpha_{ik}^{q_{jk}}$ in Equation (5.44) equal 1 will $\xi_{ij}=1$. The discrete IRFs relate the ideal latent item-score vector to the fallible real-data item-score vector by allowing masters ($\xi_{ij}=1$) to fail an item accidentally, called slipping and quantified by the slipping parameter (also, Dayton & Macready, 1980),

$$s_j = P\left(X_{ij} = 0 \mid \xi_{ij} = 1\right), \tag{5.45}$$

and non-masters ($\xi_{ij}=0$) to succeed accidentally, quantified by the guessing parameter (also, Dayton & Macready, 1980),

$$g_j = P\left(X_{ij} = 1 \mid \xi_{ij} = 0\right). \tag{5.46}$$

Acknowledging that

$$1 - s_j = P\left(X_{ij} = 1 \mid \xi_{ij} = 1\right) \tag{5.47}$$

stands for the probability that a master has the item correct, and using the definitions in Equations (5.44), (5.45), and (5.46), one defines the IRF as

$$P\left(X_{ij} = 1 \mid \alpha_i, s_j, g_j\right) = \left(1 - s_j\right)^{\xi_{ij}} g_j^{\,1 - \xi_{ij}}. \tag{5.48}$$

Equation (5.48) shows that for non-masters ($\xi_{ij}=0$), we have $P(X_{ij} = 1 \mid \alpha_i, s_j, g_j)$ $= g_j$ and for masters ($\xi_{ij}=1$), we have $P(X_{ij} = 1 \mid \alpha_i, s_j, g_j) = 1 - s_j$. Hence, the class of non-masters has a probability at the guessing level to solve the item correctly, and the class of masters has a probability reflecting non-slipping or, indeed, mastery. A feature of the IRF in Equation (5.48) is that it is coordinate-wise monotone in α_i if and only if $1 - s_j > g_j$ (Junker & Sijtsma, 2001b). Thus, the probability of obtaining a 1 score due to possessing the attributes item j requires (and not slipping) should be greater than the probability of a 1 score by guessing. One can verify this monotonicity property by checking that changing 0s in α_i into 1s can change $\xi_{ij}=0$ into $\xi_{ij}=1$ but not vice versa; hence, by adding attributes, a non-master can become a master, but this makes sense only if scoring $X_{ij}=1$ becomes more likely.

 The DINA model and other CDMs discussed later can be seen as restricted LCMs. We illustrate this point for the DINA model using an artificial example. Consider four latent attributes ($K=4$) and two items. We enumerate the different vectors of latent-attribute scores as α_w, $w=1, \ldots, 2^K$; that is, for $K=4$, we have 16 latent classes, each class defined by a unique constellation of presence or absence of latent attributes: $\alpha_1 = (0000)$, $\alpha_2 = (1000)$, $\alpha_3 = (0100)$, ..., $\alpha_6 = (1100)$, ..., $\alpha_{14} = (1011)$, ..., $\alpha_{16} = (1111)$. Item j requires the first two latent attributes for its solution, so that $\mathbf{q}_{jk} = (1100)$, and item l requires the first and the fourth latent attributes, so that $\mathbf{q}_{lk} = (1001)$. Someone belonging to the 14th latent class possesses the first, third, and fourth attribute, and for item j we compare α_{14} (replacing the person index with the class index) with $\mathbf{q}_{jk}$ and find that

$$\xi_{14j} = \prod_{k=1}^{4} \alpha_{14k}^{q_{jk}} = 1^1 \times 0^1 \times 1^0 \times 1^0 = 0,$$

and

$$P\left(X_{14j} = 1 \mid \alpha_{14}, s_j, g_j\right) = g_j.$$

Similarly, the comparison of α_{14} with $\mathbf{q}_{lk}$ yields

$$\xi_{14l} = \prod_{k=1}^{4} \alpha_{14k}^{q_{lk}} = 1^1 \times 0^0 \times 1^0 \times 1^1 = 1,$$

and

$$P\left(X_{14l}=1\,|\,\alpha_{14},s_l,g_l\right)=1-s_l.$$

Given a total of J items in the test, one can complete this exercise for the other $J-2$ items and then for the other 15 latent classes, J items each. Using these results, and using Equation (5.4) for the general LCM, we substitute $P(\theta=w)$ with $P(\alpha_w)$ and $P(X_j=x_j\,|\,\theta=w)$ with $P(X_{ij}=1\,|\,\alpha_i,s_j,g_j)=(1-s_j)^{\xi_{ij}}\,g_j^{\,1-\xi_{ij}}$ [Equation (5.48)], so that the LCM formulation of the DINA model is

$$P\left(\mathbf{X}=\mathbf{x}\right)=\sum_{w=1}^{16}P\left(\alpha_w\right)\prod_{j=1}^{J}\left(1-s_j\right)^{\xi_{wj}}\,g_j^{\,1-\xi_{wj}}.$$

For the psychological measurement of individuals, one might assign the individual to the latent class that is the most likely for her; see Equation (5.6) for LCMs. This assignment informs one about the presence and absence of latent attributes for that individual among the array of four latent attributes that together determine in a non-compensatory way the responses to the items in the test. In an educational context, information about absence of vital skills (i.e., $\hat{\alpha}_{wk}=0$) might prove useful for remedial teaching. In a clinical context, presence of particular pathologies (i.e., $\hat{\alpha}_{wk}=1$) might prove useful in the context of treatment of the individual. This information is only valid provided that the DINA model fits the data sufficiently well, a topic that we will discuss later in this chapter.

The DINA model derives its rather complex name from the following model ingredients. Equation (5.44) shows how the latent attribute vector for person i and the attributes that item j requires for its solution combine to produce person i's latent score on item j; specifically, the inputs—the $\alpha_{ik}^{q_{jk}}$s —are binary and combine to produce a binary outcome, ξ_{ij}. The way the attributes combine in the response process is conjunctive; hence, a logical "AND" operation is involved and its outcome is deterministic. The terminology of "AND" gate stems from electronic engineering, in particular, transistors, where an AND gate acts as an electronic switch, and is further found in many different application areas, such as computer science. An AND gate is a logical operation on two inputs to produce an output, and this only happens when both inputs are true. Finally, getting from the latent item score ξ_{ij} to the observable item score X_{ij} is a noisy or probabilistic process, governed by guessing and slipping probabilities.

Haertel (1989) defined a latent difficulty parameter for item j, using indicator or latent item score, ξ_{ij} [Equation (5.44)], and the class probability $P(\theta=w)$, which corresponds with the proportion of people who possess the subset of latent attributes typical of latent class w. As before, $\sum_w P\left(\theta=w\right)=1$. One may notice that each person in the same latent class has the same latent item score, ξ_{ij}; hence, to be consistent with Haertel (1989), we rewrite $\xi_{ij}=\xi_{wj}$. Further, we use shorthand $P(\theta=w)=P_w$. Haertel (1989) called the guessing parameter, g_j, the item's *false positive probability*, and the probability of not slipping, $1 - s_j$, the item's *true positive probability*. Then, using these ingredients, one defines the latent difficulty for item j as

$$\delta_j = \sum_{w=1}^{W} \xi_{wj} P_w. \tag{5.49}$$

This is the proportion of people in the whole population who have a latent score on item j equal to 1. Given the definitions discussed so far, one can also define the manifest difficulty p_j of item j as

$$p_j = g_j \left(1 - \delta_j\right) + \left(1 - s_j\right) \delta_j = \delta_j + g_j \left(1 - \delta_j\right) - s_j \delta_j. \tag{5.50}$$

Equation (5.50) shows that the discrepancy between the manifest difficulty p_j and the latent difficulty δ_j depends on two contributions. One contribution is $g_j(1 - \delta_j)$, which represents the contribution from the people who have a latent item score equal to 0 and guess correctly, and the other contribution is $-s_j \delta_j$, which represents the contribution of the people who have a latent item score equal to 1 but slip. The first group are the false positives, and the second group are the false negatives, and both represent errors according to the CDM approach, having a biasing effect on item difficulty.

Reduced Reparametrized Unified Model. The reduced reparametrized unified model (RRUM; Chung & Johnson, 2017; Hartz, 2002; Roussos, DiBello, Stout, Hartz, Henson, & Templin, 2007) resulted from previous, more complex models that suffered from technical problems such as non-identifiability and estimation complexities and thus needed simplification. The RRUM defines guessing and slipping parameters for each combination of item and attribute, that is, g_{jk} and s_{jk}. Thus, the RRUM deviates in this respect from the DINA and other CDMs discussed later, which define these parameters as item parameters, g_j and s_j.
The building blocks of the RRUM are the following. First, we define

$$\pi_j^* = \prod_{k=1}^{K} \left(1 - s_{jk}\right)^{q_{jk}}. \tag{5.51}$$

Equation (5.51) shows that, for item j, one takes the product of the probabilities of not slipping for the attributes that the item requires for its solution, that is, for which $q_{jk} = 1$. This yields the probability of having a score $X_j = 1$ given possession of all the attributes needed for solving the item, that is, $\pi_j^* = P(X_j = 1 \mid \alpha_i, s_j)$. Second, as an initial step in further developing the RRUM, we define

$$r_{jk}^* = \frac{g_{jk}}{1 - s_{jk}}. \tag{5.52}$$

Because of the constraint that $1 - s_{jk} > g_{jk}$ (following the DINA model; Junker & Sijtsma, 2001b), it follows that $0 < r_{jk}^* < 1$. When the guessing probability is greater relative to the non-slipping probability, r_{jk}^* is closer to 1. In the RRUM, r_{jk}^* plays the part of a penalty term, but to play that part effectively, first one has to identify the circumstance in which responding justifies a sensible penalty and then adjust r_{jk}^* to be effective in that circumstance but not in others. For this purpose, third, we consider the following power of r_{jk}^*,

$$R_{ijk}^* = \left(r_{jk}^{*(1-\alpha_{ik})} \right)^{q_{jk}} . \tag{5.53}$$

In Equation (5.53), if item j requires attribute k, then $q_{jk} = 1$, and if in addition person i does not possess attribute k, then $(1 - \alpha_{ik}) = 1$, so that Equation (5.53) equals $R_{ijk}^* = r_{jk}^*$. Hence, when person i does not possess attribute k, which is required for solving item j, r_{jk}^* can be considered a penalty that weighs more heavily when the guessing probability for combination (j, k) is greater relative to the non-slipping probability [Equation (5.52)]. The other three combinations of $(1 - \alpha_{ik})$ and q_{jk}, which are (1, 0), (0, 1), and (0, 0), all produce $R_{ijk}^* = r_{jk}^* = 1$; hence, they do not affect a product in which they are included as we will see next. Together, these ingredients define the RRUM. Given $\mathbf{r}_j^* = (r_{j1}^*, \ldots, r_{jK}^*)$, the RRUM is defined as

$$P\left(X_{ij} = 1 \mid \alpha_i, \pi_j^*, \mathbf{r}_j^* \right) = \pi_j^* \prod_{k=1}^{K} \left(r_{jk}^{*(1-\alpha_{ik})} \right)^{q_{jk}} , \tag{5.54}$$

in which we included the right-hand side of Equation (5.53). Thus, the probability of a correct answer to item j is the product of the probabilities of not slipping for the attributes that the item requires for its solution, also interpreted as the item difficulty, multiplied by the penalties for each of the required attributes the person does not possess. Typically, when the person misses more required attributes, her probability of a correct answer to item j reduces further.

Like the DINA model, the RRUM is a non-compensatory, conjunctive model, because the relation between a latent attribute and the item score depends on possession or absence of the other latent attributes. In particular, the relation between any attribute and the item score is smallest when the person possesses none of the other attributes required for item j.

Deterministic Input, Noisy "OR" Gate Model. We consider the non-compensatory, disjunctive deterministic input, noisy "OR" gate model (DINO model; Templin & Henson, 2006; also, see Maris, 1999) to illustrate a disjunctive process model. The DINO model may be considered the mirror image of the DINA model (Köhn & Chiu, 2016). Templin and Henson (2006) described the DINA model and other models, which are non-compensatory and conjunctive, as suited for educational abilities or skills. For such models, one needs

all the attributes that the item's solution requires to be successful, and failure at one attribute—for example, being able to divide two numbers in a
more-complex arithmetic problem—produces incorrect results. Templin and
Henson considered non-compensatory and conjunctive models, such as the
DINA model, unsuited for personality and clinical disorders (see Van der
Ark, Rossi, & Sijtsma, 2019). As an example, they discussed ten criteria or
latent attributes for the diagnosis of pathological gambling and noticed that
different persons may have different reasons, and sometimes more than one
reason, to provide an affirmative answer to an item like "I am ashamed of
the things I've done to obtain money for gambling". John may be ashamed
because he was involved in illegal acts and Mary because she relied on others to finance her gambling, whereas Peter felt ashamed because of the combination of these two reasons. Thus, persons do not have to satisfy all ten
criteria as the DINA model requires but only a subset, and different persons
can satisfy different subsets of criteria. In particular, following agreed upon
diagnostics, a person is diagnosed as a pathological gambler when she satisfies at least five out of ten of the criteria. Obviously, the DINA model is
unsuited as a measurement model in this situation, because it would require
people to satisfy all the criteria needed to answer item j positively.

The DINO model assumes that the person needs to possess only one attribute, α_{ik}, to produce a latent response equal to 1. The latent response variable
is defined as

$$\eta_{ij} = 1 - \prod_{k=1}^{K} \left(1 - \alpha_{ik}\right)^{q_{jk}} . \tag{5.55}$$

From Equation (5.55) it can be seen that the combination of the item requiring an attribute that the person possesses ($\alpha_{ik} = q_{jk} = 1$), is the only combination that produces $(1 - \alpha_{ik})^{q_{jk}} = 0$ and hence, a product equal to 0 and latent
response, $\eta_{ij} = 1$. Only if a person has none of the attributes the item requires,
that is, $\left(1 - \alpha_{ik}\right)^{q_{jk}} = 1$, all k, does the product in Equation (5.55) equal 1, and
do we have $\eta_{ij} = 0$. Thus, one needs to possess at least one attribute necessary
for item j to produce a latent response $\eta_{ij} = 1$. Latent response η_{ij} defines two
groups, one for which $\eta_{ij} = 1$ that possesses at least one attribute needed to
solve item j, and the other for which $\eta_{ij} = 0$ that possesses none of the attributes needed to solve item j. The IRF of the DINO model equals

$$P\left(X_{ij} = 1 \mid \alpha_i, s_j, g_j\right) = \left(1 - s_j\right)^{\eta_{ij}} g_j^{1 - \eta_{ij}}, \quad 1 - s_j > g_j. \tag{5.56}$$

One may notice that Equation (5.56) is identical to the IRF in the DINA model
[Equation (5.48)] but that the difference resides in how the latent item scores
originate. In both the DINA and the DINO models, the latent item scores
originate in a non-compensatory way, but in the DINA model, the attributes combine conjunctively, and in the DINO model, the attributes combine

disjunctively; hence, a logical "OR" operation is involved. The difference between the models results in different parameter estimates based on the same data.

General Diagnostic Model. We noticed previously that LCMs are mixture models (Von Davier & Rost, 2016). Von Davier (2008, 2010) discussed the general diagnostic model (GDM), which is an LCM and therefore a mixture model. First, we discuss the GDM, and then, we discuss the GDM as a mixture model. The GDM handles both dichotomous and polytomous item scores and binary, ordinal, and continuous latent attributes. Hence, item scores are defined as $x = 0, \ldots, M_j$, $j = 1, \ldots, J$, and M_j can vary across items (see Chapter 3, where we recommended equal score format across items in the context of nonparametric IRT). Von Davier (2008, 2010) defined a logistic link function between the function's argument defining the way in which observable random variables and person and item parameters combine in producing a response to the item, and the probability of the observable response to the item expressed as an item score.

As before, let the Q-matrix be of the order $J \times K$, containing binary indicator values $q_{jk} \in \{0,1\}$. Let $\alpha = (\alpha_1, \ldots, \alpha_N)$ be a vector containing the binary-attribute vectors for the N respondents taking the test. Furthermore, we define real-valued parameters for combinations of items, attributes, and item scores, denoted γ_{jkx} with $j = 1, \ldots, J$, $k = 1, \ldots, K$, and $x = 0, \ldots, M_j$, and real-valued parameters for items and item scores, denoted β_{jx} with $j = 1, \ldots, J$, and $x = 0, \ldots, M_j$. Function $h(q_{jk}, \alpha_k)$ defines how requirements that items make of attributes and attributes that persons have available combine to produce an item score and is specified later. Using these definitions, the GDM is defined as

$$P(X_j = x \mid \alpha) = P_j(\alpha) = \frac{\exp\left[\beta_{jx} + \sum_{k=1}^{K} \gamma_{jkx} h(q_{jk}, \alpha_k)\right]}{1 + \sum_{y=1}^{M_j} \exp\left[\beta_{jy} + \sum_{k=1}^{K} \gamma_{jky} h(q_{jk}, \alpha_k)\right]}. \quad (5.57)$$

Von Davier (2010) discussed choices for γ_{jkx}, in particular

$$\gamma_{jkx} = x\gamma_{jk}, \quad (5.58)$$

and the real-valued function $h(q_{jk}, \alpha_k)$, which he defined as

$$h(q_{jk}, \alpha_k) = q_{jk}\alpha_k. \quad (5.59)$$

Von Davier noticed that these choices reduce Equation (5.57) to the general diagnostic model for partial credit data (Muraki, 1992); see Chapter 4. He also noticed that the model in Equation (5.57) encompasses a large number of models from several origins, such as IRT, LCM, and CDM. Magidson and Vermunt (2001) discussed the two-factor LCM, which resembles the GDM in Equation (5.57), and showed its relation with the traditional LCM formulation in Equation (5.4). We will interpret the model after we have introduced its mixture version, which is called the discrete mixture distribution version of the GDM, acronym MGDM.

In the MGDM, the distribution of the data, $\mathbf{x} = (x_1,\ldots,x_J)$, depends on the attribute vector, $\boldsymbol{\alpha}$, and another unobservable subgrouping variable, denoted $U = u$. The underlying idea is that sometimes one can distinguish different subpopulations that vary with respect to their attribute distribution, the strategies employed to solve the items, or both. The subpopulations are the latent classes. Different subpopulations are distinguished by an unobservable group variable U with scores u_i for each individual, $i = 1,\ldots,N$. Thus, for the situation in which we assume response probabilities to vary with different but unobservable subpopulations, the data from an individual are her item-score vector $\mathbf{x}_i = (x_{i1},\ldots,x_{iJ})$, her latent-attributes vector $\alpha_i = (\alpha_{i1},\ldots,\alpha_{iK})$, and her latent-group membership $U = u_i$. In what follows, person indices are dropped for convenience. Following Von Davier (2010), we first notice that the data $\mathbf{x}$ depend only on the attribute parameters $\boldsymbol{\alpha}$ and the grouping variable $U = u$ but not on any other random variable one might denote $Z = z$; thus, local independence means that

$$P(\mathbf{x} \mid \alpha, u, z) = P(\mathbf{x} \mid \alpha, u) = \prod_{j=1}^{J} P(X_j = x_j \mid \alpha, u). \tag{5.60}$$

Again, one needs to model the marginal probability of the data,

$$P(\mathbf{x}) = \sum_u P(U = u) P(\mathbf{x} \mid u), \tag{5.61}$$

in which probability $P(U = u)$ is the class probability. To obtain the second probability, $P(\mathbf{x}|u)$, one has to sum across the different attribute patterns, so that

$$P(\mathbf{x} \mid u) = \sum_{\alpha} P(\alpha \mid u) P(\mathbf{x} \mid \alpha, u). \tag{5.62}$$

Finally, the group-dependent conditional item-score probabilities, $P(\mathbf{x} \mid \alpha, u)$, are modeled by means of Equation (5.57), using the restrictions

$$\gamma_{jkx} = x_j \gamma_{jku} \tag{5.63}$$

and the restriction on $h(q_{jk}, \alpha_k)$ in Equation (5.59), resulting in

$$P(\mathbf{x} \mid \alpha, u) = \prod_{j=1}^{J} P(X_j = x_j \mid \alpha, u)$$

$$= \prod_{j=1}^{J} \frac{\exp\left[\beta_{jxu} + \sum_{k=1}^{K} x_j \gamma_{jku} q_{jk} \alpha_k\right]}{1 + \sum_{y=1}^{M_j} \exp\left[\beta_{jyu} + \sum_{k=1}^{K} y \gamma_{jku} q_{jk} \alpha_k\right]}. \tag{5.64}$$

In Equation (5.64), parameter β_{jxu} stands for the class-specific item location or item difficulty and parameter γ_{jku} for the slope or discrimination parameter that links attribute α_k to item j in group u. It may be noted that this parameter only plays a role in the response probability when the item score $x > 0$ and $q_{jk} = \alpha_k = 1$, thus, when attribute k is required to solve item j ($q_{jk} = 1$) and the person possesses this attribute ($\alpha_k = 1$); otherwise, in Equation (5.64), the product $x_j\gamma_{jku}q_{jk}\alpha_k = 0$. Finally, using the results in Equations (5.62) and (5.64), when completely detailed, the model in Equation (5.61) equals the MGDM, that is,

$$P(\mathbf{x}) = \sum_u P(U = u)\sum_\alpha P(\alpha \mid u)\prod_{j=1}^{J} \frac{\exp\left[\beta_{jxu} + \sum_{k=1}^{K} x_j\gamma_{jku}q_{jk}\alpha_k\right]}{1 + \sum_{y=1}^{M_j}\exp\left[\beta_{jyu} + \sum_{k=1}^{K} y\,\gamma_{jku}q_{jk}\alpha_k\right]}. \quad (5.65)$$

Von Davier (2010) also discussed a hierarchical GDM with a mixture structure for the situation in which the data are nested within classrooms, schools, etc. This can be done by introducing an observable nesting variable to the model, indicating, for example, that persons are in the same or different schools. We will not discuss the hierarchical model any further, not because it is not of interest but rather because we limit our attention to the conceptual perspective on CDMs and the meaning these models have for the measurement of psychological attributes. This means that we ignore model variations and extensions, such as ones that facilitate observable subgrouping. We refer the interested reader to relevant sources, in the case of hierarchical models Von Davier and Carstensen (2007) and Vermunt (2003).

The GDM and the MGDM as we discussed them are compensatory models, because each attribute required for item j contributes to the item-score probability in Equations (5.57) and (5.65), respectively, and does this independent of the possession or the absence of the other attributes required for item j. Different model versions are possible that render models non-compensatory, but we refrain from discussing those here. Returning to the compensatory models: For example, if the person possesses none of the attributes needed for item j, based on Equation (5.63), for item j all attribute contributions equal $x_j\gamma_{jku}q_{jk}\alpha_k = 0$, and

$$P\left(X_j = x_j \mid \alpha, u\right) = \frac{\exp\left[\beta_{jxu} + \sum_{k=1}^{K} x_j\gamma_{jku}q_{jk}\alpha_k\right]}{1 + \sum_{y=1}^{M_j}\exp\left[\beta_{jyu} + \sum_{k=1}^{K} y\,\gamma_{jku}q_{jk}\alpha_k\right]} = \frac{\exp\left[\beta_{jxu}\right]}{1 + \sum_{y=1}^{M_j}\exp\left[\beta_{jyu}\right]}. \quad (5.66)$$

Equation (5.66) represents the lowest item-score probability, and each required attribute the person possesses increases the exponent in numerator and denominator by an amount $x_j\gamma_{jku}$ and thus increases the item-response probability, and does this irrespective of the possession or the absence of the other required attributes.

Generalized-DINA or G-DINA Model. De la Torre (2011; de la Torre & Minchen, 2019; Lee & Luna-Bazaldua, 2019) discussed a flexible framework, which he called the Generalized-DINA or G-DINA model, that encompasses the DINA model, the DINO model, and the GDM as special cases. As with other CDMs, the G-DINA model defines parameters for each separate *item*, in particular, slipping parameter s_j and guessing parameter g_j.

As the name suggests, the G-DINA model generalizes the DINA model. The DINA model assumes that one needs to possess all required attributes to give the correct answer to the item and that all attribute sets in which the person failed at least one attribute have equal probability; see the computational example we discussed for the DINA model. Thus, the DINA model classifies the attribute-score vectors containing 1s and 0s in two groups. The G-DINA model relaxes the demanding assumption that does not distinguish failing just one attribute from failing all. This is done as follows.

First, we define the number of attributes needed to solve item j correctly as $K_j^* = \sum_{k=1}^K q_{jk}$; this is the count of the number of indicators $q_{jk}=1$ in row j of the Q-matrix. One may notice that $K_j^* \leq K$. For example, for the set of J items, K attributes may be needed in total, but for item j, only K_j^* are relevant. One may also notice that for different items j and l, possibly $K_j^* \neq K_l^*$. In the example with respect to clinical disorders based on the MCMI-III, $K=4$ (Table 5.6), and for the items 1 and 12, $K_1^* = K_{12}^* = 2$ whereas for item 37, $K_{37}^* = 1$. It may also be noted that K_1^* counts scores equal to 1 for latent attributes α_2 (somatoform disorder) and α_4 (major depression), and K_{12}^* counts scores equal to 1 for α_1 (anxiety) and α_3 (thought disorder). Second, indexing different attribute vectors by m, we consider the reduced attribute vector α_{mj}^* that only contains the K_j^* parameters for the attributes needed to solve item j. Here, an attribute vector characterizes a particular subset of attributes but not the latent attributes that individual i possesses; thus, subscript m does not index individuals but particular vectors. For the MCMI-III example, endorsing item 1 means one suffers from the second disorder, which is somatoform, and the fourth disorder, which is major depression, so that $K_1^* = 2$. Rather than using vector $\alpha_{m1} = (0, \alpha_{m12}, 0, \alpha_{m14})$, for item 1, we define $\alpha_{m1}^* = (\alpha_{m11}^{(2)}, \alpha_{m12}^{(4)})$, with superscripts indicating the original latent attributes indexed "2" and "4" with which the "1" and "2" re-indexed latent attributes correspond. In general, for latent-attribute vector m and item j, the K_j^* enumerated latent attributes are characterized by $q_{mjk} = 1$, meaning they are required for item j. We replace index $k = 1, \ldots, K$ by index $k^* = 1, \ldots, K_j^*$, so that $\alpha_{m1k^*}^{(k)} = \alpha_{m1k}$, with $k^* \leq k$. Third, we assume that K_j^* latent attributes define $2^{K_j^*}$ different latent groups. For larger K (larger than, for example, $K=4$), $2^{K_j^*}$ can represent quite a reduction of the 2^K latent groups defined based on a vector containing all K attribute parameters, that is, often $2^{K_j^*} \ll 2^K$.

We also discuss the idea that two different vectors of the same size can be ordered, such that one is coordinate-wise ordered relative to the other. That is, consider α_{mj}^* and $\alpha_{m'j}^*$; then, the ordering $\alpha_{mj}^* \prec \alpha_{m'j}^*$ implies that at least

TABLE 5.6

Q-Matrix of 44 Items by Four Clinical Syndromes (MCMI-III)

Item	Syndrome A	H	SS	CC	Item	Syndrome A	H	SS	CC	Item	Syndrome A	H	SS	CC	Item	Syndrome A	H	SS	CC
1	0	1	0	1	12	1	0	1	0	23	1	0	0	0	34	1	0	0	0
2	0	1	0	1	13	0	0	1	0	24	1	0	0	0	35	0	1	1	1
3	0	1	0	0	14	0	0	1	0	25	0	1	0	1	36	1	0	0	1
4	0	0	1	0	15	1	1	0	1	26	0	0	1	0	37	0	0	0	1
5	0	0	1	1	16	1	1	0	0	27	1	0	0	0	38	0	0	1	1
6	0	1	0	0	17	1	0	1	0	28	0	0	0	1	39	0	0	0	1
7	1	0	0	0	18	0	0	1	0	29	0	1	0	1	40	0	1	0	0
8	0	0	0	1	19	0	0	1	0	30	0	0	1	0	41	1	0	1	0
9	0	1	0	1	20	0	0	1	0	31	1	0	0	0	42	0	0	0	0
10	0	0	1	0	21	0	0	0	1	32	0	0	1	1	43	1	0	0	0
11	1	0	0	0	22	0	1	0	1	33	1	1	0	0	44	0	0	0	1

Note: Item = item number, A = anxiety, H = somatoform disorder, SS = thought disorder, CC = major depression. The item numbers do not correspond to the original item numbers in the MCMI-III (de la Torre et al., 2018).

one pair of attribute parameters is strictly ordered, $\alpha_{mjk} < \alpha_{m'jk}$, whereas other pairs are ordered, $\alpha_{mjk} \leq \alpha_{m'jk}$, but no pair is reversely ordered, $\alpha_{mjk} \not> \alpha_{m'jk}$. Formally, this means that in a separate equation, where "attribute pair" refers to the same attribute considered in two different vectors,

$$\alpha_{mj}^{*} \prec \alpha_{m'j}^{*} \Rightarrow \tag{5.67}$$

$$\alpha_{mjk} \leq \alpha_{m'jk}, \quad \text{for at most } K_j^{*} - 1 \text{ attribute pairs,}$$

$$\alpha_{mjk} < \alpha_{m'jk}, \quad \text{for at least 1 attribute pair,}$$

and

$$\alpha_{mjk} \not> \alpha_{m'jk}.$$

Because α_{mj}^{*} is just a condensed representation of α_{mj}, the ordering $\alpha_{mj} \prec \alpha_{m'j}$ implies the ordering $\alpha_{mj}^{*} \preceq \alpha_{m'j}^{*}$, where α_{mj}^{*} and $\alpha_{m'j}^{*}$ are of the same length. In the context of nonparametric IRT, Rosenbaum (1987a) discussed an ordering property for functions of item scores that he called decreasingness in transposition and that is similar to the property shown in Equation (5.67); also, see Sijtsma and Junker (1996) and Ligtvoet et al. (2011). An example of a function decreasing in transposition is the number of attributes that two different groups possess to solve item j, because

$$\alpha_{mj}^{*} \prec \alpha_{m'j}^{*} \Rightarrow \sum_{k=1}^{K_j^{*}} \alpha_{mjk}^{*} < \sum_{k=1}^{K_j^{*}} \alpha_{m'jk}^{*}. \tag{5.68}$$

Equation (5.68) expresses that we consider attribute vectors that one can strictly order in the sense that the smaller vector cannot contain attributes that are not in the larger vector and also contains fewer attributes than the largest vector.

The probability of having item j correct is denoted $P(X_j = 1 \mid \alpha_{mj}^{*}) = P(\alpha_{mj}^{*})$. De la Torre (2011) noted that in many applications it makes sense to restrict the model to $P(\alpha_{mj}^{*}) \leq P(\alpha_{m'j}^{*})$ whenever $\alpha_{mj}^{*} \prec \alpha_{m'j}^{*}$. However, the G-DINA model also allows $P(\alpha_{mj}^{*}) > P(\alpha_{m'j}^{*})$ even if $\alpha_{mj}^{*} \prec \alpha_{m'j}^{*}$. This situation can happen when the subset of attributes that a person possesses draws her consistently toward a particular distractor in a multiple-choice item, implying small success probability, whereas someone lacking all K_j^{*} attributes may guess correctly with a higher success probability. In this case, we have that $0_{mj}^{*} \prec \alpha_{mj}^{*}$ so that $P(0_{mj}^{*}) > P(\alpha_{mj}^{*})$.

The G-DINA model originally used the identity-link function, but de la Torre (2011) also discussed the logit-link and log-link functions. Because of the restriction of the response probability to the interval between 0 and 1, and because of the practice in much of modern psychometrics to use the logit-link, we present the G-DINA model in the logit formulation. We use de la Torre's notation where λs denote regression parameters, and let

$$\text{logit}\left[P\left(X_j = 1 \mid \alpha^*_{mj}\right)\right] = \log\left[\frac{P\left(X_j = 1 \mid \alpha^*_{mj}\right)}{P\left(X_j = 0 \mid \alpha^*_{mj}\right)}\right] = \lambda_{j0} + \sum_{k=1}^{K^*_j} \lambda_{jk}\alpha_{mjk}$$

$$+ \sum_{k'=k+1}^{K^*_j}\sum_{k=1}^{K^*_j-1} \lambda_{jkk'}\alpha_{mjk}\alpha_{mjk'} + \cdots + \lambda_{j12\ldots K^*_j}\prod_{k=1}^{K^*_j}\alpha_{mjk}. \tag{5.69}$$

We abbreviate the sum on the right-hand side of Equation (5.69) by

$$f\left(\lambda, \alpha\right) = \lambda_{j0} + \sum_{k=1}^{K^*_j} \lambda_{jk}\alpha_{mjk} + \sum_{k'=k+1}^{K^*_j}\sum_{k=1}^{K^*_j-1} \lambda_{jkk'}\alpha_{mjk}\alpha_{mjk'}$$

$$+ \cdots + \lambda_{j12\ldots K^*_j}\prod_{k=1}^{K^*_j}\alpha_{mjk}, \tag{5.70}$$

and rewrite the response probability as a logistic function of $f(\lambda, \alpha)$, so that

$$P\left(X_j = 1 \mid \alpha^*_{mj}\right) = P\left(\alpha^*_{mj}\right) = \frac{\exp\left(f\left(\lambda, \alpha\right)\right)}{1 + \exp\left(f\left(\lambda, \alpha\right)\right)}. \tag{5.71}$$

Like a log-linear model (e.g., Heinen, 1993, pp. 34–38; Heinen, 1996, pp. 31–43), the G-DINA model uses an intercept, main effects of all the K^*_j attributes relevant to the item's solution, and all the interaction effects to fully explain the conditional probability of giving a positive answer to item j. In the logit-function in Equation (5.69), intercept λ_{j0} equals the logit of a conditional probability of a positive response when the person does not possess any attributes, but in the logistic model in Equation (5.71), the contribution of the intercept still is constant, in this case to the response probability but through a logit transformation. Using his notation, de la Torre's (2011) identity-link formulation of G-DINA model in which δs denote regression parameters equals

$$P\left(X_j = 1 \mid \alpha^*_{mj}\right) = P\left(\alpha^*_{mj}\right) = \delta_{j0} + \sum_{k=1}^{K^*_j} \delta_{jk}\alpha_{mjk} + \sum_{k'=k+1}^{K^*_j}\sum_{k=1}^{K^*_j-1} \delta_{jkk'}\alpha_{mjk}\alpha_{mjk'}$$

$$+ \cdots + \delta_{j12\ldots K^*_j}\prod_{k=1}^{K^*_j}\alpha_{mjk}. \tag{5.72}$$

Here, one sees that intercept δ_{j0}, which is numerically different from intercept λ_{j0}, equals the response probability when the person does not possess any of the attributes. However, as we noticed, ease of interpretation comes at the expense of the scale problems when a dependent variable in a regression model is restricted between fixed values 0 and 1; hence, our preference for the logit-link.

Next, in the logistic model in Equation (5.69), parameter λ_{jk} stands for the contribution of possessing attribute k (i.e., $\alpha_{mjk} = 1$) compared to not possessing attribute k (i.e., $\alpha_{mjk} = 0$); hence, it represents a main effect. Parameter $\lambda_{jkk'}$ represents the additive change when one possesses both attributes k and k' in addition to possessing these attributes separately; hence, the parameter represents a first-order interaction effect. Finally, $\lambda_{j12...K_j^*}$ represents the change in addition to all other main effects and lower-order interaction effects. The saturated model has $2^{K_j^*}$ parameters λ. Next, we consider how the logistic G-DINA version is related to the GDM (Von Davier, 2010). For a special case when only binary attributes are considered, the GDM is a special case of the G-DINA model.

For the GDM with restrictions [Equation (5.64)], we consider dichotomous items, scored 0 and 1 (i.e., $x = 0,1$), leave out the unobservable subgrouping variable, U, and consider attribute vector α_{mj}^*, so that $q_{jk} = 1$ for the latent attributes in α_{mj}^*; using these simplifications, the GDM with restrictions equals

$$P\left(X_j = 1 \mid \alpha_{mj}^*\right) = \frac{\exp\left[\beta_j + \sum_{k=1}^{K_j^*}\gamma_{jk}\alpha_{jk}\right]}{1+\exp\left[\beta_j + \sum_{k=1}^{K_j^*}\gamma_{jk}\alpha_{jk}\right]}. \tag{5.73}$$

Next, we take the logit formulation of Equation (5.73) and obtain

$$\operatorname{logit}\left[P\left(X_j = 1 \mid \alpha_{mj}^*\right)\right] = \log\left[\frac{P\left(X_j = 1 \mid \alpha_{mj}^*\right)}{P\left(X_j = 0 \mid \alpha_{mj}^*\right)}\right] = \beta_j + \sum_{k=1}^{K_j^*}\gamma_{jk}\alpha_k, \tag{5.74}$$

with

$$\beta_j = \lambda_{j0} \quad \text{and} \quad \gamma_{jk} = \lambda_{jk}; \tag{5.75}$$

see Equation (5.69). Equation (5.74) is identical to the linear logistic model (LLM; Hagenaars, 1990; Maris, 1999). The model equals the logit version of the G-DINA model when all the interaction terms are 0 and thus has $K_j^* +1$ parameters. We have shown that for this special case the G-DINA model encompasses the GDM. Although not completely, this special case at least sheds some light on the relation between the models.

To investigate the relation between the G-DINA model, and the DINA model (Junker & Sijtsma, 2001b) and the DINO model (Templin & Henson, 2006), we show how the latter two constrained models can be obtained by leaving out main effects or interaction effects from the G-DINA model.

For this exercise, it proves convenient to take the identity-link version of the G-DINA model [Equation (5.72)] as the point of departure.

The DINA model is a special case of the G-DINA model. Let $\mathbf{1}_{K_j^*}$ denote a vector with K_j^* elements, all equal to 1. First, we notice that in the DINA model,

$$P\left(X_j = 1 \mid \alpha_{mj}^*\right) = g_j, \quad \text{if } \alpha_{mj}^* \prec \mathbf{1}_{K_j^*}, \tag{5.76}$$

and

$$P\left(X_j = 1 \mid \alpha_{mj}^*\right) = 1 - s_j, \quad \text{otherwise.} \tag{5.77}$$

As we saw previously, one guesses for the correct answer with probability g_j if one does not possess all required attributes, that is, if $\alpha_{mj}^* \prec \mathbf{1}_{K_j^*}$, and only if $\alpha_{mj}^* = \mathbf{1}_{K_j^*}$ does one have a success probability $1 - s_j$. This means that if one does not possess all K_j^* attributes necessary to answer item j positively, one falls in one of the $2^{K_j^*} - 1$ latent classes that have equal probability of correctly answering item j. We noticed that the success probability for the item only increases if one possesses all K_j^* attributes. Using the identity link, the G-DINA model formulation of the DINA model is obtained if one defines

$$g_j = \delta_{j0} \tag{5.78}$$

and

$$1 - s_j = \delta_{j0} + \delta_{j12\ldots K_j^*} \prod_{k=1}^{K_j^*} \alpha_{mjk}. \tag{5.79}$$

The G-DINA model formulation in Equation (5.72) shows that the intercept, which defines the success probability when the person does not possess any of the K_j^* attributes, equals the guessing parameter in Equation (5.78). The non-slipping probability, $1 - s_j$, increases the guessing probability by adding the effect $\delta_{j12\ldots K_j^*}$ [Equation (5.72)] due to possessing all K_j^* attributes, implying $\prod_{k=1}^{K_j^*} \alpha_{mjk} = 1$ and a latent item score $\xi_{ij} = 1$. Thus, we have demonstrated that the G-DINA model encompasses the DINA model.

The G-DINA model also encompasses the DINO model. The DINO model equals (**0** representing a vector with only 0 entries)

$$P\left(X_j = 1 \mid \alpha_{mj}^*\right) = g_j, \quad \text{if } \alpha_{mj}^* = \mathbf{0}_{K_j^*}, \tag{5.80}$$

and

$$P\left(X_j = 1 \mid \alpha_{mj}^*\right) = 1 - s_j, \quad \text{otherwise.} \tag{5.81}$$

Equation (5.80) shows that the success probability equals g_j if one lacks all K_j^* attributes and $1 - s_j$ if one possesses at least one of the K_j^* attributes [Equation (5.81)]. To obtain the DINO model from the G-DINA model [Equation (5.72)], we consider the case of $K_j^* = 2$, and we notice that for $\alpha_{mj}^* = 0_{K_j^*}$, from the G-DINA model [Equation (5.72)] it follows that upon inserting 0 values for the two attribute parameters, we obtain

$$P\left(\alpha_{mj}^* = (00)\right) = \delta_{j0}, \tag{5.82}$$

whereas the other probabilities are

$$P\left(\alpha_{mj}^* = (10)\right) = \delta_{j0} + \delta_{j1}, \tag{5.83}$$

$$P\left(\alpha_{mj}^* = (01)\right) = \delta_{j0} + \delta_{j2}, \text{ and} \tag{5.84}$$

$$P\left(\alpha_{mj}^* = (11)\right) = \delta_{j0} + \delta_{j1} + \delta_{j2} + \delta_{j12}. \tag{5.85}$$

From Equation (5.81), we know that the latter three probabilities are equal to $(1 - s_j)$; hence, Equations (5.83) and (5.84) imply that $\delta_{j1} = \delta_{j2}$, and using this equality, Equation (5.85) implies that $\delta_{j1} = \delta_{j2} = -\delta_{j12}$. For larger values of K_j^*, one may check that all main effects and interaction effects are equal in absolute size but that their sign is positive if an odd number of attributes is involved in the effect—for example, δ_{jk}—and negative if the number is even—for example, $-\delta_{jk'k''}$, that is,

$$\delta_{jk} = -\delta_{jk'k''} = \cdots = (-1)^{K_j^* + 1} \delta_{j12\ldots K_j^*}, \tag{5.86}$$

for $k = 1, \ldots, K_j^*$, $k' = 1, \ldots, K_j^* - 1$, and $k'' > k' \ldots K_j^*$. One obtains the DINO model from the G-DINA model if one defines

$$g_j = \delta_{j0}, \tag{5.87}$$

and, arbitrarily,

$$1 - s_j = \delta_{j0} + \delta_{jk}. \tag{5.88}$$

Equation (5.88) shows that the guessing probability, which is representative of the situation in which one does not possess any of the K_j^* attributes, is increased by the effect of possessing one of the attributes. The result is the probability of non-slipping, that is, mastery. Each item is characterized by two parameters; see Equations (5.87) and (5.88). Hence, we have shown that the G-DINA model encompasses the DINO model.

The RRUM [Equation (5.54)] can be shown to be special case, not of the G-DINA model but of a second general, flexible CDM that de la Torre (2011) proposed. This second general CDM generalized a specialized CDM that Maris (1999) and Junker and Sijtsma (2001b) proposed and that is based on slipping and guessing parameters defined for each *attribute* rather than each item, so that parameters s_k and g_k are used. This model is known as the *noisy inputs, deterministic "AND" gate* (NIDA; Junker & Sijtsma, 2001b) model. The NIDA model assumes that slipping and guessing occur at the attribute level rather than the item level. Next to his G-DINA model, de la Torre (2011) discussed a flexible framework for the NIDA model, which he called the generalized-NIDA or G-NIDA model. Because the NIDA model received considerably less attention than other CDMs, such as the DINA model, and because their discussion would require much space, we refrain from also discussing the NIDA model and the G-NIDA model. However, the RRUM, which we did discuss more extensively, happens to be a special case of the G-NIDA model. Therefore, we chose to provide some information for interested readers about the relation of the RRUM and the G-NIDA model; see de la Torre (2011) for details.

BOX 5.2   The RRUM is a Special Case of the G-NIDA Model

We rewrite the RRUM so that it appears in a form comparable to de la Torre's G-NIDA model; see de la Torre (2011) for further discussion of the G-NIDA model. Hence, the log versions of both models are also equal. This result is interesting, because the log version of the G-NIDA model is a special case of the log version of the G-DINA model with only an intercept and main effects but without interaction terms. As a result, we have that the RRUM is a special case of the log version of the G-DINA model. To avoid a long derivation, we rewrite the RRUM [Equation (5.54)] for the reduced attribute vector α_{mj}^* that only contains the K_j^* attribute parameters for the attributes needed to solve item j, and notice that it equals the G-NIDA model. Then, by definition $q_{jk} = 1$ for all the attributes in α_{mj}^*, and Equation (5.54) becomes

$$P\left(X_j = 1 \mid \alpha_{mj}^*, \pi_j^*, \mathbf{r}_j^*\right) = \pi_j^* \prod_{k=1}^{K} \left(r_{jk}^{*(1-\alpha_{mjk})} \right)^{q_{jk}} = \pi_j^* \prod_{k=1}^{K} r_{jk}^{*(1-\alpha_{mjk})}. \tag{5.89}$$

We can write the right-hand side of Equation (5.89) as

$$P\left(X_j = 1 \mid \alpha_{mj}^*, \pi_j^*, \mathbf{r}_j^*\right) = \pi_j^* \prod_{k=1}^{K^*} r_{jk}^* \prod_{k=1}^{K^*} \left(\frac{1}{r_{jk}^*} \right)^{\alpha_{mjk}},$$

and inserting Equation (5.51) for π_j^* (and noticing that $q_{jk} = 1$) and Equation (5.52) for r_{jk}^*, next we write

$$\prod_{k=1}^{K^*}\left(1-s_{jk}\right)\times\frac{g_{jk}}{1-s_{jk}}\prod_{k=1}^{K^*}\left(\frac{1-s_{jk}}{g_{jk}}\right)^{\alpha_{mjk}}=\prod_{k=1}^{K^*}g_{jk}\prod_{k=1}^{K^*}\left(\frac{1-s_{jk}}{g_{jk}}\right)^{\alpha_{mjk}}. \tag{5.90}$$

The formulation on the right-hand is precisely the G-NIDA-model. Thus, the RRUM is a re-parameterization the G-NIDA model (de la Torre, 2011). Because de la Torre (2011) showed that the log version of the G-NIDA is a special case of the log version of the G-DINA (which we did not explicitly discuss here), the RRUM also is a special case of the log version of the G-DINA.

Log-Linear Cognitive Diagnostic Model. In addition to Van Davier's (2008, 2010) GDM and de la Torre's (2011) G-DINA model, Henson et al. (2009) proposed a third general framework for cognitive diagnosis based on a log-linear approach with latent variables. This is the log-linear cognitive diagnostic model (LCDM). Because the GDM and the G-DINA model have much in common with the LCDM, we wish to avoid an abundance of detail and skip the log-linear formulation of the model and transfer to the model's formulation as it appears often in the literature (e.g., Liu, Tian, & Xin, 2016; Sen & Bradshaw, 2017). This is the probability of a 1 score on the target item j conditional on the vector of latent attributes needed for responding positively to item j, that is, vs denoting regression parameters,

$$P\left(X_{ij}=1\mid\alpha_i\right)=\frac{\cdot\exp\left[v_{j,0}+v'_j\mathbf{h}\left(\alpha_i,\mathbf{q}_j\right)\right]}{1+\exp\left[v_{j,0}+v'_j\mathbf{h}\left(\alpha_i,\mathbf{q}_j\right)\right]}. \tag{5.91}$$

In Equation (5.91), the notation is straightforward and consistent with similar notation in the general diagnostic model and the G-DINA model but emphasizes that different models yield different numerical values for their parameters. Parameter $v_{j,0}$ provides a baseline probability when the person does not possess any of the latent attributes item j requires, and vector product $v'_j\mathbf{h}(\alpha_i,\mathbf{q}_j)$ provides a selection of main and interaction effects tailored to the (sub-) set of latent attributes that person i possesses for responding to item j. The target CDM determines which main and interaction effects are included in the model, and the combination of person and item determines which combinations of elements in α_i and $\mathbf{q}_j$ are effective.

One can derive restricted CDMs from Equation (5.91). For example, assume that only two latent attributes are relevant for responding positively to a particular item type, and assume that we test the DINA model as representing the mechanism that combines latent attributes to produce a response. This means that both latent attributes are needed to increase the response probability in Equation (5.91), whereas possessing only one latent attribute does not affect the response probability. Hence, Equation (5.91) becomes

$$P\left(X_{ij}=1\,|\,\alpha_i\right)=\frac{\exp\left[v_{j,0}+v_{j,12}\alpha_1\alpha_2\right]}{1+\exp\left[v_{j,0}+v_{j,12}\alpha_1\alpha_2\right]}.\qquad(5.92)$$

Only if one possesses both latent attributes will $\alpha_1\alpha_2=1$ and will parameter $v_{j,12}$ exert an influence on the response probability, whereas possessing only α_1 or α_2 has no effect, because Equation (5.92) does not contain main effects. Henson et al. (2009) discussed how other restricted CDMs are obtained from the general model in Equation (5.92). For more information, see the monograph by Rupp, Templin, and Henson (2010).

Estimation

First, for the DINA model, the joint distribution of the data conditional on the latent variables, here the K binary attributes, is the product of the marginal distributions of the item scores; that is, assuming local independence, we have that

$$P\left(\mathbf{X}_i\,|\,\alpha_i\right)=\prod_{j=1}^{J}P\left(X_{ij}\,|\,\alpha_i\right).\qquad(5.93)$$

Also assuming that the data records of different persons are independent, we write the distribution of the data in matrix $\mathbf{X}$ as

$$P\left(\mathbf{X}\,|\,\alpha\right)=\prod_{i=1}^{N}P\left(\mathbf{X}_i\,|\,\alpha_i\right).\qquad(5.94)$$

This joint likelihood can be maximized for the parameters $\mathbf{g}=(g_1,\dots,g_J)$ and $\mathbf{s}=(s_1,\dots,s_J)$, but because the estimators are known to have unfavorable statistical properties, alternatively one rather uses the marginal likelihood approach,

$$L(\mathbf{X})=\prod_{i=1}^{N}P(\mathbf{X}_i)=\prod_{i=1}^{N}\sum_{w=1}^{W}P(\mathbf{X}_i\,|\,\alpha_w)P(\alpha_w),\qquad(5.95)$$

where the number of possible attribute patterns equals $W=2^K$. The latent class structure is apparent on the right-hand side of Equation (5.95). De la Torre (2009) discussed an estimation method based on Equation (5.95) that uses EM, and noticed that this method is labor-intensive due to the huge number of different latent attribute vectors α_w. A variation on this method assumes that the elements of vector α are locally independent given a continuous higher-order latent variable θ, having the structure of Equation (5.1),

$$P(\alpha \mid \theta) = \prod_{k=1}^{K} P(\alpha_k \mid \theta), \tag{5.96}$$

and $P(\alpha_k \mid \theta)$ is modeled as a Rasch model,

$$P(\alpha_k \mid \theta) = \frac{\exp(\lambda_{0k} + \lambda_1\theta)}{1 + \exp(\lambda_{0k} + \lambda_1\theta)}, \tag{5.97}$$

in which λ_{0k} is the intercept, $\lambda_1 > 0$ is the slope, and $\theta \sim \mathcal{N}(0,1)$ by assumption. Equation (5.97) renders the probability monotone in θ and dependent on an intercept parameter and a slope parameter that is equal across attributes. This is the higher-order DINA (de la Torre & Douglas, 2004), and an MCMC algorithm can be used to estimate the $K-1$ intercept parameters and 1 slope parameter. Yang and Embretson (2007) discussed an equation similar to Equation (5.97) for inferring the posterior probability of a person's most likely latent class α_w given her item-score pattern $\mathbf{X}_i$, the item parameters $\mathbf{g}$ and $\mathbf{s}$, and design matrix $\mathbf{Q}$.

Goodness-of-Fit Methods

Several goodness-of-fit methods for CDMs are available (e.g., Chen, de la Torre, & Zhang, 2013; de la Torre & Lee, 2013; Hansen, Cai, Monroe, & Li, 2016; Kunina-Habenicht, Rupp, & Wilhelm, 2012; Liu et al., 2016; Ma, Iaconangelo, & de la Torre, 2016; Sen & Bradshaw, 2017; Sorrel, Abad, Olea, de la Torre, & Barrada, 2017). The literature shows that the investigation of these methods is still underway, and a definitive choice of which methods to prefer may be premature. Several methods are also appropriate for LCM and IRT goodness-of-fit assessment, and authors discussing some methods in the context of CDMs borrowed the methods from the context of other models. We do not aspire to provide one toolkit filled with methods that are universally applicable but discuss in this section methods that are used in the context of CDMs irrespective of their origin.

As with LCMs, absolute and relative fit methods are available that can be used in conjunction to assess the degree to which the model fits the data. Absolute fit measures assess the goodness of fit of the hypothesized model to the data, and relative fit measures assess the fit of a model relative to another model. In both cases, the model consists of two parts, which are the particular CDM the researcher used and the Q-matrix that relates the attributes to the solution of each of the items that together constitute the test (Chen et al., 2013). Fit measures that are powerful identify a CDM that misrepresents how particular latent attributes combine, a Q-matrix that assigns the wrong subset of latent attributes to the items, or both causes of model misspecification. One additional source of misfit relates to the identification of the correct attributes, meaning that latent attributes irrelevant to the problem at hand may have been included in the Q-matrix, or relevant latent attributes may have been excluded (e.g., Culpepper, 2019). Another source of misfit relates to the way in which a set of attributes combine in producing a response to an item, that is, the correct CDM. These two potential sources of misfit belong to the realm of theory formation, which is essential to constructing a scale for an attribute, and precede the psychometric evaluation of a collected data set from which to infer a scale. Thus, fit assessment can reveal whether the chosen CDM and the chosen Q-matrix together explain the data well, but fit assessment cannot reveal the true theoretical model including the set of latent attributes that govern a particular problem. One sometimes sees endless trial and error attempts following a misfitting model, resulting, for example, in an approximately fitting model for a limited number of items, but we rather urge the reader to reconsider the theoretical underpinnings of the item set. Chen et al. (2013) noticed that absolute fit measures are rarer than relative fit measures, such as AIC [Equation (5.41)] and BIC [Equation (5.42)].

The order in which to assess absolute and relative misfit can be a topic for discussion. For example, Sen and Bradshaw (2017) suggested first assessing absolute fit of a couple of CDMs and then comparing the best fitting CDMs by means of relative fit methods. We think this strategy makes sense only if one follows an exploratory approach, trying different CDMs reflecting different ways in which attributes combine to produce an outcome without an *a priori* hypothesis about the solution process of interest. Such a strategy may be useful if one knows very little of the target attribute but also raises the question of whether one should not first try to obtain more information about the attribute through other channels, prior to collecting a sample of data, and test CDMs more or less haphazardly. The other channels through which to collect information can involve research to establish a draft of a theory. In contrast to an exploratory approach, if one has constructed the test based on theoretical considerations, one likely hypothesizes one CDM and focuses on absolute fit assessment. Relative fit assessment comes in only if the hypothesized model is rejected and well-informed adaptations of the Q-matrix are tried to find suggestions for follow-up studies in which the adapted and possibly improved theory is tested. De la Torre and Chui (2016; also, de la Torre, 2008) proposed an index that can be used to empirically validate the Q-matrix entries by identifying and replacing initially incorrectly specified entries in the Q-matrix. Next, we first focus on absolute fit assessment and then have a brief look at relative fit assessment.

Absolute Fit Assessment. Absolute fit assessment in the CDM context may use two kinds of methods, known as full information methods and limited information methods. Full information methods, such as statistic G^2 [Equation (5.24)], use the complete contingency table for the J items and incorporate each pattern of J item scores, but they encounter problems when cells are empty or nearly so (Hansen et al., 2016). This happens unavoidably, because as J grows, the number of cells soon is enormous relative to a realistic sample size, a phenomenon we previously discussed as the *curse of dimensionality*. For example, for $J = 10$ dichotomous items, we have $2^J = 1,024$ cells, and for greater J, with real data it is almost unavoidable that many cells are empty or nearly so. The result of the badly filled cell problem is Type I error rates that differ from the nominal significance level and thus have little to say about the fit of the hypothesized model (Liu et al., 2016; Maydeu-Olivares, 2013). Limited information methods (Chen et al., 2013; Liu et al., 2016) do not use the complete crosstable for the J items but assess marginal quantities obtained by collapsing the cross table in certain ways (Liu et al., 2016). This way, fewer cells are realized among which the observations are distributed, resulting in better-filled cells; see Table 5.7 for an example involving three items j, k, and l, each scored 1, 0, and proportions $\pi_{x_j x_k x_l}$ as cell entries. For example, π_{101} is the proportion of people scoring 1 on items j and l and 0 on item k. One obtains a 2×2 cross table for the bivariate proportions for items j and k by collapsing the three-dimensional table across the two values for item l; similarly, one obtains cross tables for items j and l and items k and l by collapsing across the two values for items k and j, respectively. For example, for items j and k, bivariate proportion $\pi_{x_j x_k} = \pi_{11} = \pi_{111} + \pi_{110}$, and likewise one obtains π_{10}, π_{01}, and π_{00}; see Table 5.7, lower panel. Next, one can collapse this bivariate cross table to obtain the two univariate proportions $\pi_{x_j} = \pi_1 = \pi_{111} + \pi_{110} + \pi_{101} + \pi_{100}$ and $\pi_{x_k} = \pi_1 = \pi_{111} + \pi_{110} + \pi_{011} + \pi_{010}$. Clearly, higher-order proportions are based on frequencies that are at least as large as the frequencies on which lower-order proportions are based, which alleviates the (near) empty cell problem.

By choosing a higher aggregation level for item-score pattern frequencies, limited information methods tend to avoid the problems that full information methods encounter when cells are empty or nearly empty. However, by collapsing the cross table, limited information methods lose information that the original cross table contained at the finer-grained level. For example, limited information methods assess the discrepancy between empirical first-order proportions of positive answers to the J items, defined by proportions

TABLE 5.7

Complete Three-Way Contingency Table (Upper panel) and its Three Two-Way Marginal Contingency Tables (Lower Panel)

			$X_3 = 0$					$X_3 = 1$	
			X_2					X_2	
			0	1				0	1
X_1	0		π_{000}	π_{010}		X_1	0	π_{001}	π_{011}
	1		π_{100}	π_{110}			1	π_{101}	π_{111}

		X_2					X_3					X_3	
		0	1				0	1				0	1
X_1	0	π_{00+}	π_{01+}		X_1	0	π_{0+0}	π_{0+1}		X_2	0	π_{+00}	π_{+01}
	1	π_{10+}	π_{11+}			1	π_{1+0}	π_{1+1}			1	π_{+10}	π_{+11}

Note: $\pi_{00+} = \pi_{000} + \pi_{001}$, etc.

π_{j}, and the corresponding proportions reconstructed by means of estimated model parameters and probabilities (Chen et al., 2013). Another possibility is the comparison between empirical and reconstructed second-order proportions of people giving positive responses to pairs of items (Liu et al., 2016), denoted π_{jk}. In addition, Chen et al. (2013) considered a statistic based on log-odds ratios. To avoid discussing too much detail, we refrain from explaining the log-odds ratio statistic and have a look at one particular kind of limited information statistic.

Because Chen et al. (2013) found that the first-order residuals were ineffective at identifying CDMs or Q-matrices that were misspecified relative to the model used to generate simulated data, we only briefly discuss the second-order residual, denoted $r_{jj'}$. Statistic $r_{jj'}$ is the absolute residual of two quantities. First, the Fisher-Z transformation of the correlation, r_{jk}, between the vector, X_{j}, of observed scores on item j, and vector X_{k} for item k, involving proportion π_{jk}, and second, the Fisher-Z transformation of the correlation, $\tilde{r}_{jk}$, between the vector of model-predicted item scores, $\tilde{X}_{j}$, based on the estimated attribute parameters, $\hat{\alpha}$, and model parameters, such as guessing and slipping parameters in the DINA model, and vector $\tilde{X}_{k}$. Statistic $r_{jj'}$ is defined as

$$r_{jj'} = \left| Z\left(r_{jk}\right) - Z\left(\tilde{r}_{jk}\right) \right|. \tag{5.98}$$

Residuals $r_{jj'}$ should be close to 0 when the CDM of interest fits the data well. Chen et al. (2013) suggested approximating the standard error of $r_{jj'}$ by $(N-3)^{-1/2}$ and using a formal test procedure for item selection that controls the familywise Type I error rate. However, there are two problems here. First, ignoring the vertical bars in Equation (5.98), it seems that the standard error of the resulting linear combination rather equals

$$SE\left(r_{jj'}\right) = \left\{ 2\left(N-3\right)^{-1} - \sigma\left[Z\left(r_{jk}\right), Z\left(\tilde{r}_{jk}\right) \right] \right\}^{\frac{1}{2}}. \tag{5.99}$$

There is no reason for the covariance to be 0. Moreover, considering absolute values, the standard error will be different, but we will not pursue this road any further here and use the difference between $Z(r_{jk})$ and $Z(\tilde{r}_{jk})$ descriptively. Second, not statistically testing rules out a familywise approach to testing. In Chapter 4, we pointed out that we consider such an approach undesirable when the purpose of the research is to identify items inconsistent with the measurement model, because a corrected, hence a smaller significance level makes it more difficult to reject the null-hypothesis of a fitting model, and hence, identify such deviant items. We conclude that the general picture for practical use of this statistic, and the log-odds ratio statistic that we did not discuss, is quite complex.

We briefly discuss a few additional methods to give the reader an impression of the available tools, but notice once more that the topic of assessment methodology has not yet fully crystallized (Hansen et al., 2016). Liu et al. (2016) discussed statistic M_2, a global limited-information fit statistic comparable with the second-order residuals just discussed and based on theory discussed by Maydeu-Olivares and Joe (2006), which addresses bivariate proportions π_{jk} and compares residuals based on (estimated) model-based bivariate proportions $\tilde{\pi}_{jk}$ and observed counterparts p_{jk}. Statistic M_2 can be generalized to proportions of order r, denoted M_r. Based on simulated data, statistic M_2 was found to have good Type I error rate and good power to detect model-data misfit, results that Hansen et al. (2016) also found in another series of simulation studies. Because Hansen et al. (2016) found global statistic M_2 to be insensitive to misspecification of the distribution of latent attributes, these authors also studied a local X^2 statistic (Chen & Thissen, 1997) for studying local independence and found it satisfying to identify sources of misfit. Liu et al. (2016) proposed the root mean square error of approximation (RMSEA), which is derived from statistic M_2, as a measure of effect size and found that it was successful in identifying important cases of misclassification and ignoring unimportant cases.

Sorrel et al. (2017) compared four (limited information) item-fit statistics that assessed the IRF of an item. The IRF provides the probability of a particular score, in this chapter a 1 score, as a function of the pattern of the attributes collected in α, that is, the latent class. The authors concluded that the likelihood-ratio statistic and the Wald statistic, which compare nested models and thus are relative fit measures, had the best Type I error and the best power. They also found that the application of these statistics requires items of good quality, defined as discrimination power (ibid., p. 620), or large samples, however, without guarantee that the statistical properties of the methods are always satisfying.

Typically, even if limited information methods give an impression of the goodness of fit of the model to the data, both fit and misfit can be difficult to understand when the methods do not directly address particular model assumptions. For example, second-order residual, $r_{jj'}$, and statistic M_2 address bivariate proportions π_{jk}, but by doing this these methods do not explicitly address one or more of the model assumptions. Hence, fit based on these statistics supports a part of the model that does not coincide with one or more assumptions, but it is unknown which part of the model the statistics support and thus which part remains unassessed. This circumstance also happens with the fit assessment of other measurement models, such as assessing IRT model-data fit by means of non-negativity of inter-item correlations; see Equation (3.23). Finding positive inter-item correlations supports an unknown part of the IRT model but does not explicitly support particular model assumptions, such as unidimensionality, monotonicity, and local independence. The likelihood-ratio statistic and the Wald statistic that Sorrel et al. (2017) studied explicitly targeted the IRFs, and fit thus supports the hypothesized IRF features. Similarly, the interpretation of model-data misfit can be problematic, because often misfit points to an item involved in one or more deviant correlations but not to the assumptions violated. This renders the diagnosis of what precisely is wrong with one's model difficult.

Relative Fit Assessment. Here, we pay less attention to relative fit statistics than absolute fit statistics, because we discussed relative fit statistics already at some length in the section on LCMs. Moreover, formally relative fit statistics remain very much the same when they are used to compare a different class of models, whereas absolute fit statistics are much more tailored to the specific features of the model under consideration and tend to be different across different models. For example, whereas the non-negativity of inter-item correlations is of interest in nonparametric IRT and, in addition, plays a role in the definition and use of the scalability coefficients, this model feature does not generalize well to CDMs in the sense that it is also used there for assessing absolute fit or other item and scale features. Researchers studying relative fit statistics focus on alternatives for AIC and BIC and investigate whether the alternatives have greater power to identify the data-generating model, misspecification of this model, and misspecification of Q-matrices.

Several studies addressed AIC and BIC (Kunina-Habenicht et al., 2012), and other studies added variations of AIC and BIC. For example, Sen and Bradshaw (2017) added SABIC, which stands for sample-size adjusted BIC (Sclove, 1987) and which is also used in the LCM context. SABIC uses an adapted multiplication term depending on sample size that reduces the influence of sample size on the fit statistic and is defined as $N^* = (N + 2)/24$. In the definition of BIC, N^* replaces N. In their simulation study, Sen and Bradshaw (2017) used items of medium and high quality, defined by medium and high item discrimination, where discrimination is quantified by the difference between probabilities of a positive response to an item between groups that do and do not possess a particular latent attribute. They found that neither AIC, BIC, nor SABIC had large power to detect the correct CDM that generated the simulated data when item quality was below excellent, but results were also disappointing albeit better when item quality was better. In addition, they distinguished between data-generating models assuming discrete latent variables (i.e., CDMs) and data-generating models assuming continuous latent variables (i.e., IRT models) and found that IRT models were often identified as the true model even when the data had been generated using a CDM. Chen et al. (2016) concluded from their simulation study that BIC was more powerful than AIC to detect misspecification of the CDM, the Q-matrix, or both.

Relationship to Nonparametric IRT

Junker and Sijtsma (2001b) studied the properties of the DINA model from the perspective of the monotone homogeneity model and focused on monotonicity properties. Before we consider their result, we first notice the stochastic ordering result different from SOL, which reverses the roles of latent and manifest variables and therefore is called stochastic ordering of the manifest variable by the latent variable (SOM; also discussed in Chapter 4, but repeated here for the sake of clarity). Starting from unidimensionality, monotonicity, and local independence, for any pair of persons v and w with $\theta_v < \theta_w$, Hemker et al. (1997) defined SOM as

$$P(X \geq x \mid \theta_v) \leq P(X \geq x \mid \theta_w). \tag{5.100}$$

The monotone homogeneity model thus supplies a latent structure justifying ordering people on the observable X total score. In Chapter 2, we saw that older approaches, such as CTT, did not supply such a justification but simply recommended the use of X. With the exception of the Rasch model (Chapter 4), modern IRT approaches based on unidimensionality, monotonicity, and local independence do not mention that they also justify SOM and even the more useful SOL, which allows one to make inferences about latent, explanatory structures—an ordinal latent scale—from observable data. The reason probably is that they rely solely on the latent-variable scale, but we can only speculate about this.

Holland and Rosenbaum (1986) introduced the notion of summaries of the item scores, denoted $g(\mathbf{X})$, that are non-decreasing coordinate-wise in X_j ($j = 1, \ldots, J$), and Junker and Sijtsma (2001b) noticed that in the DINA model,

$$P\left[g(\mathbf{X}_i) \mid \alpha_i\right] \text{ is coordinate-wise non-decreasing in } \alpha_i. \tag{5.101}$$

Obviously, this is an SOM property, meaning that the mastery of more attributes yields a higher summary score. Junker and Sijtsma (2001b) were unable to derive similar SOL properties for the DINA model but succeeded for the NIDA model that we mentioned in this chapter but did not discuss extensively.

Data Example: Identifying Patients' Disorder Profiles Using the MCMI-III

We started the section on CDMs by discussing an application of a CDM to clinical data aimed at identifying patients' disorder profiles based on their responses to 44 items measuring anxiety, somatoform disorder, thought disorder, and major depression. The items were part of a clinical inventory, which is the MCMI-III (Millon et al., 2009). In this subsection, we use CDMs to analyze this data set (also, see de la Torre et al., 2018, and Van der Ark et al., 2019). The sample consisted of 1,210 Caucasian patients and inmates in Belgium (61% males; for more details, see Rossi, Elklit, & Simonsen, 2010) who responded to the Dutch version of the MCMI-III (Rossi, Sloore, & Derksen, 2008). In the MCMI-III, the abbreviations A (anxiety), H (somatoform disorder), SS (thought disorder), and CC (major depression) indicate the scales of the four clinical disorders. As the Q-matrix (Table 5.6) shows, due to comorbidity of symptoms, several items are associated with more than one clinical disorder.

For each of the 44 dichotomous items, the item score was 1, if the respondent endorsed the item, and the item score was 0 otherwise. The proportion of 1 scores on an item is denoted π_j. The higher the proportion π_j, the more popular the item in the sense that more people give a positive response. In an item pair, a Guttman error occurs when a person produces a score of 1 on the more unpopular item in the pair and a score of 0 on the more popular item (Chapter 4). In a pattern of J item scores, the total number of Guttman errors counts the frequency of such (10) scores in the $\frac{1}{2}J(J-1)$ unique item pairs. For a particular person, the number of Guttman errors indicates the inconsistency of her item-score pattern and can be used as an outlier score (Zijlstra, Van der Ark, & Sijtsma, 2007). Two persons had an unexpectedly large number of Guttman errors, well beyond the cutoff value suggested by the adjusted boxplot (Hubert & VanderVieren, 2008). Their item-score patterns were removed from the data, yielding a sample size of $N = 1,208$. The data contained no missing values.

Characteristic of clinical disorders is that having a single disorder is sufficient to increase the probability of endorsing a multi-disorder item. Hence, starting from the DINO model seemed to be a plausible choice. As having more disorders increases the probability of endorsing an item, the G-DINA model [Equation (5.69)], which generalizes the DINO model, may show a better fit. The data were analyzed by using the R package GDINA (Ma & de la Torre, 2018).

As expected, the G-DINA model showed a better fit than the DINO model (Table 5.8). Hence, further analysis concentrated on the G-DINA model. Figure 5.2 shows a histogram of the $\binom{44}{2} = 946$ residuals $d_{jj'} = Z(r_{jk}) - Z(\tilde{r}_{jk})$. If the G-DINA model fits well, one expects that $\bar{d}_{jj'} = 0$. Most values in Figure 5.2

TABLE 5.8

Comparing the Fit of the DINO Model and the GDINA Model

	Model	
	DINO	**GDINA**
# parameters	103	139
AIC	53,038.25	52,335.31
BIC	53,563.21	53,043.76
M2	3,353.8	3,378.8
	$df = 887, p < .0001$	$df = 851, p < .0001$
Difference		
$\chi^2 = 774.94, df = 36, p < .0001$		

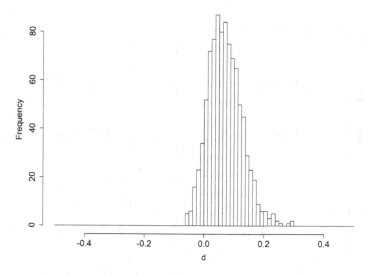

FIGURE 5.2

Histogram of $d_{jj'} = Z(r_{jk}) - Z(\tilde{r}_{jk})$. If the G-DINA model fits well, residuals $d_{jj'}$ are scattered symmetrically around 0 with small variance. Here, value 0 is almost in the left tail of the distribution.

are positive, which shows that the correlational structure under the G-DINA model is weaker than it is in the data, and this may be taken as evidence that the G-DINA model does not fit well. Compared to the number of degrees of freedom, the value of the limited information statistic M_2 (Liu et al., 2016; Table 5.8) corroborates this result.

For each of the 44 items, Table 5.9 shows the endorsement probabilities for each cluster of disorders hypothesized in the Q-matrix (Table 5.6). As expected, when the disorder is present the endorsement probabilities are larger than when the disorder is absent. Some items may not function well. For example, item 3 (Having a hard time keeping balance when walking; somatoform disorder) and item 18 (Not noticing nearby people while awake; thought disorder) have a low endorsement probability, even when the disorder is present. An explanation for this low endorsement probability is that the symptoms these items describe are rather extreme and are only endorsed by those who suffer severely from the clinical disorder. Also, item 19 (Mood changing from day to day; thought disorder) and item 34 (Certain thoughts keep coming back; anxiety) have a rather high endorsement probability, even when the disorder is absent. An explanation is that the symptoms these items describe are so ordinary that people not suffering from clinical disorders regularly experience them as well.

Table 5.10 shows the probability that a person suffers from a clinical disorder, given that the person gave a positive response to an item. Boldface probabilities correspond to an entry $q_{jk} = 1$ in the Q-matrix. One can use these probabilities to select items for predictive purposes. All probabilities are

TABLE 5.9

Item Endorsement Probabilities for the Hypothesized Combinations of Clinical Syndromes

Item	A = 0	A = 1	Item	H = 0	H = 1	Item	SS = 0	SS = 1	Item	CC = 0	CC = 1
7	.21	.77	3	.03	.21	4	.08	.48	8	.19	.91
11	.27	.78	6	.04	.29	10	.24	.82	21	.09	.51
23	.13	.61				13	.18	.77	28	.03	.50
24	.28	.80				14	.16	.80	37	.02	.57
27	.05	.45				18	.01	.22	39	.23	.54
31	.19	.70				19	.32	.81	44	.07	.60
34	.36	.94				20	.06	.32			
41	.11	.65				26	.07	.30			
43	.07	.48				30	.07	.48			
						40	.18	.73			
						42	.10	.41			

	H = 0		H = 1	
Item	CC = 0	CC = 1	CC = 0	CC = 1
1	.03	.58	.56	.92
2	.05	.47	.57	.90
9	.04	.40	.45	.48
15	.05	.32	.49	.74
22	.03	.22	.21	.47
25	.08	.40	.33	.67
29	.01	.32	.40	.81

	A = 0		A = 1	
Item	H = 0	H = 1	H = 0	H = 1
16	.10	.33	.34	.73
33	.14	.46	.71	.88

	A = 0		A = 1	
Item	SS = 0	SS = 1	SS = 0	SS = 1
12	.16	.64	.73	.95
17	.04	.24	.42	.74

(Continued)

TABLE 5.9 (CONTINUED)

Item Endorsement Probabilities for the Hypothesized Combinations of Clinical Syndromes

	A = 0		A = 1	
Item	CC = 0	CC = 1	CC = 0	CC = 1
36	.05	.16	.33	.54

	SS = 0		SS = 1	
Item	CC = 0	CC = 1	CC = 0	CC = 1
5	.19	.61	.54	.85
32	.04	.30	.37	.82
38	.06	.23	.19	.59

	H = 0				H = 1			
	SS = 0		SS = 1		SS = 0		SS = 1	
Item	CC = 0	CC = 1	CC = 0	CC = 1	CC = 0	CC = 1	CC = 0	CC = 1
35	.09	.35	.24	.59	.23	.37	.40	.79

Note: A = anxiety, H = somatoform disorder, SS = thought disorder, CC= major depression.

TABLE 5.10

Probability of Having a Clinical Syndrome Given the Endorsement of an Item, and the Marginal Probabilities of Having a Clinical Syndrome (Last Row) for the G-DINA Model

Item	Syndrome				Item	Syndrome				Item	Syndrome				Item	Syndrome			
	A	H	SS	CC		A	H	SS	CC		A	H	SS	CC		A	H	SS	CC
1	.82	**.84**	.85	.80	12	**.82**	.69	**.80**	.66	23	**.84**	.65	.74	.65	34	.75	.60	.67	.59
2	.81	.85	.84	**.87**	13	.75	.69	**.83**	.64	24	.77	.61	.69	.60	35	.80	.76	**.84**	.78
3	.74	**.88**	.80	.64	14	.76	.71	**.85**	.66	25	.78	**.78**	.80	.77	36	**.91**	.71	.82	.78
4	.77	.72	**.87**	.67	15	.80	**.85**	.84	.76	26	.74	.68	**.82**	.64	37	.87	.74	.86	**.96**
5	.75	.67	**.78**	.71	16	.83	.79	.79	.68	27	**.92**	.69	.79	.70	38	**.83**	.73	.87	.83
6	.73	**.86**	.78	.63	17	**.92**	.74	**.87**	.73	28	.85	.72	.84	**.92**	39	.67	.59	.66	.66
7	**.81**	.63	.72	.63	18	.83	.78	**.95**	.72	29	.85	**.90**	**.88**	.84	40	.74	.69	**.82**	.64
8	.76	.66	.75	**.80**	19	.68	.63	.73	.58	30	.80	.75	**.91**	.70	41	**.88**	.66	.76	.67
9	**.82**	**.85**	**.85**	.80	20	.76	.71	**.85**	.66	31	**.81**	.63	.72	.63	42	.74	.68	**.81**	.63
10	.72	.67	**.79**	.62	21	.78	.67	.77	**.82**	32	.86	.77	**.92**	**.85**	43	**.89**	.67	.77	.68
11	**.77**	.61	.69	.60	22	.82	**.84**	**.84**	.81	33	**.82**	.72	.77	.65	44	.82	.70	.81	**.88**
MP	.53	.48	.52	.45		.53	.48	.52	.45		.53	.48	.52	.45		.53	.48	.52	.45

Note: A = anxiety, H = somatoform disorder, SS = thought disorder, CC = major depression, MP = marginal probability. Bold face probabilities correspond to a 1 entry in the Q matrix (Table 5.6).

high, which is partly due to the high prevalence of the four clinical disorders in the sample. Some items are excellent predictors of clinical disorders. For example, endorsing item 18 (Not noticing nearby people while awake; see previous paragraph) appears to be an excellent predictor of thought disorder; that is, when we let $SS = 1$ indicate presence of thought disorder, given a positive score on item 18, we have $P(SS = 1 \mid X_{18} = 1) = .95$. Several items show high probabilities for all disorders and do not distinguish well between clinical disorders. For example, item 39 (Having tried to commit suicide; major depression) has a relatively low predictive value for each of the four clinical disorders, whereas item 9 (Feeling worn out for no special reason; somatoform disorder and major depression) has a relatively high predictive value for the four clinical disorders.

The final goal of a CDM analysis is predicting a person's attributes from the data to diagnose the person's clinical disorders, if any are present. For example, Table 5.11 shows the item-score patterns and the diagnosis for the first ten persons tested. In addition, Table 5.12 shows the prevalence of all possible combinations of clinical disorders. Approximately two-thirds of the persons have either no disorder (33.8% of the complete sample) or all four disorders (30.7%). For the remaining third, 15.8% has three disorders, 8.7% has two disorders, and 10.7% has one disorder, anxiety being the most prevalent. Using the DINO model, Van der Ark et al. (2019) found rather different prevalence rates (e.g., no disorder 45.3%, and all four disorders 6.0%). Their results illustrate that the type of model and the goodness of fit of the model may have a large effect on the outcomes.

TABLE 5.11

Item-Score Pattern and Diagnosis for the First Ten Respondents

Resp.	Item-score pattern	Diagnosis
1	011110111111011010110010101110100110001110101	All disorders
2	000010000000000000000001010000001000000000000	No disorders
3	110000110011011110110011010100000010000111000	All disorders but major depression
4	000000011010001100110100000011000000000110000	Somatoform
5	000111101111111101001111111011111101011111011	All disorders but somatoform
6	000000000000000000000001000000100100000010010	No disorders
7	001100000001001010001011000011011010100101100	Anxiety and somatoform
8	000010001000100000000000000001000000010000100010	No disorders
9	000000000000000000000000000000000000000000000	No disorders
10	000000000011000010100011010000000110000010100	Anxiety

Note: Resp. = respondent.

TABLE 5.12

Class Probabilities According to the G-DINA
Model Expressed as Percentages

Class	Prevalence	Class	Prevalence
No disorder	33.8%	H and SS	4.1%
CC	2.4%	A and SS	2.0%
SS	.6%	A and H	1.0%
H	3.4%	All but A	1.7%
A	4.3%	All but H	6.8%
CC and SS	.4%	All but SS	1.4%
CC and H	.0%	All but CC	5.9%
CC and A	1.3%	All disorders	30.7%

Note: A = anxiety, H = somatoform disorder, SS = thought disorder, CC = major depression.

Discussion

CDMs are an interesting class of models for producing nominal scales that measure psychological and possibly other attributes. Von Davier and Lee (2019) provide an up-to-date anthology of models, special topics such as CDM adaptive testing, classification consistency, and differential item functioning, and applications and software. Of the models discussed in this monograph, together with the LCM of which they constitute a special class, CDMs aspire to classify people based on their possession of a set of latent attributes that are discrete. We consider this a welcome addition to the group of available measurement models for ordinal scales, let alone the models that aspire metric scales without too much theoretical underpinning, either by assumption (e.g., CTT, factor analysis) or implication (e.g., parametric IRT). CDMs focus explicitly on how a set of latent attributes combine to produce the correct probability of giving the positive response to an item. This is the attractive feature of CDMs as much as it is their Achilles' heel, because with the exception of IRT models such as the linear logistic test model and multidimensional IRT models, CDMs implicitly require that psychologists are able to define and operationalize a well-developed attribute theory as the basis of a CDM application. This should be a challenge to psychometricians and psychologists alike, and the success of CDMs very much depends on the availability of well-developed attribute theory.

CDMs are relatively new and much theory development is still underway, which is witnessed by the existence of three general frameworks—the GDM, the G-DINA model, and the LCDM—that show more similarities than differences, suggesting that in due time the field may convergence to one general framework that encompasses all the specific models as special cases. The discussion about goodness-of-fit methods clarified that several methods were proposed, but that best practices have yet to be developed. This requires experience that has to accumulate in many analyses of real data. Another feature of CDMs that may render goodness-of-fit research difficult is the complexity of the models that also include the Q-matrix specifying how latent attributes relate to specific items whereas the CDM specifies how the latent attributes combine to produce item scores. So far, it appears as if absolute fit methods are more powerful in assessing model-data fit than relative fit methods. Chiu and Douglas (2013) proposed a nonparametric approach that uses the binary-valued Q-matrix in combination with each possible binary-valued latent-attribute vector α_m, $m = 1, \ldots, 2^K$, to determine the corresponding latent item-score vector ξ_m [based on Equation (5.44)], and then uses a distance measure to find the latent-attribute vector α_m^* that minimizes the distance between the corresponding latent item-score vector ξ_m^* and person i's observed item-score vector x_i. Latent-attribute vector α_m^* represents the latent class for person i. Their approach avoids model estimation and model-fit assessment and focuses on classification only, can be used irrespective of sample size, and may be useful when particular CDMs are misspecified. Chiu, Douglas, and Li (2009) discussed other classification or clustering approaches.

Finally, we have discussed CDMs together with the LCM as a restricted LCM that is of special interest to psychological measurement and possibly other research areas as well. However, CDMs may as well be conceived as discrete parametric IRT models, where the discreteness refers to the latent attributes. This discreteness results in discrete IRFs; see the equations we presented for each of the CDMs discussed in this chapter. Because CDMs stand out for their ambition to classify people based on the latent attributes they do and do not possess, we preferred to discuss them together with LCMs in one separate chapter on classification or nominal measurement.

General Discussion

Latent class models and cognitive diagnostic models, which are special cases of latent class models as we demonstrated for the DINA model, provide the tools for constructing nominal scales for psychological measurement. Such scales are rather unusual in psychology, where the habit of counting numbers of correct or affirmative answers, or counting total numbers of credit points has become the norm. Even though such counts have proven valuable for measurement and in many formal measurement models may replace the technically useful but practically awkward latent variable scales, we believe that other possibilities have received too little attention for two reasons. One is that psychology has been oriented for a long time toward the measurement of quantities, and the other is that theory about attributes that drives scale construction, in general, is underdeveloped. The work of Jansen and Van der Maas (1997, 2002) in proportional reasoning has shown that classifying people by means of their typical item-score patterns may do more justice to the attribute than counting the number of correct solutions. For transitive reasoning, Bouwmeester and Sijtsma (2007) have shown that ordering classes based on item-score patterns provides an intelligible ordinal scale, consistent with the theory of transitive reasoning. The fact that such scales are rare in psychology and elsewhere where individuals are assessed with respect to particular attributes, in our view witnesses too much reliance on old habits and too little on the systematic development of theories supporting particular attributes. This is not to say that we should do away with cumulative scales, only that they may not be the panacea for the measurement of just any attribute.

Like cognitive IRT models (Chapter 4), cognitive diagnostic models provide formal structures for the way in which hypothesized sets of sub-attributes, skills, or characteristics of cognitive processes combine to produce a particular item score. If only one could show particular attributes to be dissectible into an organized series of steps that people take to arrive at an answer to a task that is either right or wrong, then these models could be considered formal tests of the hypothesis that contribute to a better foundation of the attributes and a well-founded scale for the attribute. In particular, in psychometrics and educational measurement, researchers attempted to found tests on a meticulous analysis of the thing they intended to measure (Embretson, 2016). Janssen (2016) provided an overview of applications of the linear logistic test model (Chapter 4) to attribute measurement in the areas of mathematics, science, reasoning, reading, personality, and emotion. Their work is vital for the development of better scales for the key attributes of psychology, but as far as we know this work has had little resonance in psychology and related fields. On the contrary, in psychology and health research alike, researchers aim at using short scales, consisting only of a few items (Kruyen et al., 2012). This preference is fueled by the little time available to question the participants in their research, the desire to reduce the burden on patients when answering questions about anxiety and depression, and examples of short scales—in fact, a limited number of indicators of a construct—that work well in the social sciences when individual differences are of no interest. While all of this is understandable and defendable from mostly practical perspectives, it works against the interest of the scientific development of the human sciences and has the effect of not spending enough energy on developing attribute theory and well-founded scales for the key attributes of the fields.

6

Pairwise Comparison, Proximity,
Response Time, and Network Models

The diversity of the phenomena of nature is so great, and the treasures hidden in the heavens so rich, precisely in order that the human mind shall never be lacking in fresh nourishment

—Johannes Kepler

Introduction

In this chapter, we briefly discuss four approaches to measurement that are conceptually highly interesting but used less frequently for measurement than the methods we discussed in the previous chapters. Pairwise comparison models date back to the 1920s and proximity models originated in the 1960s, but despite the time they have been around, they have never become popular in psychological measurement to a degree that even comes close to that of CTT and IRT. The pairwise comparison model scales items in a first data collection round and then presents the scaled items to persons to determine their scale values in a second data collection round. The proximity models that we discuss model dichotomous and polytomous item scores. Proximity models are IRT models which have unimodal rather than monotone IRFs, and this is because they model a response process based on the proximity of a person's scale location to that of the item. Items and persons are scaled based on one data collection round. Response time models and network psychometrics are of a more recent date and in the short period of their existence have attracted quite some attention. Response time models seem to be gaining popularity in educational measurement, but we see no barriers to using them in measurement in other research areas. Response time models analyze dichotomous item scores together with the response times that people take to answer items. In particular, network psychometrics seems to be quickly growing popular, especially for the assessment of psychopathologies. Network models for psychometrics study networks of symptoms and identify the relationships between symptoms, assuming that symptoms affect one another. Pairwise comparison, proximity, and response time models are latent variable models much comparable in principle to the models we discussed in Chapters 3, 4, and 5, but network models only contain the observable symptoms acting as causal agents upon one another meanwhile lacking one or more latent variables serving that purpose.

Pairwise Comparison Models

The point of departure of models for pairwise comparison is a set of stimuli that judges compare with respect to an attribute. Originally, the stimuli pertained to aspects of attitudes (Thurstone, 1928), but in fact, they can be anything that can be compared with respect to a well-defined attribute. For example, stimuli can be written texts and the attribute can be the quality of the authors' language use, but stimuli can also be politicians who are compared with respect to trustworthiness, children's drawings that are compared with respect to developmental level, and sounds that are compared with respect to their audibility. Central in Thurstone's theorizing about the comparison of stimuli is the discriminal process. Discriminal processes refer to the various, usually unidentified mechanisms people use to react to different stimuli and from which comparisons of different stimuli on a psychological dimension can be accomplished. Thus, a discriminal process is not well defined with respect to its content and operation, and only serves as a generic term for the cognitive or other processes that a stimulus activates and that produce a response or a distribution of responses. The terminology has become well-known, but content of the discriminal process is of no further importance for the formal model Thurstone proposed. Thurstone (1927a, b) assumed that judges accomplished comparisons based on one or more discriminal processes. Each stimulus elicits a discriminal process in people typical of that stimulus, but when people assess a stimulus, they may not only use the typical discriminal process but also discriminal processes similar to the discriminal process typical of the target stimulus. On a hypothetical attribute continuum, the discriminal processes that people use for assessing a stimulus are adjacent to one another and represent different levels of the attribute. Consequently, sets of discriminal processes used for assessing different stimuli often overlap. Thurstone did not explicitly mention the idea of a propensity distribution as we discussed in Chapter 2 in the context of CTT, but he did consider a hypothetical frequency distribution of the discriminal processes used for assessing a particular stimulus. Because people use different sets containing different numbers of adjacent, similar discriminal processes, these frequency distributions vary with respect to mode and dispersion. When the different frequency distributions show overlap, across different occasions—replications, no matter how they are defined—one does not expect the discriminal processes to produce the same result when comparing two stimuli, and the inconsistency is greater the closer the stimuli are with respect to the psychological dimension on which one compares them.

The discriminal process that a particular stimulus typically elicits, and which is typical of a particular level of the attribute, is the modal discriminal process for the given stimulus. Thurstone assumed the frequency distribution of discriminal processes for a given stimulus to be normal. Discriminal processes on any occasion at which stimuli are compared deviate from the modal discriminal process for that stimulus. Deviations above the mode indicate an occasional judgment that indicates, for example, a relatively high level of trustworthiness for a particular politician, and a deviation below the mode indicates a level of trustworthiness that is relatively low. The standard deviation of the distribution is the discriminal dispersion. Discriminal dispersions can vary across stimuli. The difference between scale values of two stimuli on a particular occasion is the discriminal difference. Discriminal differences vary across occasions, and so does their sign. The distance between two stimuli is the difference of their modal discriminal processes.

Interestingly, in specific cases, knowing the discriminal process in detail could shed light on the kind of measurement model one needs, but Thurstone clearly focused on the model he wished to propose for analyzing pairwise comparison data. The discriminal process only served as a summary for everything that happens in a person's mind in preparation of the response she provides when she compares two stimuli, but precisely what happens remained a black box. We now focus on Thurstone's model for paired comparisons. On a historical note, Thurstone's ideas go back to nineteenth-century psychophysicists such as Fechner and Weber (e.g., Stigler, 1986, Chap. 7), and Thurstone (1927a) provided a detailed account of the relations of his model and those of his predecessors. Böckenholt (2006) and Böckenholt and Tsai (2006) provided extensive reviews of models for pairwise comparison, which not only span a long period of interest and study but also span a wide variety of research areas, among which is economics, including consumer behavior and marketing; see Catellan (2012) for additional model developments and application areas. Andrich (1978b) and Jansen (1984) discussed the relation between Thurstone's model and the Rasch model. Here, we briefly discuss the original model (Thurstone, 1927a, b) and the model that became the most popular (Bradley & Terry, 1952; Luce, 1959).

Thurstone Model

Thurstone (1927a, b) used a notation and a conceptual framework that deviated considerably from what we use today and in particular in this monograph. Of course, this is what one finds in many writings published before, say, 1960, and what we tried to do here (and elsewhere) is to adapt the notation and the concepts used in older days to the nomenclature we use today and throughout this monograph. This is to take away potential barriers for understanding the ideas expressed in the older sources, to clarify how particular ideas and models can be understood from today's perspective, and, simply, to improve the readability of the monograph. Having said this, Thurstone must be applauded for the meticulous way in which he laid out his ideas, which helped considerably to translate his ideas to a modern-day text, possible small translation errors aside.

The model for pairwise comparison that Thurstone (1927a, b) proposed assumes that each stimulus or item has a location on a psychological continuum. What Thurstone called a continuum, we called a scale throughout this monograph. Of course, one can only establish whether one has constructed a scale for an attribute based on the data, but to keep the discussion simple, we speak of a scale as if we already knew one exists. Then, each item has a location on the scale, and the responses that a person provides to the item across hypothetical replications form a distribution of responses of which the mode determines the scale location. When a person compares two items, then two discriminal processes are involved, one for each item, and whether item j is preferred to item k or the other way around depends on which of the two discriminal processes dominates the other in the particular confrontation of the person with the two items. If the discriminal process for item j dominates the discriminal process for item k, the person says that item j possesses more of the attribute than item k, and reversely, if the discriminal process for item k dominates the discriminal process for item j, the person says item k possesses more of the attribute than item j. Ties are not allowed (but were allowed in later proposals; see Böckenholt, 2006, and Catellan, 2012). We assume that with every item corresponds a distribution of discriminal processes and that these discriminal processes produce responses quantified as scores following a normal distribution with the mean equal to the item's scale location, which, as always, we denote δ_j, and dispersion σ_j due to variation of discriminal processes that the item elicits in the person's mind. If we let the observable response of person i to item j based on the discriminal process be u_{ij}, then $u_{ij} \sim \mathcal{N}\left(\delta_j, \sigma_j^2\right)$. The distribution is considered across independent replications comparable to the person's propensity distribution as in CTT (Chapter 2), but unlike in CTT, the distribution is the same for all persons.

For person i's comparison of two items, for simplicity, we ignore subscript i and consider the distribution of the differences $u_{jk} = u_j - u_k$. Let the mean of the distribution of u_{jk} be equal to

$$\mu_{u_{jk}} = \delta_j - \delta_k, \tag{6.1}$$

and let the variance be equal to

$$\sigma_{u_{jk}}^2 = \sigma_j^2 + \sigma_k^2 - 2\rho_{jk}\sigma_j\sigma_k; \tag{6.2}$$

then, we have

$$u_{jk} \sim \mathcal{N}\left(\mu_{u_{jk}}, \sigma_{u_{jk}}^2\right). \tag{6.3}$$

Figure 6.1a shows the situation and clarifies how a person compares two items. In both panels of Figure 6.1, the location of the difference $u_{jk}=0$ corresponds with the value 0 on the abscissa, and the difference $\delta_j - \delta_k$ corresponds with the mean of the distribution, $\mu_{u_{jk}}$. For example, if $\delta_j=2$ and $\delta_k=-1$, then $\mu_{u_{jk}}=3$, and the shaded area in Figure 6.1a shows the proportion of the replications in which item k is ranked higher than item j, $p_{k>j}$. If $p_{k>j}>.5$, then one concludes that item k is ranked higher than item j; otherwise,

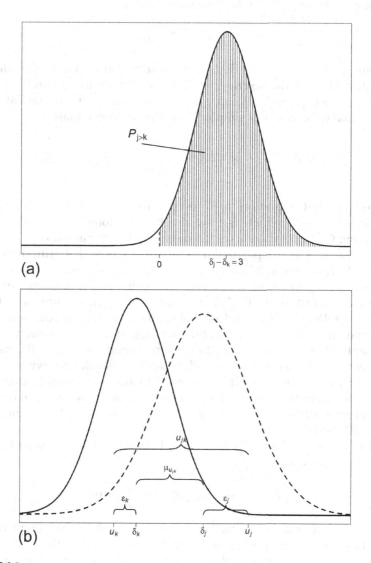

(a)

(b)

FIGURE 6.1

(a) Distribution of differences u_{jk} with $\mu_{u_{jk}}=3$ ($\sigma_{u_{jk}}^2=1.79$), resulting in $P_{j>k}$. (b) Discriminal process for item j ($\delta_j=2$, $\sigma_j^2=2$) and discriminal process for item k ($\delta_k=-1$, $\sigma_k^2=1.55$) with $\rho_{jk}=.5$.

item j is ranked higher than item k. Thus, to decide which of two items should be ranked higher, one compares the outcomes of the discriminal processes associated with the two items.

Following Sijtsma and Junker (2006), one may formulate Thurstone's model of pairwise comparison as an IRT model that closely resembles Lord's normal ogive model in Equation (4.4). Figure 6.1b shows that the observable difference u_{jk} may be written as the sum of its distribution mean, $\mu_{u_{jk}}$, and the normal deviate that we denote ε_{jk}, so that

$$u_{jk} = \mu_{u_{jk}} + \varepsilon_{jk} = \delta_j - \delta_k + \varepsilon_{jk}. \tag{6.4}$$

We assume that this normal deviate is the sum of the separate deviates for the distributions of the discriminal processes for the items j and k; that is, $\varepsilon_{jk} = \varepsilon_j + \varepsilon_k$, see Figure 6.1b. Because the distribution of ε_{jk} is normal, say, standard normal to keep things simple, Thurstone's model equals

$$P\left(u_{jk} > 0\right) = P\left[\varepsilon_{ij} > -\left(\delta_j - \delta_k\right)\right] = \frac{1}{\sqrt{2\pi}} \int_{-(\delta_j - \delta_k)}^{\infty} \exp\left(-u^2/2\right) du. \tag{6.5}$$

We believe that Equation (6.5) points to an interesting feature of Thurstone's model, which is how much Thurstone's model starts out resembling CTT, and then takes it a huge step further as a model for attribute measurement. In particular, Thurstone's model distinguishes an observable response u_j (CTT: item score X_j) from its mean $\mu_{u_j} = \delta_j$ (CTT: true item score T_j) by means of a random error ε_j (CTT: random measurement error E_j), but unlike CTT, Thurstone's model then assumes that random error follows a normal distribution whereas CTT assumes random measurement error has mean 0 and correlates 0 with any other variable; see Chapter 2. As a result, estimation methods and goodness-of-fit assessment methods for Thurstone's model could be developed, whereas this remained problematic for CTT. In the CTT case, later model extensions, such as those made possible by factor models, were necessary to develop a model that allowed the estimation and fit assessment of a modified "classical" model.

Back to the original model formulation, Thurstone's *law of comparative judgment* is formulated as

$$\mu_{u_{jk}} = \delta_j - \delta_k = Z_{jk}\sqrt{\sigma_j^2 + \sigma_k^2 - 2\rho_{jk}\sigma_j\sigma_k}, \tag{6.6}$$

where Z_{jk} is a normal deviate. The law of comparative judgment relates the scale locations of the two items to the dispersion of the discriminal processes and the items' correlation. For each item pair, an equation such as Equation (6.6) holds, and for J items one has $\frac{1}{2}J(J-1)$ such pairs. Each item is characterized by an unknown location parameter and an unknown dispersion parameter, and each item pair is characterized by a correlation. Hence, $\frac{1}{2}J(J-1)$ equations contain $J+J+\frac{1}{2}J(J-1)$ parameters, which are too many to solve from the set of equations. Therefore, Thurstone (1927b) discussed five cases, each case simplified with respect to Equation (6.6) and each case facilitating the estimation of the parameters or a subset of the parameters or commenting on particular aspects of the estimation problem (also, see Tsai, 2000, 2003, for critical comments with respect to identification problems). To provide an impression, Case I entails setting for one item, say, item 1, the scale value equal to 0, that is, $\delta_1 = 0$, and the standard deviation of the discriminal processes equal to 1, that is, $\sigma_1^2 = 1$. In addition, one assumes that all correlations are equal, that is, $\rho_{jk} = \rho$, all j and k, $j \neq k$. These simplifications leave $2J - 1$ unknowns, and one can check that one needs at least five items to have at least as many equations as there are unknowns and have a solvable problem. Case II argues that the model for one person, assuming the availability of hypothetical replications for this person, in several cases can be generalized to a group of persons, each individual producing one comparison per item pair and all persons together producing a distribution of observations comparable to the hypothetical distribution one person allegedly produces. Case III argues in favor of assuming $\rho_{jk} = 0$, all j and k, $j \neq k$; also, see Tsai (2000). Case IV assumes that the standard deviations in each pair differ only little, say, by a quantity that is a fraction of the smaller of the two standard deviations. This opens up possibilities for further simplification of Equation (6.6), leaving $2J - 2$ unknowns and the necessity to have at least four items to have a solvable problem. Finally, Case V assumes that, in addition to $\rho_{jk} = 0$, all j and k, $j \neq k$, all standard deviations are equal, that is, $\sigma_j^2 = \sigma^2$, all j. Consequently, the law of comparative judgment simplifies to

$$\mu_{u_{jk}} = \delta_j - \delta_k = Z_{jk}\sqrt{2\sigma^2}. \tag{6.7}$$

If one assumes $\sigma^2 = 1$ as a unit, then the law becomes even simpler. Other publications often refer to Case V, but it obviously provides a coarse approximation to the real scale; also, see Mosteller (1951a) and Tsai (2000). Goodness-of-fit methods were discussed by, for example, Mosteller (1951b) and Maydeu-Olivares (2001).

The method of pairwise comparison produces a scale for the items but not a scale for persons. To measure persons' scale values, one administers the set of scaled items to the target group one at a time and then asks a person for each item whether she endorses the item. It is assumed that people endorse only the items that are located close to their subjective scale location. Translated to IRFs, one assumes a unimodal IRF. The person's scale value is the mean or the median scale value of the item she endorsed (Johnson & Junker, 2005).

Bradley–Terry–Luce Model

Here we follow the "modern psychometric" notation and conceptual framework Fischer (1974, pp. 186–193) used to explain the exponential model for
pairwise comparison that Bradley and Terry (1952) and Luce (1959) discussed
and which has become known as the Bradley–Terry–Luce model. Their work
builds on that of Zermelo (1929). Johnson and Junker (2005) and Tsukida and
Gupta (2011) argued that the Bradley–Terry–Luce model provides an exponential alternative to the Gaussian model that Thurstone proposed and thus
facilitates the estimation of the model. One may compare this simplification
to IRT where, from the viewpoint of estimation convenience, exponential
models provided an improvement on the older normal-ogive models; see
Chapter 4.

The point of departure is that of Thurstone's model: One presents a set
of J items in pairs to a set of N judges and asks each judge to determine
which of the two items in a pair, j or k, is ranked highest on a dimension that
is explicitly mentioned and defined. Thus, the model assumes that persons
judge items with respect to the same attribute. The comparison task yields
a data matrix with N rows and $\frac{1}{2}J(J-1)$ columns, and entries X_{ijk} that have
one of two values: $X_{ijk}=1$ if person i prefers item j to item k, and $X_{ijk}=0$ if
person i prefers item k to item j. Ties are not allowed (but see Böckenholt,
2006; Catellan, 2012). The column totals in the data matrix are denoted s_{jk}
and count the frequency with which item j is preferred to item k at the group
level. The attribute on which persons compare items is denoted θ, and items
have locations ϵ_j and ϵ_k, which indicate the degree to which items possess
the attribute. Unlike the previous section, we use item location parameters
ϵ (to be distinguished from error components we define in the context of
response time models, which we discuss later), because this choice facilitates
comparison with the Rasch model (Chapter 4), as we will see shortly. The use
of item locations denoted δ in the previous section served to make a connection to present-day treatment of measurement models but not to relate the
Thurstone model to IRT analogs.

The Bradley–Terry–Luce model assumes further that judges are interchangeable in the sense that their individual response models are equal; see
Thurstone's Case II. One finds this assumption throughout this monograph.
Then, the response probability that person i ranks item j higher than item k
equals

$$P\left(X_{ijk}=1\,|\,\epsilon_j,\epsilon_k\right)=\frac{\epsilon_j}{\epsilon_j+\epsilon_k}. \tag{6.8}$$

If $\epsilon_j > \epsilon_k$, then $P\left(X_{ijk}=1\,|\,\epsilon_j,\epsilon_k\right)>.5$, and if $\epsilon_j < \epsilon_k$, then $P\left(X_{ijk}=1\,|\,\epsilon_j,\epsilon_k\right)<.5$. If
$\epsilon_j = \epsilon_k$, then $P\left(X_{ijk}=1\,|\,\epsilon_j,\epsilon_k\right)=.5$. Equation (6.8) is similar to the Rasch model;

this is obvious when we insert the monotone transformations $\theta = \ln \xi$ and $\delta_j = \ln \epsilon_j$ in Equation (4.11) and obtain

$$P\left(X_{ij} = 1 \mid \xi_i, \epsilon_j\right) = \frac{\xi_i}{\xi_i + \epsilon_j}. \tag{6.9}$$

In Equation (6.9), one compares the scale location of person i with the scale location of item j, and an interpretation similar to that for the Bradley–Terry–Luce model, in which a person dominates an item (when $\xi_i > \epsilon_j$) or the other way around ($\xi_i < \epsilon_j$), or when they are tied ($\xi_i = \epsilon_j$), is equally well feasible. Mathematically, it makes no difference how one interprets the parameters, so that the models are simply identical. Statistically, when estimating parameters, the Bradley–Terry–Luce model has only J item parameters to estimate whereas the Rasch model is faced with the circumstance that in addition to J item parameters one has to estimate N person parameters; see Chapter 4.

Because it is so convenient, we outline the estimation of the maximum likelihood estimation of the J item parameters of the Bradley–Terry–Luce model.

BOX 6.1   Estimation of Parameters in the Bradley–Terry–Luce Model

The estimation of the parameters is straightforward using maximum likelihood. First, one writes for one person the probability of choosing one of the two rankings as

$$P\left(X_{ijk} = x_{ijk} \mid \epsilon_j, \epsilon_k\right) = \frac{\epsilon_j^{x_{ijk}} \epsilon_k^{1-x_{ijk}}}{\epsilon_j + \epsilon_k}. \tag{6.10}$$

Equation (6.10) is the basis for the likelihood of the complete data matrix. Let $\mathbf{X}$ denote the data matrix of N rows and $\frac{1}{2}J(J-1)$ columns with scores x_{ijk} as entries, and let ϵ denote the vector of item parameters; then, the likelihood is

$$L\left(\mathbf{X} \mid \epsilon\right) = \prod_{j<k} \prod_i \frac{\epsilon_j^{x_{ijk}} \epsilon_k^{1-x_{ijk}}}{\epsilon_j + \epsilon_k} = \prod_{j<k} \frac{\epsilon_j^{s_{jk}} \epsilon_k^{N-s_{jk}}}{\left(\epsilon_j + \epsilon_k\right)^N}. \tag{6.11}$$

One may notice that in Equation (6.11), individual item scores x_{ijk} have been replaced by column totals s_{jk}. These column totals are sufficient statistics, which can be shown by dividing the likelihood in Equation (6.11), $L(\mathbf{X} \mid \epsilon)$, by the probability of the column totals collected in vector $\mathbf{s}$ conditional on the model parameters collected in vector ϵ. We provide the outcome and refer the interested reader to Fischer (1974, p. 189) for more details; that is, the outcome equals

$$\frac{L(\mathbf{X}|\epsilon)}{P(\mathbf{s}|\epsilon)} = \prod_{j<k}\binom{N}{s_{jk}}^{-1}. \tag{6.12}$$

The point we wish to make is that the outcome in Equation (6.12) does not contain the parameters in ϵ anymore; hence, the column totals are sufficient statistics for the estimation of the model parameters.

One may have noticed that we have $\frac{1}{2}J(J-1)$ column totals for estimating J parameters. A more efficient approach is obtained by noticing that the total frequency of item j being ranked higher than any of the other $J-1$ items, denoted s_j, equals

$$s_j = \sum_{k=1}^{j-1}\left(N - s_{jk}\right) + \sum_{k=j+1}^{J} s_{jk}, \tag{6.13}$$

and then rewriting the numerator in the likelihood in Equation (6.11) as

$$\prod_{j<k}\epsilon_j^{s_{jk}}\epsilon_k^{N-s_{jk}} = \prod_j \epsilon_j^{s_j},$$

yields an alternative equation for the likelihood,

$$L(\mathbf{X}|\epsilon) = \frac{\prod_j \epsilon_j^{s_j}}{\prod_{j<k}\left(\epsilon_j + \epsilon_k\right)^N}. \tag{6.14}$$

Now, the J sum statistics s_j replace the $\frac{1}{2}J(J-1)$ sufficient statistics s_{jk}, and are minimally sufficient for the estimation of the J model parameters ϵ_j. This means that they are less informative for the estimation of the model parameters, which is due to their higher aggregation level, but they do contain the minimum amount of information necessary for estimating the parameters (Fischer, 1974, pp. 190–191). Further elaboration that we skip here shows that the likelihood in Equation (6.14) yields a system of equations,

$$\frac{s_j}{N} = \sum_{k:k\neq j}\frac{\epsilon_j}{\epsilon_j + \epsilon_k}, \tag{6.15}$$

from which the J model parameters can be estimated.

Discussion

Thurstonian pairwise comparison models did not gain a strong foothold in the behavioral sciences, probably because collecting pairwise comparison data costs a lot of effort when the number of items is greater than, say, eight (nine implying already 36 comparisons), and the measurement of individuals requires a second data collection round. Procedures were developed for not having to execute all pairwise comparisons, but the laboriousness of the two-step procedure soon inspired Thurstone and Chave (1929) to develop the scaling methods of equal-appearing intervals and successive intervals that asked people to sort items into scale intervals without having to execute all the separate pairwise comparisons. We ignore these methods that as far as we know are out of use and move to a consecutive class of measurement methods that primarily became known as unfolding models but that we shall discuss as proximity models.

Proximity Models

Two response processes produce data that constitute the basis for constructing scales for psychological attributes (Andrich, 1988a; Andrich & Luo, 1993) and, we add, social science and health attributes as well. One response process involves the principle that as a person's scale value is higher, the probability that she gives a positive response also is higher. The data the response process produces are scores expressing whether the response was positive or not with respect to the target attribute, meaning that depending on the attribute, scores express the degree to which a response to a cognitive problem was correct or the person endorsed a particular item assessing a trait or an attitude. The scale that one constructs based on data like this is cumulative; hence, the scale is a cumulative scale. The more items the person answered positively, the higher her scale score. The CTT model and the factor model (Chapter 2), the IRT models (Chapters 3 and 4), ordinal versions of LCMs and CDMs (Chapter 5), and the pairwise comparison model (this chapter) all study cumulative scales. Readers will recognize that cumulative scales are dominant in the social, behavioral, and health sciences. The other process involves the principle that as the person's scale location is closer to the item's scale location, the greater the probability that she will express preference for the item. Here, it is not the number of preferred or otherwise selected items that determines a person's scale location but which items the person picked that determine her location. The person's scale location is close to the locations of the items she picked and remote from the locations of the items she did not pick. Hence, the scale is based on the proximity between person and item and may be called a proximity scale.

Proximity scales are sometimes called preference scales, and originally, they were called unfolding scales. The metaphor of unfolding expresses the idea that if one imagines a person's scale location as a hinge and one folds the scale parts from both sides to form one common scale, one obtains the item ordering for this person (Coombs, 1964, Chap. 5). The general scale on which persons and items are located is the J-scale (J for joint), and personal scales are called I-scales (I for individual). For example, if the J-scale for person i and items A, B, C, D, and E looks like

$$A\text{-----}B\text{---}i\text{------}C\text{-----}D\text{--------}E,$$

then *folding* the scale at location i yields the I-scale for person i, equal to

$$i\text{---}B\text{---}C\text{--}A\text{---}D\text{--------}E.$$

The item ordering may vary, depending on the person's position. Clearly, based on a known J-scale, every person has an I-scale but, reversely, if one only knows the data from asking people, for example, to simultaneously rank J items according to their preference, the question is whether a J-scale is consistent with the data. *Unfolding* the I-scales is a technique for finding a J-scale, provided one exists. This is a deterministic method but, for example, Bechtel (1968), Sixtl (1973), Heiser (1981), Van Schuur (1984, 1988), and Van Blokland-Vogelesang (1991) proposed probabilistic versions. We prefer to use the intuitively appealing name of proximity scale and focus on what we consider an elaboration of the older methods that may be considered part of IRT for unimodal IRFs. The reader will have noticed that the second phase of Thurstone scaling, in which people receive scale scores, was based on the proximity principle.

Proximity models aim at scaling a set of items the content of which matches people's preferences together with a set of persons on one dimension (see DeSarbo & Hoffman, 1986, for a multidimensional approach). The data come from a sample of people indicating their preference for each of J stimuli or items. The items may describe viewpoints concerning various political issues, bitterness of brands of beer, or statements with respect to immigration. For example, people are asked to indicate whether they agree with viewpoint j concerning a flight tax in relation to fighting consequences of climate change, whether they think the bitterness of brand j is just right, or whether they agree with statement j putting a quota on the number of immigrants entering the country per annum. The questionnaires consist of J viewpoints, beer brands, and statements. In each example, the answer people provide is either yes or no, but different answers are possible, depending on the task. The scales' interpretations are political conservatism, bitterness (as a taste), and attitude with respect to immigration. The idea is that a person implicitly compares the item's scale position to her own ideal scale position and answers yes with higher probability the smaller the perceived distance between the two scale positions. The scaling method does not model the process that leads to the expression of a preference, only the outcome of such a process. Hence, whether people really compare the items to their own positions serves as a thought model, but the real process is unknown and not a topic of further research, as far as we know.

On a unidimensional scale, the perceived distance between one's ideal scale location and the items' scale locations can extend in two directions. For example, one may assess one's ideal scale location for political conservatism below (i.e., less conservative) or above (i.e., more conservative) the scale location of viewpoint j. Similarly, one's ideal preference for bitterness may be either below (i.e., less bitter) or above (i.e., more bitter) beer j's bitterness, and one's acceptance of immigration may be below or above statement j's location. For example, let the statement be "The government should allow no more than 1,000 immigrants from the Middle East each year into the country", then one may disagree, because one is of the opinion that 1,000 immigrants is too many (personal location below item location) or because one thinks 1,000 is not enough (personal location above item location).

Deterministic Model

Following Post (1992, p. 6) and based on Coombs' parallelogram model (Coombs, 1964, Chap. 4), we assume that people (1) differ with respect to their ideal scale locations but agree with respect to the ordering of the items' scale locations; and (2) prefer only those items of which the locations are close to their own ideal scale locations. Unlike Coombs, we use the latent variable concept from the IRT context to be consistent with the treatment of other topics in this monograph. Thus, let θ denote the scale and let δ_j be the scale location of item j. Notation d_j expresses the length of the scale interval in which a person who is located there prefers item j. We assume that δ_j lies halfway along the interval. The interval sometimes is called the "latitude of acceptance" (Coombs, 1964, pp. 196, 298). The location of intervals for different items varies across items and so does the interval length. Then, mathematically, the deterministic proximity model with uniform IRFs (Post, 1992, p. 18) equals

$$|\theta-\delta_j|\le\frac{d_j}{2} \Leftrightarrow P(X_j=1|\theta)=1,$$
(6.16a)

and

$$|\theta-\delta_j|>\frac{d_j}{2} \Leftrightarrow P(X_j=1|\theta)=0.$$
(6.16b)

To have unique orderings of people and items, the IRFs have to be restricted such that

(1) If persons a and c both prefer item j and person b is located between persons a and c, so that $\theta_a < \theta_b < \theta_c$, then person b also prefers item j (Figure 6.2a); and

(2) If three items are ordered such that $\delta_j < \delta_k < \delta_l$, then the three lower (L) bounds and the three upper (U) bounds must exhibit the same ordering: $\delta_j^L < \delta_k^L < \delta_l^L$ and $\delta_j^U < \delta_k^U < \delta_l^U$ (Figure 6.2b).

The restriction on lower and upper bounds limits the interval length d_j of the different items. Figure 6.2b shows for this example that the interval lower bounds, the item-location parameters, and the interval upper bounds divide the latent variable scale into seven adjacent intervals, each of which corresponds with a unique pattern of three item scores. Two patterns are missing in this example, which are (010) and (101). Figure 6.2c shows a different set of three IRFs that satisfy the restrictions with respect to lower and upper bounds and that allow pattern (010) as one of the five possible patterns for this IRF configuration. Only pattern (101) is impossible to obtain with the model in Equations (6.16a) and (6.16b) including the order restrictions on the

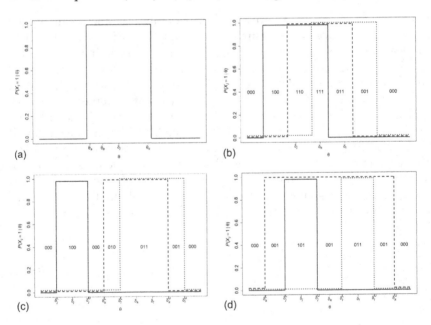

FIGURE 6.2
(a) Deterministic proximity models showing general response principle for one item. (b) Order restrictions on parameters of different items. (c) Three items allowing five possible item-score patterns. (d) Three items violating order restrictions on item parameters.

lower and upper bounds. This is an error pattern, comparable to the error pattern in the Guttman (1944, 1950) scalogram model defined by (01) whenever $\delta_j < \delta_k$; that is, one cannot fail an easier item when one has succeeded a more difficult item (Figure 4.2). Figure 6.2d shows three IRFs that violate the order restrictions on the upper bounds: That is, $\delta_j < \delta_k < \delta_l$, but lower and upper bounds with respect to items k and l have been interchanged, so that $\delta_k^L < \delta_j^L < \delta_l^L$ and $\delta_j^U < \delta_l^U < \delta_k^U$ (Post, 1992, p. 19). The third interval in Figure 6.2d corresponds with item-score pattern (101), which is not allowed in the model we defined in Equations (6.16a) and (6.16b), because it stands in the way of an unambiguous item ordering. Thus, item-score vectors consistent with the model show the following strings of 0s and 1s: (1) first a string of 1s and then a string of 0s; (2) first a string of 0s, then a string of 1s, and finally a string of 0s; or (3) first a string of 0s and then a string of 1s. Table 6.1 shows the typical parallelogram structure that emerges when items (columns) are ordered by increasing item location and persons (rows) by increasing person location.

An interesting feature of the deterministic model is that the width of the latitude of acceptance corresponds to the discrimination power in cumulative IRT models, where a steeper IRF slope corresponds to a higher item discrimination (Figure 4.1). Higher item discrimination means a sharper distinction of people to the left and to the right of the IRF's inflexion point: That is, more people have either a low or a high response probability, and fewer people have an intermediate response probability. For the deterministic Guttman model [Equations (4.2a) and (4.2b)], the IRF sharply divides the distribution of latent variable θ in two parts, one part having response probability 0 and the other part having response probability 1 (Figure 4.2).

TABLE 6.1

Parallelogram Structure That Emerges When Nine Items (Columns) Are Ordered by Increasing Item Location and Persons (Rows) by Increasing Person Location

0	0	0	0	0	0	0	0	0
1	0	0	0	0	0	0	0	0
1	1	0	0	0	0	0	0	0
1	1	1	0	0	0	0	0	0
1	1	1	1	0	0	0	0	0
0	1	1	1	1	0	0	0	0
0	0	1	1	1	1	0	0	0
0	0	0	1	1	1	1	0	0
0	0	0	0	1	1	1	1	0
0	0	0	0	0	1	1	1	1
0	0	0	0	0	0	1	1	1
0	0	0	0	0	0	0	1	1
0	0	0	0	0	0	0	0	1
0	0	0	0	0	0	0	0	0

The IRFs in Figure 6.2c,d show varying latitudes of acceptance, and it is clear that given a fixed item location, a narrower latitude of acceptance assigns fewer people a preference probability 1 and on both sides of the latitude of acceptance, more people a preference probability 0. Hence, the item better distinguishes people who prefer the item from the other people, and thus it has greater discrimination power.

Probabilistic Models

Each of the models we discuss makes three assumptions of which the first two are well-known. First, the assumption of unidimensionality entails a latent variable θ on which persons and items are located and that serves as a scale for the target attribute. Second, local independence means that the joint distribution of the J item scores is the product of the J marginal distributions conditional on θ. Third, the IRF $P(X_j = 1|\theta, \delta_j)$ is assumed weakly unimodal, meaning that there exists a value δ_j for which $P(X_j = 1|\theta, \delta_j)$ reaches its maximum, so that for $\theta < \delta_j$ the IRF is nondecreasing and for $\theta > \delta_j$ the IRF is nonincreasing. We primarily consider models for dichotomous item scores, which are the most frequently studied proximity models, but also discuss a model for polytomous items.

Andrich (1988a) suggested that proximity models are especially suited for attributes that go through a developmental sequence where skills mastered at previous stages fade away as soon as the person progresses to the next stages where more advanced skills are learned. Similarly, one may imagine asking volunteers of a political party whether they are prepared to use particular means to reach potential voters, ranging from putting up posters over town, handing out flyers at a shopping mall, sending people emails, calling them up at home, and so on, up to visiting them at home. Possibly, someone who is prepared to visit people at their home is not prepared to engage in weaker means of communication when they consider them less effective and therefore a waste of time. Andrich proposed a logistic model in which the probability of possessing a particular skill or preferring a particular means of communication, both represented as items, depends on the squared distance between the person parameter θ_i and the item location, such that

$$P(X_j = 1|\theta_i, \delta_j) = \frac{\exp\left[-(\theta_i - \delta_j)^2\right]}{1 + \exp\left[-(\theta_i - \delta_j)^2\right]}. \tag{6.17}$$

This is the *squared simple logistic model*. Figure 6.3 provides the IRFs of two items, denoted $P(X_j = 1|\theta)$ and $P(X_k = 1|\theta)$. The probability of preferring item j is maximal when $\theta = \delta_j$; then, $P(X_j = 1|\theta) = .5$. On both sides of this maximum, the response probability decreases as the distance of the person location to the item location increases. The IRF is symmetrical in δ_j. One

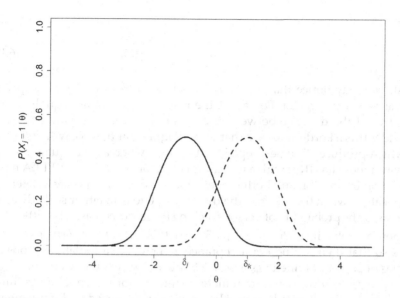

FIGURE 6.3
The IRFs of item j (solid curve) and item k (dashed curve) under a squared simple logistic model.

may notice that the maximum response probability is a technical property of the model but that there is no compelling empirical reason to expect that response probabilities for items cannot exceed .5. In addition, there is no *a priori* reason why the response probability at equal distances on either side of the IRF's maximum should be the same. Whether the model provides a good description of the data has to be investigated in concrete cases, as always.

The squared simple logistic model resembles the Rasch model (Chapter 4). Unlike the Rasch model, the squared simple logistic model does not have simple sufficient statistics, and person and item parameters are not separable; hence, conditional maximum likelihood is unfeasible here. Instead, Andrich (1988a) discussed a joint maximum likelihood estimation procedure for the parameters, characterized by inconsistency of estimates, and derived standard errors and Fisher's information function, which is bimodal.

Comparable to the squared simple logistic model, the *PARELLA model* (Hoijtink, 1990, 1991) has a single-peaked IRF dependent on the squared difference between the person and the item parameters, but also dependent upon a power parameter denoted ϕ, so that

$$P\left(X_j = 1 | \theta_i\right) = \frac{1}{1 + \left[\left(\theta - \delta_j\right)^2\right]^\phi},$$ (6.18a)

and

$$P(X_j = 0 \mid \theta_i) = 1 - P(X_j = 1 \mid \theta_i). \qquad (6.18b)$$

First, one may notice that for $\theta = \delta_j$, probability $P(X_j = 1 \mid \theta) = 1$, irrespective of the power parameter. For $\phi = 0$, the response probability equals .5, irrespective of the distance between the person and the item parameters. We consider this a borderline case that we will ignore, and so we will ignore $\phi < 0$, which produces "single-dipped" IRFs. Thus, whereas the squared simple logistic model has IRFs with a maximum of .5, for $\phi > 0$, the PARELLA model has a maximum of 1, and both models thus seem to incorporate a deterministic feature when $\theta = \delta_j$. When the distance of θ to δ_j in either scale direction increases, the probability of person i picking the item decreases. For $|\theta - \delta_j| = 1$, response probability $P(X_j = 1 \mid \theta) = .5$, and this model property fixates the IRFs so that their latitude of acceptance, which is manipulated by positive values of power parameter ϕ, is restricted; see Figure 6.4 for examples.

To avoid the inconsistent parameter estimates for which joint maximum likelihood estimation is known, Hoijtink (1990) proposed and implemented marginal maximum likelihood estimation in combination with the expectation maximization (EM) algorithm (Dempster et al., 1977). A simulation study showed that the estimation procedure was able to recover the data-generating parameter values with sufficient precision. Hoijtink (1991) discussed

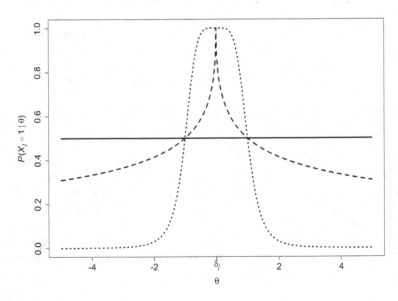

FIGURE 6.4
Three IRFs under the PARELLA model with $\delta_j = 0$ ($j = 1,2,3$): $\phi = 0$ (solid curve), $\phi = 0.25$ (dashed curve), and $\phi = 2.5$ (dotted curve).

a descriptive method that compares observed and expected IRFs and thus provided a tool for assessing the fit of the PARELLA model to the data.

Andrich and Luo (1993) discussed an alternative proximity model based on the cumulative Rasch model for polytomous items (Andersen, 1977), and Verhelst and Verstralen (1993) came up with the same idea based on the partial credit model [Masters, 1982; also, see Chapter 4, Equation (4.76)]. The cumulative Rasch model for polytomous items and the partial credit model are different parameterizations of one another and thus are the same model (see Chapter 4), rendering the approaches by Andrich and Luo (1993) and Verhelst and Verstralen (1993) equivalent. Arbitrarily, we take the approach by Andrich and Luo (1993) as our point of departure. The logic for their choice was to recognize that the single-peaked IRF, $P(X_j = 1|\theta)$, typical of the response process in proximity models implies a second response function, $P(X_j = 0|\theta)$, such that for each latent variable value the two probabilities add to 1. This means that as θ increases, $P(X_j = 0|\theta)$ decreases until $\theta = \delta_j$, and then increases as θ increases further; see Figure 6.5a, which shows that both response functions are one another's mirror image. More important, Figure 6.5a shows that when a person picks item j, her scale value likely is close to the item location, but if she does not pick item j, her scale value may be on either side of the item location; one simply does not know. Cumulative scaling models do not share this ambiguity.

The Rasch model for polytomous items can be formulated for three ordered response categories, that is, $M = 2$, and $X_j = x$, with $x = 0,1,2$. Figure 6.5b shows the response functions for three ordered item scores, and the similarity with the response functions for the proximity model in Figure 6.5a is obvious, but an important difference is that each of the three item scores is governed by its own response probability with three kinds of item parameters. The item location parameter δ_j corresponds with the maximum probability of obtaining item score $x = 1$. Threshold parameter τ_{1j} corresponds to the intersection point of response functions $P(X_j = 0|\theta)$ and $P(X_j = 1|\theta)$, and threshold parameter τ_{2j} corresponds to the intersection point of $P(X_j = 1|\theta)$ and $P(X_j = 2|\theta)$. In $\theta = \tau_{1j}$, the probabilities of obtaining either item scores $x = 0$ or $x = 1$ are equal, and in $\theta = \tau_{2j}$, the probabilities of obtaining either item scores $x = 1$ or $x = 2$ are equal. Hence, parameter $\varphi_j = \frac{1}{2}(\tau_{2j} - \tau_{1j})$, lies halfway between the two thresholds and represents the distance from the item location to each of the two thresholds. Beyond a distance φ_j on either side of midpoint δ_j, the response probability of an item score 0 or 2 exceeds the response probability of item score 1. Parameter φ_j is the unit parameter of item j; a greater unit means a larger scale area in which score 1 is dominant, suggesting weaker discrimination. In a probabilistic context, the item's latitude of acceptance is the interval between the two threshold parameters, and unit parameter φ_j is the latitude of acceptance parameter (Andrich, 2016).

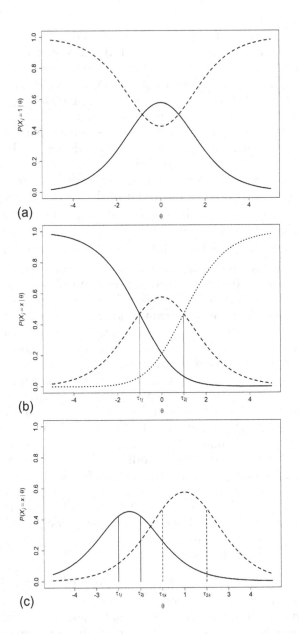

FIGURE 6.5

(a) $P(X=1|\theta)$ (solid curve) and $P(X=0|\theta)$ (dashed curve) for a dichotomous proximity model. (b) Category characteristic curves $P(X_j=0|\theta)$ (solid curve), $P(X_j=1|\theta)$ (dashed curve), and $P(X_j=2|\theta)$ (dotted curve) for the polytomous Rasch model with $\delta_j=0$, $\tau_{1j}=-1$, $\tau_{2j}=1$, and $\varphi_j=\frac{1}{2}(\tau_{2j}-\tau_{1j})=1$. (c) Hyperbolic cosine model for item j (solid curve; $\delta_j=-1.5,\tau_{j1}=-2,\tau_{j2}=-1,\varphi_j=0.5$) and item k (dashed curve; $\delta_k=1,\tau_{1k}=0,\tau_{2k}=2,\varphi_k=1$).

In the next step, given the resemblance of the response process for binary proximity data and the cumulative Rasch model for items having three ordered scores, it makes sense to adapt the latter model to binary proximity data. This can be done by maintaining the response function for $x=1$ representing "picking item j", $P(X_j=1|\theta)$, and combining the two response functions for $x=0$ and $x=2$ together representing "not picking item j", with probability equal to $1-P(X_j=1|\theta)$. Skipping some algebraic developments, but noticing that the hyperbolic cosine function, denoted cosh, equals $\cosh(a)=[\exp(-a)+\exp(a)]/2$, the *hyperbolic cosine model* for proximity data is defined as

$$P(X_j=1|\theta)=\frac{1}{\gamma_j}\exp(\varphi), \tag{6.19a}$$

and

$$P(X_j=0|\theta)=\frac{1}{\gamma_j}2\cosh(\theta-\delta_j), \tag{6.19b}$$

with normalizing constant

$$\gamma_j = \exp(\varphi_j)+2\cosh(\theta-\delta_j),$$

and

$$\cosh(\theta-\delta_j)=[\exp(\delta_j-\theta)+\exp(\theta-\delta_j)]/2.$$

Figure 6.5c shows response functions for the hyperbolic cosine model for two items with two different unit or discrimination parameters, φ_j and φ_k. The item with the smaller unit parameter, which is item j, discriminates better. The maximum value for $P(X_j=1|\theta)$ equals $\dfrac{1}{1+2\exp(-\varphi_j)}$ (Johnson & Junker, 2003), and as φ_j increases, the maximum probability of picking item j increases. Thus, unlike the squared simple logistic model and the PARELLA model, the hyperbolic cosine model does not define the maximum probability of picking item j at a fixed value, such as .5 or 1. By fixing unit parameter $\varphi_j=\varphi=\ln 2$, and inserting this value into Equations 6.19a and 6.19b for the hyperbolic cosine model, one obtains the *simple hyperbolic cosine model*. This model assumes equal discrimination for all items. Comparable to the squared simple logistic model [Equation (6.17)], the maximum value of $P(X_j=1|\theta)$ in the simple hyperbolic cosine model equals .5.

The hyperbolic cosine model and the identical model version that Verhelst and Verstralen (1993) proposed allow any maximum probability between 0 and 1 of picking the target item. This maximum varies across items and is determined by the data, not by a structural limitation imposed by the model, and thus provides a more flexible approach than any of the other models we discussed. The simple hyperbolic cosine model resembles the Rasch model, but unlike the Rasch model, it is not an exponential model and it does not have simple sufficient statistics for parameter estimation. Thus, conditional maximum likelihood estimation (Chapter 4) is unfeasible. Andrich and Luo (1993) used joint maximum likelihood estimation to derive an (theoretically inconsistent) estimation procedure for the hyperbolic cosine model, together with standard errors for parameter estimates and information functions. The information functions are bimodal, as in the squared simple logistic model [Equation (6.17)]. Verhelst and Verstralen (1993) discussed marginal maximum likelihood estimation in combination with expectation maximization (EM: Dempster et al., 1977). Maximum likelihood estimation is problematic in all proximity models, because the models' likelihood can have several local maxima, and thus estimation may run into problems. Verhelst and Verstralen (1993) noticed this problem for the hyperbolic cosine model, and Johnson and Junker (2003) pointed the problem out in general for proximity models. The latter authors proposed a Bayesian estimation procedure using a Markov chain Monte Carlo technique to estimate the model parameters for the hyperbolic cosine model and suggested that the method is applicable to other proximity models as well.

Andrich (2016) discussed a Pearson chi-square statistic that assesses whether the observed IRF is consistent with the IRF based on the model. Specifically, in a class of people denoted g, the statistic assesses the squared difference between the observed frequency of endorsement with the target item j and the frequency expected by the estimated model, and then adds the squared differences across adjacent and non-overlapping classes, $g = 1,...,G$, and across items, $j = 1,...,J$. The statistic has $(G-1)(J-1)$ degrees of freedom. The test has more power as the dispersion of the item locations is greater and the sample size is greater. Verhelst and Verstralen (1993) proposed using an asymptotic chi-square G^2 test statistic that also facilitates the testing of nested models against one another.

With their graded unfolding model, Roberts and Laughlin (1996) took up the generalization of the hyperbolic cosine model to polytomous item scores (also, see Roberts, 2016). The authors modeled the response process of interest by means of the rating scale model [Andrich, 1978a; also, see Chapter 4, Equation (4.77)], but one might use other cumulative IRT models as well. If an item has $M+1$ ordered response categories scored $x = 0,...,M$, then a rating scale model formulation with $C=2M+1$ hypothetical, subjective response options, "scored" $y = 0,...,C$, in pairs reflecting opposite reasons to rate the item, for example, as disagree, describes the response process based on a person proximity from either below or above. For example, four $(M=3)$ observable item scores may correspond to *strongly disagree* $(x=0)$, *disagree* $(x=1)$, *agree* $(x=2)$, and *strongly agree* $(x=3)$, and the eight $(C=2M+1=7)$ subjective item scores reflect *strongly disagree from below* $(y=0)$, *disagree from* below ($y=1$), *agree from below* ($y=2$), *strongly agree from below* ($y=3$), *strongly agree from above* ($y=4$), *agree from above* ($y=5$), *disagree from above* ($y=6$), and *strongly disagree from above* ($y=7$). Thus, an observed item score such as $x=0$ corresponds to two subjective item scores, $y=0$ and $y=7$. Likewise, $x=1$ corresponds with $y=1$ and $y=6$, and so on. Hence, $X=x$ corresponds with $Y=x$ and $Y=C-x$. The graded unfolding model is derived from

$$P(X_j = x \mid \theta) = P(Y_j = x \mid \theta) + P(Y_j = C - x \mid \theta), \tag{6.20}$$

where the response probabilities are rating scale models and the derivation takes into account that the threshold parameters τ_x and τ_{C-x} have opposite signs, so that $\tau_x = -\tau_{C-x}$; see Equation (4.77). The graded unfolding model equals

$$
\begin{aligned}
&P(X_j = x \mid \theta) \\
&= \frac{\exp\left[x(\theta - \delta_j) - \sum_{k=0}^{x}\tau_k\right] + \exp\left[(C-x)(\theta - \delta_j) - \sum_{k=0}^{x}\tau_k\right]}{\sum_{w=1}^{M}\left\{\exp\left[w(\theta - \delta_j) - \sum_{k=0}^{w}\tau_k\right] + \exp\left[(C-w)(\theta - \delta_j) - \sum_{k=0}^{w}\tau_k\right]\right\}}.
\end{aligned} \tag{6.21}
$$

The operation in Equation (6.21) causes the "threshold" parameter on the left-hand side to lose the interpretation it has in the rating scale model. For four item scores, Figure 6.6 shows the response functions. De la Torre, Stark,

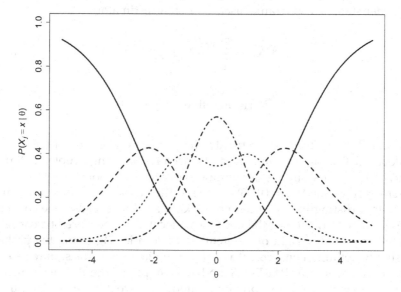

FIGURE 6.6
Polytomous unfolding model for item j, with $\delta_j=0$, eight threshold parameters $\tau_k = -3.5 + k$ for $k = 0,\ldots,7$), yielding four category response curves: $P(X_j = 0 \mid \theta)$ (solid curve), $P(X_j = 1 \mid \theta)$ (dashed line), $P(X_j = 2 \mid \theta)$ (dotted line), and $P(X_j = 3 \mid \theta)$ (dashed dotted line).

and Chernyshenko (2006) proposed Bayesian estimation of the parameters in the generalized graded unfolding model, including item discrimination parameters (also, see Johnson & Junker, 2003).

Discussion

Based on work by Post (1992) with respect to a nonparametric proximity model (also, see Van Schuur, 1984, 1988), Johnson (2006) proved ordering properties for proximity models. For example, Post (1992, p. 33) assumed uni-dimensionality, local independence, and weakly unimodal IRFs such that if the IRF has its maximum for a range of θ values, the item location δ_j is the midpoint of this range (Johnson, 2006) so that δ_j is uniquely defined, and adds two assumptions about stochastic ordering based on observable item scores, of which we only discuss the fourth. Arbitrarily define an item order-ing, such that $\delta_1 < \delta_2 < \cdots < \delta_J$. The fourth assumption is that as the item loca-tion shifts further to the right on the scale, the probability that, given $X_j = 1$, $\theta > \theta_0$, with θ_0 fixed, is nondecreasing; that is, $P(\theta > \theta_0 \mid X_j = 1)$ is nondecreas-ing in j. Johnson (2006) showed that several but not all parametric proximity models satisfy these assumptions and then showed that items and persons can be ordered consistently using simple observable statistics. For items this is the item rank number r_j $(r_j = 1, \ldots, J)$ or the bounded item quantile $q_j = \dfrac{r_j}{J}$ $(q_j = \dfrac{1}{J}, \ldots, 1)$. Let $\mathbf{X}_i$ be the vector with J item scores for person i, then the per-son order statistic is the rank-based Thurstone estimator,

$$T_j^q(\mathbf{X}_i) = \frac{\sum_{j=1}^J q_j X_{ij}}{\sum_{j=1}^J X_{ij}} \text{ if } \sum_{j=1}^J X_{ij} > 0; \qquad (6.22a)$$

$$T_j^q(\mathbf{X}_i) \text{ is undefined if } \sum_{j=1}^J X_{ij} = 0. \qquad (6.22b)$$

The rank-based Thurstone estimator has values on the unit interval. The rank-based Thurstone estimator can be used for ordering people in all prox-imity models satisfying the assumptions outlined previously.

Verhelst and Verstralen (1993) discussed the problem that proximity mod-els cannot distinguish between not picking an item because one considers oneself to be to the left of the item's scale location or because one considers oneself to be to the right of the item's scale location. Ignoring the problem by simply defining item score 0 as not picking item j (see the squared simple logistic model and the PARELLA model) or adapting cumulative IRT models for ordered item scores to a proximity model for two item scores by combin-ing the response functions for item scores 0 and 2 (hyperbolic cosine model and graded unfolding model) invites problems with respect to model iden-tification and parameter estimation. Collecting data asking people whether they agree below or above the target item could provide a solution to the problem of response ambiguity, but, as far as we know, whether this is a viable solution has not been investigated thus far.

Despite the effort that several researchers put into the topic (e.g., Polak, 2011), proximity models are not very popular techniques for scale construction. The main reason is that it proves difficult to construct items that elicit responses consistent with the typical unimodal IRF. First, one has to formulate the item content such that people with opposite scale locations have the same probability of picking the item (in case of binary-scored items), while people located in between these groups have a greater probability. This simply is a difficult exercise, and, except for a couple of exceptions, the result sometimes may look rather unwieldy. Second, even if one succeeds in formulating a set of typical "unimodal items" by appearance, one may find that people produce responses that are best modeled using a cumulative IRT model. Let us consider once more the example asking volunteers of a political party whether they are prepared to use particular means to reach potential voters. We suggested that someone who is prepared to visit people at their home may not be prepared to engage in weaker means of communication, but one could also argue that someone whose ideal point coincides with visiting people at home also endorses weaker means of action, simply because such means may also contribute to raising votes. We do not argue that unimodal IRFs are unrealistic, only that for many applications they may be not as realistic as cumulative IRFs. How people from a particular population respond to particular stimuli is an empirical issue.

Some authors argued in favor of proximity models in case of personality trait measurement. Stark, Chernyshenko, Drasgow, and Williams (2006) compared what they called ideal-point models (Coombs, 1964, p. 80), a terminology synonymous to proximity models, to IRT models with monotone IRFs and found that the former models better fitted a couple of personality scales than their traditional IRT counterparts. The authors argued that, generally, the usefulness of ideal-point models is difficult to assess using traditional personality inventories, because the items have been constructed with monotone IRFs in mind, that is, items that have a location that is rather extreme, whereas ideal-point models show their merit best when items have a rather neutral scale location. Carter, Dalal, Boyce, O'Connell, Kung, and Delgado (2014) adopted a similar position and argued that a curvilinear relation may exist between items measuring the Big Five trait of conscientiousness and performance; being too conscientious may result in worse performance due to overconcentration and too much concern. CTT, factor analysis, and IRT obscure this single-peaked relationship by assuming a monotone relation. Based on a literature review, the authors concluded that ideal-point models fit non-cognitive data better than models assuming a monotone relation between item and performance scale, because they are consistent with the internal process in respondents when they ask themselves whether their location fits the item's location or ideal point. Liu and Chalmers (2018) discussed a large number of unidimensional and multidimensional proximity models and provided software to analyze data.

Response Time Models

The idea that the time it takes a person to complete a task holds information about the person's mode and skill to tackle particular tasks has been around for quite some time. In the older literature (Gulliksen, 1950; Lord & Novick, 1968; Thurstone, 1937), completion time often appeared as time pressure put upon a person when she is required to complete as many tasks as she can within a fixed amount of time, and it is recorded how many tasks the person completes. Thus, "response time" is fixed and the number of completed tasks is the individual-difference variable across people. Tasks performed under a fixed time limit are quite simple and easy, and examples are simple administrative job skills such as counting, categorizing, and storing of information. Such tasks are assumed to be of trivial difficulty, so that errors, if any, can be assumed coincidental and then ignored, meaning that the number of tasks completed is the test score irrespective of the correctness of the task outcome. However, this clearly represents an extreme case, because most tasks do not have a difficulty level up to a point that one can consider failure to come up with the correct task outcome a matter of coincidence. For example, counting the number of hits on a keyboard that a person produces within a fixed and restrictive time limit tells us something about her working speed, but ignoring the number of errors arguably is not a good idea, because the number of errors, or reversely, the number of correct hits, also is informative. One would not want to hire someone who types at the speed of light and only produces errors. Obviously, because the vast majority of tests consist of items addressing more-demanding cognitive processes, ignoring the correctness of a response is out of the question. On the other hand, one might as well ask whether ignoring the speed at which a person solves a non-trivial cognitive problem and finds an answer, right or wrong, is such a good idea and whether the information about speed should not be part of the performance assessment in addition to the number-correct score or other scores indicating quality of performance. These considerations already give a hint that models for performance speed and performance quality deserve a place in measurement.

Gulliksen (1950, p. 230) spoke of a speed test if the items were of trivial difficulty, thus allowing counts of completed items only and ignoring the issue of correctness of responses. He (ibid., pp. 230–231) distinguished speed tests from power tests that consist of items addressing non-trivial cognitive processes, skills, or other attributes, and that allow assessing test performance by means of number-correct and related scores, thus ignoring speed issues. In principle, unlimited time is allotted to solve the problems. However, also in the case of pure power tests one may ask if it does not make sense to distinguish between the performances of two persons both solving the same problem but one much faster than the other. In addition, what to think of two persons having the same ability level but one working at a much slower pace than the other does, even if the test administrator puts some time pressure on both? Without implying that the slower person delivers the weaker performance, it makes sense to record their time performances and study how to use them in, for example, decision making based on test performances.

Thus, it seems reasonable to assume that with almost every test performance it makes sense to consider a person's working speed in addition to her task achievement and consider these two inputs as complementary and interesting sources of information for the assessment of an attribute. Based on the literature (also, see Molenaar & Visser, 2017), we discern two approaches to response time modeling. The first approach is typical psychometric, aimed at measuring individual differences. We discuss the lognormal model in more detail, which produces results that can be used in three ways. First, response times are used to increase measurement precision of the latent variable that is usually assessed only using number-correct scores and related scores (e.g., Bolsinova & Tijmstra, 2018; Molenaar, Tuerlinckx, & Van der Maas, 2015). Second, response times are used to identify aberrant responding (Marianti, Fox, Avetisyan, Veldkamp, & Tijmstra, 2014), for example, in computerized adaptive testing (Van der Linden & Van Krimpen-Stoop, 2003) and for detecting differences between response strategies (Van der Maas & Jansen, 2003). Third, response times are used for constructing tests (Van der Linden, Scrams, & Schnipke, 1999). The second approach is typical cognitive and recently made its appearance in psychometrics. Within this approach, we discuss the diffusion model, which may be considered appropriate primarily for studying the cognitive processes eliciting responses to items. We continue by discussing some preliminaries for the first approach to response time modeling and return to the second approach later.

Several proposals have been made for models incorporating both parameters for speed and achievement and combinations of separate models for either speed or achievement (e.g., Jansen, 1997; Maris, 1993; Rasch, 1960; Thissen, 1983; Van der Linden et al., 1999; Verhelst, Verstralen, & Jansen, 1997). Van der Linden (2009) discussed several basic issues with respect to choosing the proper quantification that does justice to both speed and achievement features of both persons and items. We briefly follow some aspects of his discussion.

To have a sensible model for speed and achievement, one needs to assume that the response time of person i to item j is a random variable, just as one assumes that the item score is a random variable. That is, we assume that across independent replications of the test administration procedure, the time it takes person i to solve an item varies and that one measurement represents a random draw from a distribution of response times for item j. We denote this random variable T_{ij} with realizations $T_{ij} = t_{ij}$. In addition to this random variable, random variables for the item score, X_{ij} with realization $X_{ij} = x_{ij}$ and completion of the item, denoted D_{ij} with realization $D_{ij} = d_{ij}$ are defined. The item score is well-known from this and the previous chapters. The item completion variable is new and has the following meaning. Let $D_{ij} = 1$ denote person i completing item j irrespective of the correctness of the response, and let $D_{ij} = 0$ denote otherwise; then random variable D_i denotes the number of completed tasks,

$$D_i = \sum_{j=1}^{J} D_{ij}. \tag{6.23}$$

As before, the test score is defined as

$$X_i = \sum_{j=1}^{J} X_{ij}. \tag{1.1}$$

Only if the items are trivially easy for the person under consideration will a person answer as many items correctly as she completes, that is, $X_i = D_i$, and the two scores contain the same information about speed. This situation represents the pure speed test, but such tests clearly are not representative of tests in general. A pure power test does not impose time limitations, but this does not mean that a person will answer an item correctly simply because she takes forever to come up with an answer. If someone does not or only imperfectly masters the subattributes to solve a problem, taking enough time may create the optimal time conditions to produce the best result for this person, but the probability of a correct answer is smaller than 1. Like pure speed tests, pure power tests only exist hypothetically.

The idea is that, without obvious time pressure, the time it takes a person to complete a higher cognitive task also may be of interest, and the three random variables for response time, T_{ij}, item completion, D_{ij} and item achievement, X_{ij} are the input for estimating interesting parameters from response time models to be discussed shortly. Van der Linden (2009) posited that each test can be conceived as a hybrid in terms of speed and power, and that the speed aspect can be captured by response time T_{ij} and completion variable D_{ij} and the power aspect by item achievement score X_{ij}.

Another issue is that of the dissimilarity of response time and speed, two concepts we have used rather informally so far but now need to explicate further. Put simply, one may spend different amounts of time on two items, but if the item that takes the most time to complete also requires the most complex cognitive operations or the largest number of operations similar to those needed for the other item, then taking more time to solve the item does not imply that working speed was slower. Physically, average speed is defined by the ratio of the distance one traveled and the time this took. For example, if hours are the time unit and one traveled 500 km in 4 hours, then the average speed of traveling was 500/4 = 125 km/hr. If one translates these concepts to cognitive operations, one replaces average speed by person i's speed of labor, denoted τ_i^*, distance traveled by amount of labor the item requires, denoted β_j^*, whereas realized time needed to complete the item is denoted by t_{ij}; then, we have

$$\tau_i^* = \frac{\beta_j^*}{t_{ij}}. \tag{6.24}$$

Rewriting yields

$$t_{ij} = \frac{\beta_j^*}{\tau_i^*}. \tag{6.25}$$

Equation (6.25) shows that the response time is the ratio of an item parameter and a person parameter, which can be interpreted as the labor intensity or the time intensity of the item and the working speed of the person.

Response time is positively skewed; hence, taking the logarithm yields a distribution that better approaches symmetry,

$$\ln t_{ij} = \ln \beta_j^* - \ln \tau_i^* = \beta_j - \tau_i. \tag{6.26}$$

Because T_{ij} is a random variable, $\ln t_{ij}$ is a realization on the log scale, whereas the right-hand side of Equation (6.26) is the difference between two parameters, which are scalars, the left-hand side of the equation actually is an expected value,

$$\mathcal{E}(\ln t_{ij}) = \beta_j - \tau_i. \tag{6.27}$$

Van der Linden (2009) named Equation (6.27) the *fundamental equation of RT* (i.e., response time) *modeling* and argued that, given the definition of average physical speed as a point of departure, this equation is basic to any model containing a person speed parameter. Such models must also have a time intensity parameter for each item.

For a combination of a person and an item, one records one response time t_{ij} and one item achievement score x_{ij}, so that one cannot estimate the correlation between time spent to produce a response and correctness of the response. Research investigating correlations between response time and item achievement typically focuses on groups or item sets and thus finds nonnegative associations produced by an aggregation effect. Van der Linden (2009) explained how, for example, in a group, speed parameters τ and ability parameters θ show variation and thus tend to correlate, which causes variation of response time T_j and item achievement scores X_j, and thus correlation between T_j and X_j. The sign and the size of the correlation is an empirical phenomenon and varies among groups and items. For example, in a group in which more able people are also faster, one expects a positive correlation between T_j and X_j. However, it is important to notice that this and other correlations are due to aggregation but do not demonstrate a within-person effect. Along these and other lines, Van der Linden (2009) argued that it is reasonable to assume three kinds of local independence: First, between item achievement scores for items j and k; that is, following Equation (3.1) where we used covariance, $\sigma(X_j, X_k \mid \theta) = 0$, all pairs, $j \neq k$; second, between response time scores, $\sigma(T_j, T_k \mid \tau) = 0$, all pairs, $j \neq k$; and third, between item achievement scores and response time scores on item j, $\sigma(X_j, T_j \mid \theta, \tau) = 0$, all items. One may notice that, just as IRT models assume constant ability for individuals during test taking, likewise response time models assume constant speed. Varying speed will produce a violation of local independence. We emphasize that speed is reflected by person parameter τ_i, not by response time t_{ij}; hence, a person's varying response time does not necessarily indicate a violation of local independence.

Lognormal Model

Van der Linden (2006, 2007, 2009, 2016b) proposed an approach to response time modeling by using a hierarchical framework consisting of two lower-level models and two higher-level models. We start discussing the lower-level models. An IRT model explains item achievement scores by incorporating the regular person ability parameter and item parameters (Chapter 4), and a lognormal model based on the fundamental equation for RT modeling [Equation (6.27)] explains response times by incorporating a person speed parameter and an item intensity parameter. For the item achievement scores, the three-parameter logistic model [Equation (4.51)], with lower asymptote γ_j, IRF slope or item discrimination parameter α_j, and IRF location or item difficulty parameter δ_j, was chosen,

$$P_j(\theta) = \gamma_j + \frac{(1-\gamma_j)\exp\left[\alpha_j(\theta-\delta_j)\right]}{1+\exp\left[\alpha_j(\theta-\delta_j)\right]}. \tag{4.51}$$

For the response times, based on the fundamental equation for RT modeling [Equation (6.27)], a lognormal model was defined. First, we add a normally distributed error term to Equation (6.26), so that

$$\ln T_{ij} = \beta_j - \tau_i + \epsilon_{ij}, \ \epsilon_{ij} \sim \mathcal{N}\left(0, \sigma_j^2\right). \tag{6.28}$$

The probability density function of a lognormal model for some random variable X of which the logarithm is normally distributed,

$$\ln(X) \sim \mathcal{N}(\mu, \sigma^2),$$
(6.29)

equals

$$f(x; \mu, \sigma^2) = \frac{1}{x} \cdot \frac{1}{\sigma\sqrt{2\pi}} \cdot \exp\left(-\frac{(\ln x - \mu)^2}{2\sigma^2}\right)$$
(6.30)

(Casella & Berger, 1990, pp. 110–111). A corresponding lognormal model for response time then equals

$$f\left(t_{ij}; \mathcal{E}(\ln t_{ij}), \sigma_j^2\right) = \frac{1}{t_{ij}} \cdot \frac{1}{\sigma_j\sqrt{2\pi}} \exp\left\{-\frac{\left[\ln t_{ij} - \mathcal{E}(\ln t_{ij})\right]^2}{2\sigma_j^2}\right\}.$$
(6.31)

To reparametrize Equation (6.31) to the conceptual framework and the corresponding notation we discussed, we define a parameter that is the reciprocal of the standard deviation of the error, σ_j, as $\varsigma_j = \sigma_j^{-1}$. In Chapter 4, we argued in the context of normal ogive IRT models that the inverse standard deviation can be interpreted as a discrimination parameter for the IRF. Fox, Klein Entink, and Van der Linden (2007) discussed a model version with an explicit discrimination parameter. In addition, we replace $\mathcal{E}(\ln t_{ij})$ by the difference of the item intensity parameter and the person speed parameter on the logarithmic scale, $\beta_j - \tau_i$. The response time model then is defined as

$$f(t_{ij}; \beta_j, \tau_i, \varsigma_j) = \frac{\varsigma_j}{t_{ij}\sqrt{2\pi}} \exp\left\{-\frac{1}{2}\left[\varsigma_j\left(\ln t_{ij} - (\beta_j - \tau_i)\right)\right]^2\right\}.$$
(6.32)

Interestingly, the lognormal model for response time has a firm basis in the conception of average speed (person i's speed of labor), distance travelled (amount of labor the item requires), and time travelled (response time). One can also find this conceptual framework in other research areas.

Next, one needs higher-order structures that allow dependencies between parameters from the two models to explain observed correlations between different item achievement scores and response time scores from the lower-level models for item achievement scores and response time scores. The two higher-level models concern joint distributions of the two person parameters from the lower-level models and the item parameters from the lower-level models. The first higher-order model is a bivariate normal model for person scores from the IRT model, θ, and the lognormal model, τ, with distribution parameters collected in vectors $\mu = (\mu_\theta, \mu_\tau)$ and $\sigma = (\sigma_\theta, \sigma_\tau)$, and correlation $\rho_{\theta\tau}$.

The second higher-order model is a multivariate normal model for the item parameters from the IRT model, α, δ, and γ, and from the lognormal model, ς and β, with distribution parameters collected in vector $\mu = (\mu_\alpha, \mu_\delta, \mu_\gamma, \mu_\varsigma, \mu_\beta)$ and covariance matrix Σ, which contains the five variances as diagonal entries and all covariances as off-diagonal entries. Estimation of the models is beyond the scope of this discussion; see Van der Linden (2006, 2007; also Fox et al., 2007; Glas & Van der Linden, 2010). Goodness-of-fit investigation involves separately assessing the IRT model and the lognormal model (Van der Linden, 2006).

Diffusion Model

Another approach that is recent in psychometrics, but already known in mathematical psychology for some time (Ratcliff, 1978), studies the underlying process that produces both the response time and the item achievement. The model used for this purpose is the diffusion model, and Tuerlinckx et al. (2016; also, see Tuerlinckx & De Boeck, 2005) discussed a general class of IRT models based on the logic of the diffusion model. The point of departure is the idea that when responding to an item, people start by collecting noisy information about the problem presented or the question asked that they think they need to provide a response, and when the balance finally tips to one side, they provide the corresponding response. Four parameters govern the process, and Figure 6.7 helps in understanding the discussion that follows. From left to right in the figure, response time progresses.

First, parameter v is the drift rate, which represents the quality of the collected information. Parameter v is a real number, and as $|v|$ is larger, the quality of the information is greater. If $v > 0$, then the person tends to provide a positive response; otherwise she tends to provide a negative response. As the process progresses, evidence for either response alternates until evidence in favor of one of the two response options gains the upper hand and is convincing enough to cross one of the two response boundaries, either the one for the positive or the one for the negative response. The steepness of the winding road that leads to boundary crossing may vary, and $|v|$ quantifies the steepness, whereas the sign of v tells us the direction. If $|v|$ is large, the road is steep toward one of the response boundaries. In this sense, drift rate is reminiscent of the mean gradient of a road running either uphill or downhill.

Second, parameter a ($a > 0$) is the boundary separation, representing the distance between the two response boundaries. Boundaries further apart require more information before one of the boundaries is crossed. This is different from saying that the boundaries represent greater cognitive distance; the more the two response options are dissimilar, the smaller the amount of information needed to reach a decision, because the options are easily distinguishable. Here, one may think of having to choose between two dishes one likes and having to choose between a tasty dish and a distasteful dish. The first choice requires more information than the second choice, because in the first choice problem the alternatives are closer. In the diffusion model

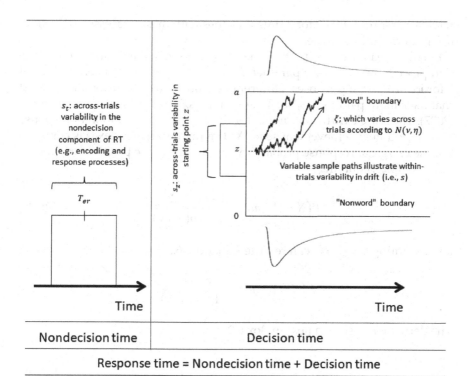

FIGURE 6.7
The drift diffusion model of two choice decisions. This example concerns the lexical decision about words and nonwords. The random walk, representing the noisy accumulation of evidence, starts at z and continues until the word or nonword boundary is hit at decision time DT. Response time RT is the sum of decision time and the time (T_{er}) required for other processes, such as the motor part of the response. Starting position z and drift rate ξ may vary over trials according to a uniform and a normal distribution, respectively. Note. Based on Figure 1 in Van der Maas et al. (2011)

this is represented by a greater boundary separation parameter *a*, quantifying a larger amount of information or greater caution to be exercised before reaching a decision. In Figure 6.7, the response boundaries are further apart.

Third, the third parameter z lies between the two response boundaries, such that $0 < z < a$, and is interpreted as the starting point of the information collection process. If $z \neq \frac{a}{2}$, then the process starts with a bias toward one of the response options before new information is collected. Often, one assumes that $z = \frac{a}{2}$. Hence, the starting point depends on boundary separation *a* and need not be estimated once boundary separation is estimated.

Fourth, response time consists of other ingredients than information collection, such as encoding the stimulus and time needed to press a button that finalizes the choice the person makes. Parameter T_{er} (*e* for encoding and *r* for response) represents the total time that goes with these processes other than information collection; hence, $T_{er} > 0$.

Finally, one may quantify the variability or noise of the information collection process by a variance parameter σ^2, but this parameter is often fixed at 1 for identification purposes. Different versions of the diffusion model exist that have more parameters, see Wagenmakers, Van der Maas, and Grasman (2007). The simple diffusion model for the bivariate distribution of response time T and item achievement score X (leaving out item index j) is complex, but the marginal response probability for providing the positive response is much simpler, equaling

$$P(X=1|v,a,z) = \frac{1-\exp(-2zv)}{1-\exp(-2av)} \tag{6.33}$$

and assuming $z = \dfrac{a}{2}$, one can rewrite Equation (6.33) as

$$P(X=1|v,a,z) = \frac{\exp(av)}{1+\exp(av)}. \tag{6.34}$$

The derivation is shown next in Box 6.2.

BOX 6.2   Rewriting Equation (6.33)

First, one may notice that after insertion of $z = \dfrac{a}{2}$, one obtains

$$P(X=1|v,a,z) = \frac{1-\exp(-av)}{1-\exp(-2av)}. \tag{6.35}$$

Second, for simplicity, we write $y = av$ and rewrite

$$P(X=1|v,a,z) = \frac{1-\exp(-y)}{1-\exp(-2y)} = \frac{\dfrac{\exp(y)-1}{\exp(y)}}{\dfrac{\exp(2y)-1}{\exp(2y)}}. \tag{6.36}$$

Next, multiplying the numerator and the denominator in the upper fraction on the right-hand side in Equation (6.36) by exp(y), one can rewrite Equation (6.36) as

$$P(X=1|v,a,z) = \frac{\exp(2y)-\exp(y)}{\exp(2y)-1}. \tag{6.37}$$

Dividing numerator and denominator by $\left[\exp(y)-1\right]$ and replacing y with av yields

$$P\left(X=1|v,a,z\right)=\frac{\exp\left(av\right)}{1+\exp\left(av\right)}. \tag{6.34}$$

The attraction of Equation (6.34) is that it provides an exponential model, thus opening possibilities for estimation and goodness-of-fit assessment.

It would lead too far to discuss results for response time as well, but considering only Equation (6.34) gives one a flavor of the kind of information a fitting model could provide. First, as drift rate v is greater negative or positive, the choice probability $P\left(X=1|v,a,z\right)$ is closer to 0 or 1, respectively. That is, a drift rate that runs steeply to the lower response boundary for $X=0$, that is, $v<0$, points to the availability of evidence of the choice for $X=0$, and similarly for $v>0$ and $X=1$. As drift rate is close to 0, the choice probability $P\left(X=1|v,a,z\right)$ is close to .5, expressing uncertainty based on the accumulated information. What is not visible here is that as drift rate v is greater negative or positive and choice probability $P\left(X=1|v,a,z\right)$ is closer to 0 or 1, respectively, response time T is shorter; that is, it takes less time to reach a decision. Second, as the boundary separation a grows, choice probabilities are either lower or higher, depending on the sign of drift rate v. However, we saw that the interpretation of a greater boundary separation a is that of exercising greater caution, hence, although invisible here, a longer response time T. Because parameter z is fixed, it does not affect choice probability. Parameter T_{er} is absent in the choice probability, and thus it only affects response time by quantifying minimum response time needed before one can actually start the response process related to the item content.

Van der Maas, Molenaar, Maris, Kievit, and Borsboom (2011; also, Tuerlinckx & De Boeck, 2005) discussed different parameterizations of the diffusion model in terms of person and item parameters, such as, $v_{ij}=\theta_i/\delta_j$, with $\theta_i>0$ and $\delta_j>0$, person parameter θ_i being the attribute parameter and item parameter δ_j being the item location or difficulty parameter, both as in IRT models, albeit on a positive scale. These properties give the model attractive properties (Van der Maas et al., 2011). Parameter estimation of the different models is done by means of the marginal maximum likelihood method (Tuerlinckx & De Boeck, 2005) and a Bayesian approach (Van der Maas et al., 2011). Several goodness-of-fit strategies are available (Van der Maas et al., 2011; Wagenmakers et al., 2007).

Discussion

Originally, response time models were more popular in cognitive experimental psychology than in psychometrics, where the conviction was dominant that the outcome of a thought process is the most informative of the target attribute. Undoubtedly, other sources provide information in addition to the outcome, and response times are a fine example. Ample applications on real data are needed to show the value that response-time data can add to the traditional outcome data. However, other data sources may be considered as well, such as information about the solution process that people follow when they try to solve cognitive problems or try to determine their best response to trait and attitude items. Thinking-aloud data are sometimes mistrusted as valid data sources, but well-thought-out constructed-response items invite the tested person to formulate her own response within well-defined formats and may provide rich information improving measurement quality. A natural example comes from the math teacher in high school, who considered only the outcome insufficient evidence for problem comprehension and used to require a detailed written account of the path leading to the final answer, because this would tell her whether you had really understood the problem. Excessive response times were considered a bad sign unless the right solution and answer were given, but a practical time limit made shorter response time preferable and longer response time ineffective.

Network Psychometrics

A recent approach to measurement views some psychological attributes as networks of symptoms that induce one another to become active. We use the terminology of symptom to emphasize that networks as we discuss them here focus on observable variables. As time progresses, symptoms that have become active reinforce particular symptoms and may weaken others. A key characteristic of the approach is that it does not assume a common cause that generates activity of these symptoms represented by one or more latent attribute variables the way we encountered them thus far in this monograph. The symptoms rather take the place of latent attribute variables and may be either cause or effect in the emerging network. It is instructive to compare this mechanism to cognitive diagnostic modeling, where a set of discrete latent attribute variables cooperate to produce a latent item score, or multidimensional IRT, where the set of latent attributes is continuous, producing an item-score probability. However, network psychometrics, as the approach is nowadays called and as we discuss it here, does not focus on latent variables inducing responses to observable variables, but entirely on observable variables or, synonymously, symptoms (e.g., Epskamp et al., 2018; Epskamp, Maris, Waldorp, & Borsboom, 2018; Marsman et al., 2018). In network psychometrics, all one has are observable variables influencing one another.

For some readers, the terminology of symptom may readily refer to psychopathologies, but in addition to psychopathologies (Borsboom & Cramer, 2013), network psychometrics is also used to model attitudes (Dalege, Borsboom, Van Harreveld, & Van der Maas, 2017) and intelligence (Van der Maas, Kan, Marsman, & Stevenson, 2017). An example of an attribute that can be conceived as a network of symptoms is depression (Borsboom & Cramer, 2013; also, see Van der Maas et al., 2006). In the network approach, depression is not a cause represented by one or more latent variables in a measurement model, but a set of symptoms that may or may not appear simultaneously and reinforce or weaken one another, while certain relationships may involve loops to affect the strength of the relationship further. Networks are often activated from outside. Activating external events may be biological (e.g., hormonal imbalance), psychological (e.g., attention deficit), or environmental (e.g., loss of someone dear). For example, an environmental event, abbreviated E, such as the loss of a dear one, may cause symptoms, abbreviated S, such as an individual starting to worry (S_1) about the future, not only during daytime but also at night. Worry may then cause sleeplessness (S_2), and sleeplessness may cause excessive fatigue (S_3). Fatigue may cause loss of concentration (S_4) during daytime, and loss of concentration may cause an increase of errors (S_5) made at one's job and, in addition, affect one's attitude toward colleagues (S_6) negatively, and so on. Borsboom (2017) argued that the difference between a normal syndrome (i.e., a set of associated symptoms) and a pathological syndrome is represented by what the waning of the external event does to the network's activation. In case of normal functioning, the waning of the external cause de-activates the symptom activity, and the network gradually returns to a base level that does not pathologically intervene with the person's daily functioning. Here, a real-world event activates the network, and the disappearance of the event causes the de-activation of the network. In case of psychopathology, the same real-world event activates the network, but the network remains active when the cause that triggered it has disappeared. Hence, psychopathologies emerge when a network of symptoms has detached from a cause external to the network and becomes self-sustained, feeding its survival through continuous feedback cycles between the symptoms. In the example, one could imagine well that when excessive worry, sleeplessness, fatigue, loss of concentration, and so on become a habit more than a reaction to an external cause, this may also induce feelings of self-doubt, deep insecurity, and so on, perhaps best summarized as depression. The phenomenon that symptoms keep activating one another, even in the absence of the external cause, is known as hysteresis (e.g., Cramer, 2013).

In a graph, the symptoms that constitute a network are represented by nodes and their relationships by edges. Figure 6.8 shows two example

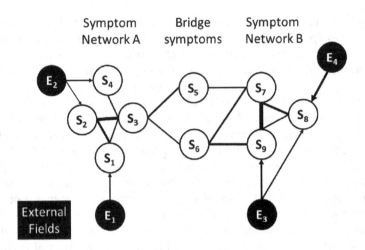

FIGURE 6.8
Two disorders (A and B) that are connected through bridge symptoms (S_5 and S_6) which play a role in both networks. Although the association of symptoms will be strongest within each network, structural overlap between the disorders is unavoidable, and as a result comorbidity will arise. Note. Based on Figure 2 in Borsboom (2017).

hypothetical networks, network A containing six nodes and network B containing five nodes, and both networks are embedded in an external field that includes four external events or causes. The presence of edges represents causality, but the causal direction is left unspecified, and the thickness of an edge represents the strength of the association between nodes. An edge can be absent, meaning that the symptom on either side of the abyss does not influence the other. Nodes they share can connect different networks, in which case they are called bridge nodes. Figure 6.8 shows two bridge nodes. For example, sleeplessness is a bridge node between networks representing major depression and generalized anxiety disorder (Borsboom & Cramer, 2013). Figure 6.8 refers to a network valid at the group level, but in principle one could also consider networks for individuals, thus entering the field of measurement. Individuals differ with respect to the strength of the edges in a network, and the strength determines the risk one runs of developing the psychopathology. Weak strength implies weakly connected networks, meaning that the mutual activation of the nodes occurs at a low level and fades out quickly when the external cause disappears, while strongly connected networks tend to continue being active when the external cause has vanished. An individual whose disposition tends to particular networks that are strongly connected is more liable to developing the target psychopathology than someone whose networks are weakly connected.

We saw that nodes conceptually represent symptoms, and we quantify nodes by random variables that we denote by notation X_j, $j = 1, ..., J$, as if they were items in a test. Data for a network analysis need not be item scores indicating whether a person reports the degree to which a symptom is present or whether a symptom is present but can also originate as longitudinal data for individuals from experience sampling using, for example, smartwatches, facilitating the estimation of networks for individuals. For simplicity, we assume we have J item scores, each representing degrees to which symptoms are present or the presence or absence of symptoms, and aim at estimating a network of the items. We also assume that the edges in a network are undirected; that is, edges express the strength or the presence or absence of a relation but not the direction. For continuous data or at least discrete ordered data, edges can be quantified as the degree to which two nodes are related. In a graph, the thickness of the line that connects two nodes visualizes the strength of the relation. For binary data that indicate whether two nodes are connected, presence or absence of a line between two nodes visualizes whether two nodes influence one another. The two kinds of data correspond with two kinds of statistical network models. As with many statistical models, one uses the covariance matrix Σ_X we discussed in Chapter 2 [Equation (2.24)] to estimate the edges in the network. Thus, for studying networks of symptoms, one uses the relations between the items. Before explaining the basics of how one uses the covariance matrix for this purpose, we discuss the concept of conditional independence that we already encountered in this and previous chapters.

What one wishes to know in psychopathology assessment (but not necessarily in other measurement contexts, a subtlety we ignore here) is whether two nodes in a graph are independent when one knows the state represented by scores on other nodes that are also connected to the two target nodes. For example, if a graph shows an edge between the symptoms "sleep problems" (say, quantified by X_j) and "concentration problems" (X_k), and both symptoms are connected to "fatigue" (X_l), then it is of interest to know whether conditioning on "fatigue" explains the connection between "sleep problems" and "concentration problems" (Van Bork, Van Borkulo, Waldorp, Cramer, & Borsboom, 2018). If "fatigue" is the only node to which the other two symptoms are connected, then, assuming $\perp$ denotes independence, variables X_j and X_k are conditionally independent when

$$X_j \perp X_k \mid X_l \text{ iff } \{j,k\} \notin E, \text{ and } l \neq j,k, \tag{6.38}$$

where E stands for the set of all edges in the network. That is, given X_l, variables X_j and X_k are independent if and only if (iff) items j and k do not share an edge. If Equation (6.38) holds, then the node represented by variable X_l explains the relation between the nodes represented by variables X_j and X_k, or in everyday parlance, fatigue explains the relationship between sleep problems and concentration problems. In a group of people varying on fatigue, one will also observe varying degrees of sleep problems and concentration problems. For example, the more one is tired, the graver the sleep and concentration problems, thus causing a positive relation between the two problems. However, if one considers a subgroup of people being equally tired and assuming there are no other causes of varying sleep and concentration problems active, one will find no relationship anymore. All the variation one observes is random variation. In this case, the network representation does not contain an edge between X_j and X_k, but these items both share an edge with X_l, because there are causal relations between either item and fatigue.

Networks can be more complex and as we already suggested, there may be more, even a whole set of variables that, when conditioned on, together explain the relation between two other variables. In particular, when one does not know how the different nodes in the network relate to one another, based on the data, one has to find out which nodes share edges and which nodes do not. A way to do this is by conditioning on all other $J - 2$ nodes in the network that are collected in a vector $\mathbf{X}_{(jk)}$, so that Equation (6.38) becomes

$$X_j \perp X_k \mid \forall X_l \in \mathbf{X}_{(jk)} \text{ iff } \{j,k\} \notin \mathrm{E}, \text{ and } l \neq j,k, \tag{6.39}$$

where the conditioning is on all variables X_l that are in vector $\mathbf{X}_{(jk)}$. Obviously, when one or more explanatory variables are not included in the network, one cannot ascertain conditional independence as in Equation (6.39) and one cannot be sure that the equation holds. Van Bork et al. (2018) mention the degree to which the environment in which people live is noisy as another explanatory variable for the connection between sleep problems and concentration problems. When this variable is absent from the network, that is, when it is not included in vector $\mathbf{X}_{(jk)}$, Equation (6.39) does not hold.

Three remarks are in order. First, conditional independence is formally identical to the assumption of local independence that we discussed in the context of IRT and LCMs. The difference with network models is that in IRT models and LCMs the explanatory agent is a latent variable. In the network approach to attribute measurement, one refrains from a latent variable and relies on the items (or other observable information) to explain the relations between pairs of items. Second, the networks we discuss are undirected in the sense that they do not indicate which node causes another node. The graphical connections are line pieces, not arrows that indicate the direction of a relation. The literature that we discuss in this chapter assumes that such networks are Markov random fields (e.g., Kinderman & Snell, 1980). To keep things simple, we discuss only Markov random fields and no other networks such as Bayesian networks that are directed and do not allow loops. By discussing only Markov random fields, we acknowledge that social science, psychological, and health research collect data often in such a way that causality cannot be ascertained beyond reasonable doubt. For example, an ineffective design is used, relevant variables are not recorded, and the available data are full of noise and beset with unidentified signals. Hence, the best one can hope for is to identify an intelligible network that sheds light on the target phenomenon. Third, the network approach uses different models for continuous or Gaussian data and binary data, but both use the covariance matrix, albeit in different ways. Next, we will briefly discuss these two approaches to identifying a network from one's data.

Network Approach for Gaussian Data

Prerequisites for Gaussian Data Networks

The number of different item pairs among J items equals $\frac{1}{2}J(J-1)$, and hence, this is the maximum number of edges in a network. We first consider correlation structures. Zero product-moment correlations suggest the absence of an edge between the two nodes involved, and non-zero correlations suggest the presence of edges. Clusters of zero-order correlations (i.e., correlations not corrected for other variables) suggest subsets of nodes that are strongly connected. However, a zero-order correlation is not very informative of whether a relation between two variables involved is direct or can be explained by other variables that act as mediators, and the two variables could also correlate because they share a common cause or because they share a common effect on another variable (Van Bork et al., 2018). Partial correlations are a viable alternative to the ambiguous zero-order product-moment correlations and provide an approximation method to study whether the property of conditional independence exists for the pairs of nodes in a network and whether one can assume presence or absence of edges between particular node pairs.

We already studied partial correlations in Chapter 3 [Equation (3.41)] in the context of modified bounds for scalability coefficients. There, we considered the correlation between two item scores, X_j and X_k, when both are corrected for their linear regression on a third item score, say, X_l. The corrected correlation equals the correlation between the residuals of the two item scores that remain after their linear regressions on the third item score have been subtracted from the scores X_j and X_k. The result is the partial correlation, denoted $\rho_{jk.l}$. The partial correlation for items j and k is the correlation between the parts of X_j and X_k that are not linearly explained by X_l. In the context of networks, the partial correlation can be determined for X_j and X_k corrected for the other $J-2$ items in the network (Epskamp, Waldorp, Mõttus, & Borsboom, 2018). This is done by subtracting the multiple regression of X_j on the other $J-2$ items from X_j and subtracting the multiple regression of X_k on the other $J-2$ items from X_k, and correlating the resulting residuals; this is the partial correlation $\rho_{jk.R_{(jk)}}$, where $R_{(jk)}$ is the set of $J-2$ items not including items j and k.

The reader may have noticed the resemblance of a partial correlation equal to zero, that is, $\rho_{jk.R_{(jk)}} = 0$, and conditional independence [Equation (6.39)], and have wondered whether a zero partial correlation is sufficient evidence for conditional independence. It is not. Next, we show that, in general, zero correlation and statistical independence are different concepts. We spend space doing this, because the distinction is so important in statistics and network psychometrics alike.

BOX 6.3   Statistical Independence Implies Zero
Correlation but Not the Other Way Around

The product-moment correlation equals $\rho_{XY} = \sigma_{XY} / \sigma_X \sigma_Y$, so that $\rho_{XY}=0$ if and only if $\sigma_{XY}=0$. Hence, we further consider the covariance and notice that it equals

$$\sigma_{XY} = \mathcal{E}(XY) - \mathcal{E}(X)\mathcal{E}(Y),$$

from which it follows that

$$\sigma_{XY} = 0 \Leftrightarrow \mathcal{E}(XY) = \mathcal{E}(X)\mathcal{E}(Y). \tag{6.40}$$

We use this result in what follows.

First, we show that statistical independence implies $\sigma_{XY}=0$. Recall from Chapter 3, Equation (3.2), that statistical independence means that the multivariate distribution of J random variables equals the product of the J marginal distributions; that is, for two variables, $g(x,y) = g_1(x)g_2(y)$ for all values (x, y). The link with zero correlation comes from

$$\mathcal{E}(XY) = \int_{-\infty}^{\infty} \int_{-\infty}^{\infty} xy\, g(x,y)\, dx dy$$

$$= \int_{-\infty}^{\infty} \int_{-\infty}^{\infty} xy\, g_1(x)g_2(y)\, dx dy \tag{6.41}$$

$$= \int_{-\infty}^{\infty} x g_1(x) dx \int_{-\infty}^{\infty} y g_2(y) dy = \mathcal{E}(X)\mathcal{E}(Y).$$

Because statistical independence implies $\mathcal{E}(XY) = \mathcal{E}(X)\mathcal{E}(Y)$, it follows that $\sigma_{XY}=0$. Hence, statistical independence implies covariance and correlation equal to 0.

Second, we show that correlation equal to 0 does not necessarily imply statistical independence (example based on https://davegiles.blogspo t.com/2013/09/on-zero-correlation-and-statistical.html). We assume that X is uniform between -1 and 1, that is, $X \sim \text{Un}(-1,1)$. It follows that $P(X < x) = \frac{1}{2}x + \frac{1}{2}$ and the probability density function $g(x) = \frac{1}{2}$. Further, X is recorded in deviation-score form, so that its mean equals $\mathcal{E}(X)=0$. Finally, we define $Y = X^2$. Using these results, we have that

$$\sigma_{XY} = \mathcal{E}(XY) = \mathcal{E}(X^3)$$

$$= \int_{-1}^{1} x^3 g(x)dx = \int_{-1}^{1} x^3 \cdot \frac{1}{2}dx = \frac{1}{8}x^4 \bigg|_{-1}^{1}$$

$$= \frac{1}{8}1^4 - \frac{1}{8}(-1)^4 = 0$$

Hence, $\rho_{XY} = 0$; yet it is obvious that X completely determines Y, thus X and Y are completely dependent.

We have shown that statistical independence implies zero correlation but that zero correlation need not imply statistical independence. This is an important result, given that the network approach to measuring attributes aims at determining which nodes in the network are connected by edges and which are not. The intention is to say something about conditional independence, and the tools for doing this are partial correlations, but partial correlations are not sufficient to prove conditional independence.

Networks for Gaussian Data

An interesting case occurs when two variables X and Y follow a bivariate normal distribution and correlate 0; then, they are also statistically independent. It is not enough that the univariate distributions of the two variables are normal; their bivariate distribution also has to be normal. The same result holds when one considers J random variables; their multivariate distribution must be normal for zero correlation to imply statistical independence. Multivariate normality is the assumption for the data that are used to estimate a network, hence the phrase Gaussian data. Given multivariate normality, a zero partial correlation suggests conditional independence as in a Markov field (Van Bork et al., 2018). To investigate zero partial correlation, one considers the standardized version of the precision matrix denoted $\mathbf{P}$, which equals the inverse of the covariance matrix, $\mathbf{P} = \Sigma_X^{-1}$ (Epskamp et al., 2018). Let an unstandardized off-diagonal element of $\mathbf{P}$ be p_{jk} and let p_{jj} and p_{kk} be different diagonal elements, then $p_{jk} / \left(p_{jj}p_{kk}\right)^{1/2}$ is a standardized off-diagonal element, and the partial correlation equals

$$\rho_{jk.R_{(jk)}} = -\frac{p_{jk}}{\left(p_{jj}p_{kk}\right)^{\frac{1}{2}}}, j \neq k. \tag{6.42}$$

From Equation (6.42), it follows that $\rho_{jk.R_{(jk)}} = 0$ if and only if $p_{jk} = 0$, and multivariate normality of the data means that $p_{jk} = 0$ implies independence of variables X_j and X_k given all other $J - 2$ variables; see Equation (6.39). Hence, one can conclude that nodes j and k in the network do not share an edge. Nonzero p_{jk} values suggest the presence of edges, provided all explanatory variables are included in the network. Van Bork et al. (2018) discussed methods for estimating the partial correlation network when the sample size is small and random noise in the data produces spurious relationships. Epskamp et al. (2018) noticed that the so-called Gaussian graphical model, which is the model we discuss here, is saturated when one does not place restrictions on the edges in the network. For example, based on theory one may hypothesize that particular edges are 0 and fix the partial correlation involved at 0.

Epskamp et al. (2018) pointed out that to estimate networks for individuals one needs data from $n = 1$ time series, in which data on an individual are collected repeatedly within a short period to prevent the model from becoming liable to change. The network can be estimated using the Gaussian graphical model, but this means that all data are treated as if they were all collected at one occasion, and interesting information about, for example, causal influence of one variable on itself and other variables at a later occasion, is lost. The authors discussed methods to model temporal effects as well, using regression models that include measurements at one or more previous occasions, called lagged measurements, to predict later measurements. These are vector autoregressive models that allow the construction of networks in which an edge suggests that a node predicts itself or another node at the next measurement occasion when one controls for all other variables observed at the previous measurement occasion or occasions. Time series data collected from multiple individuals can be modeled using a multilevel approach, allowing the estimation of individual networks and the comparison of networks between people (e.g., Bringmann et al., 2013). Epskamp et al. (2018) discussed multiple limitations with respect to vector autoregressive models, such as assumptions concerning time between measurement occasions and invariance of model parameters over time, and limitations with respect to the estimation of the multilevel approach, such as size of the data set and maximum number of variables. Violation of the multivariate normality assumption implying linear relations between all variables underlying all of what we have discussed is another threat to the usefulness of the Gaussian graphical model and the vector autoregressive model.

Network Approach for Binary Data

Multivariate normality of the data is attractive for the obvious reason that the use of partial correlations is informative of conditional independence, rendering the interpretation of networks convenient. However, in psychology, most variables are neither continuous nor normal, and one need only think of the rating-scale items used in personality inventories, which produce a limited number of ordered scores. For factor analysis assuming multivariate normality to estimate the model, using rating-scale data, Dolan (1994) concluded that the parameter estimates are robust against violation of multivariate normality when the number of ordered scores is at least five. As far as we know, the robustness of network analysis for violations of multivariate normality has not been studied yet.

Several authors proposed the Ising model (Ising, 1925) to estimate networks for dichotomous data, based on the model's origin in physics coded −1 and 1 (Dalege et al., 2017; Epskamp et al., 2018; Marsman et al., 2018; Marsman, Maris, Bechger, & Glas, 2015; Van Borkulo et al., 2014). These values indicate the state that the model tends to be in, and in the context of networks for psychological or other attributes, the states refer to symptoms represented by nodes and quantified by variables X_j with values $x_j \in \{-1,1\}$. The Ising model has two sets of parameters. First, parameter τ_j is a threshold parameter, of which the value $\tau_j = 0$ indicates that the model does not tend to either of the two states that X_j represents, which are that the symptom is absent or present. Values $\tau_j < 0$ indicate preference for state $x_j = -1$ (i.e., absence of symptom), and greater negative values indicate stronger preference, whereas values $\tau_j > 0$ indicate preference for state $x_j = 1$ (i.e., presence of symptom). Second, parameters ω_{jk} are network parameters for second-order interactions in the network between nodes represented by X_j and X_k. Value $\omega_{jk} = 0$ indicates that the interaction between the two nodes is absent; hence, the two nodes do not share an edge, and $\omega_{jk} \neq 0$ indicates that the nodes share an edge. Higher negative ω_{jk} values indicate that the two nodes prefer to be in different states (i.e., one absent and the other present), and higher positive ω_{jk} values indicate that the two nodes prefer to be in the same state (i.e., both absent or both present). The Ising model describes the joint distribution of the J item scores by main effects and second-order interaction effects as

$$P(\mathbf{X} = \mathbf{x}) = \frac{\exp\left(\sum_{j=1}^{J}\tau_j x_j + \sum_{\langle jk \rangle}\omega_{jk} x_j x_k\right)}{\sum_{\mathbf{x}}\exp\left(\sum_{j=1}^{J}\tau_j x_j + \sum_{\langle jk \rangle}\omega_{jk} x_j x_k\right)}. \tag{6.43}$$

The model norms the numerator against the denominator, which adds the possible values of the numerator across all 2^J possible item-score vectors (which may be problematic if 2^J is large). The summation of interaction effect across $\langle jk \rangle$ refers to taking into account only the interaction effects of the nodes that are adjacent in the network and thus share edges. This is possible only in confirmatory research in which one knows the structure of the network in advance and estimates the strength of the nodes that one knows to be connected. In exploratory research, one has to consider all possibilities and estimates the network parameters ω_{jk} for all pairs of nodes.

Estimating the Ising model in an exploratory context runs into numerical problems as soon as J is not small anymore, and this happens readily. One can see this from the log-likelihood of the model. Let τ denote the vector containing the parameters τ_j and let Ω denote the matrix containing parameters ω_{jk}. Let the denominator of the Ising model [Equation (6.43)] be denoted Z, then the log-likelihood equals

$$L(\tau,\Omega;x) = \ln P(\mathbf{X} = \mathbf{x}) = \sum_{j=1}^{J} \tau_j x_j + \sum_{\langle jk \rangle} \omega_{jk} x_j x_k - \ln Z. \qquad (6.44)$$

In Equation (6.44), ln Z has to be evaluated when maximizing the log-likelihood. Several solutions that circumvent the problem of large 2^J were proposed, and given the ambition of this chapter to introduce its topics briefly, we pick out one solution, which seems to be rather simple and practical and is frequently used (also, see Epskamp et al., 2018; Van Bork et al., 2018b; Van Borkulo et al., 2014).

One starts by deriving the distribution of the scores on X_j conditional on the other $J - 1$ nodes collected in vector $\mathbf{X}_{(j)} = \mathbf{x}_{(j)}$. Noticing that merging the score for item j, X_j, with the scores for the other $J - 1$ items, collected in $\mathbf{X}_{(j)}$, one obtains the vector containing all J item scores, $\mathbf{X}_j$; that is, $(X_j \wedge \mathbf{X}_{(j)}) = \mathbf{X}$, we have that (Epskamp et al., 2018; Van Bork et al., 2018b)

$$P(X_j \mid \mathbf{X}_{(j)} = \mathbf{x}_{(j)}) = \frac{P(\mathbf{X} = \mathbf{x})}{P(\mathbf{X}_{(j)} = \mathbf{x}_{(j)})}$$

$$= \frac{P(\mathbf{X} = \mathbf{x})}{\sum_{x_j} P(X_j = x_j, \mathbf{X}_{(j)} = \mathbf{x}_{(j)})}$$

$$= \frac{\exp\left[x_j \left(\tau_j + \sum_k \omega_{jk} x_k \right) \right]}{\sum_{x_j} \exp\left[x_j \left(\tau_j + \sum_k \omega_{jk} x_k \right) \right]}. \qquad (6.45)$$

Equation (6.45) is a logistic regression model (e.g., Agresti, 1990, pp. 84–91) in which item score X_j is predicted from $J - 1$ item scores X_k, $k \neq j$, parameter τ_j is an intercept parameter, and $J - 1$ parameters ω_{jk}, $k \neq j$, are the regression parameters. The model is often written as the log-odds of a positive response (provided $x_j = 0,1$),

$$\ln \frac{P(X_j = 1 \mid \mathbf{X}_{(j)} = \mathbf{x}_{(j)})}{P(X_j = 0 \mid \mathbf{X}_{(j)} = \mathbf{x}_{(j)})} = \tau_j + \sum_{k: k \neq j} \omega_{jk} x_k, \qquad (6.46)$$

(Agresti, 1990, p. 86). The estimation procedure for logistic regression models is repeated for each of the J item scores, which are separately regressed on each of the $(J - 1)$-item sets $\mathbf{X}_{(j)}$, $j = 1,\ldots,J$, not including target item j. These J regression models yield one estimate for each of the J intercepts (logistic regression) or thresholds (Ising model) τ_j, and two estimates for each of the $\frac{1}{2}J(J-1)$ regression or network parameters ω_{jk}, of which one takes the mean (Van Borkulo et al., 2014). Using ℓ_1-regularized logistic regression, also known as least absolute shrinkage and selection operator (LASSO), can help to select the best fitting of a couple of models while satisfying the sparsity principle by means of a penalty function.

Discussion

In this monograph, we are particularly interested in the construction of scales for measuring attributes of individuals. Network psychometrics is another contribution to nominal measurement of psychological and other attributes. If longitudinal data are available, the network of an individual's pathology can be estimated, allowing the psychologist to intervene effectively at the individual level. Network psychometrics shares similarities with other measurement models for nominal measurement, such as the LCM and the CDM. The network approach provides for each individual an estimate of the edges in the network that shows which symptoms are particularly dominant because they are strongly connected. LCMs provide an estimate of the most likely latent class to which an individual belongs along with the response probabilities of each symptom typical of the most likely class. CDMs provide an estimate of the set of latent attributes for each individual, suggesting which symptoms (i.e., latent attributes) the person possesses and which ones she misses.

Network models for individuals seem to require more data for their estimation than LCMs and CDMs. Experience has to accumulate in order to know which model is the most appropriate for studying individual categorization and whether a model's appropriateness depends on particular applications more than others. Epskamp et al. (2018) studied similarities and differences of network models for Gaussian data and regression analysis, factor analysis, and structural equation models. Marsman et al. (2018) studied the relationship of the Ising model and similar models for binary data with item response models, such as the Rasch model and Reckase's multidimensional item response model (both models discussed in Chapter 4). The next years will have to show the utility of this new approach to measurement.

References

Adams, R. J., Wilson, M. R., & Wu, M. L. (2016). Mixed-coefficients multinomial logit models. In W. J. van der Linden (Ed.), *Handbook of item response theory. Volume One. Models* (pp. 541–564). Boca Raton, FL: Chapman & Hall/CRC.

Agresti, A. (1990). *Categorical data analysis*. New York, NY: Wiley.

Akaike, H. (1974). A new look at the statistical model identification. *IEEE Transactions on Automatic Control, 19*, 716–723.

American Psychiatric Association. (2013). *Diagnostic and statistical manual of mental disorders* (5th edition). Washington, DC: Author.

Andersen, E. B. (1973a). A goodness of fit test for the Rasch model. *Psychometrika, 38*, 123–140.

Andersen, E. B. (1973b). Conditional inference for multiple choice questionnaires. *British Journal of Mathematical and Statistical Psychology, 26*, 31–44.

Andersen, E. B. (1977). Sufficient statistics and latent trait models. *Psychometrika, 42*, 69–81.

Andrich, D. (1978a). A rating formulation for ordered response categories. *Psychometrika, 43*, 561–573.

Andrich, D. (1978b). Relationships between the Thurstone and the Rasch approaches to item scaling. *Applied Psychological Measurement, 2*, 451–462.

Andrich, D. (1988a). The application of an unfolding model of the PIRT type to the measurement of attitude. *Applied Psychological Measurement, 12*, 33–51.

Andrich, D. (1988b). *Rasch models for measurement*. Thousand Oaks, CA: Sage.

Andrich, D. (2016). Hyperbolic cosine model for unfolding responses. In W. J. van der Linden (Ed.), *Handbook of item response theory. Volume One. Models* (pp. 353–367). Boca Raton, FL: Chapman & Hall/CRC.

Andrich, D., & Luo, G. (1993). A hyperbolic cosine latent trait model for unfolding dichotomous single-stimulus responses. *Applied Psychological Measurement, 17*, 253–276.

Atkins, D. C., Bedics, J. D., McGlinchey, J. B., & Beauchaine, T. P. (2005). Assessing clinical significance: Does it matter which method we use? *Journal of Consulting and Clinical Psychology, 73*, 982–989.

Baker, F. B. (1992). *Item response theory. Parameter estimation techniques*. New York, NY: Marcel Dekker.

Baker, F. B., & Kim, S.-H. (2004). *Item response theory. Parameter estimation techniques* (2nd edition). New York, NY: Marcel Dekker.

Baldwin, J. M., Cattell, J. M., & Jastrow, J. (1898). Physical and mental tests. *Psychological Review, 5*, 172–179.

Bartholomew, D. J., & Schuessler, K. F. (1991). Reliability of attitude scores based on a latent trait model. *Sociological Methodology, 21*, 97–123.

Bartolucci, F., Bacci, S., & Gnaldi, M. (2016). *Statistical analysis of questionnaires: A unified approach based on R and Stata*. Boca Raton, FL: Chapman & Hall/CRC.

Bauer, S., Lambert, M. J., & Nielsen, S. L. (2004). Clinical significance methods: A comparison of statistical techniques. *Journal of Personality Assessment, 82*, 60–70.

Bechtel, G. G. (1968). Folded and unfolded scaling from preferential paired comparisons. *Journal of Mathematical Psychology, 5,* 333–357.

Bejar, I. I. (2008). Standard setting: What is it? Why is it important? *R&D Connections.* Princeton, NJ: Educational Testing Service. Downloaded from https://www.ets.org/Media/Research/pdf/RD_Connections7.pdf.

Bentler, P. M. (2009). Alpha, dimension-free, and model-based internal consistency reliability. *Psychometrika, 74,* 137–143.

Bentler, P. M., & Woodward, J. A. (1980). Inequalities among lower bounds to reliability: With applications to test construction and factor analysis. *Psychometrika, 45,* 249–267.

Benton, T. (2015). An empirical assessment of Guttman's lambda 4 reliability coefficient. In R. E. Millsap, D. M. Bolt, L. A. van der Ark, & W.-C. Wang (Eds.), *Quantitative psychology research: The 78th Annual Meeting of the Psychometric Society* (pp. 301–310). New York, NY: Springer.

Bergqvist, L., & Rossiter, J. R. (2007). The predictive validity of multiple-item versus single-item measures of the same constructs. *Journal of Marketing Research, 44,* 175–184.

Bergsma, W., Croon, M., & Hagenaars, J. A. (2009). *Marginal models for dependent, clustered, and longitudinal categorical data.* New York, NY: Springer.

Binet, A., & Simon, Th. A. (1905). Méthodes nouvelles pour le diagnostic du niveau intellectuel des anormaux. *L'Année Psychologique, 11,* 191–244.

Birnbaum, A. (1968). Some latent trait models and their use in inferring an examinee's ability. In F. M. Lord & M. R. Novick (Eds.), *Statistical theories of mental test scores* (pp. 396–479). Reading, MA: Addison-Wesley.

Bock, R. D. (1972). Estimating item parameters and latent ability when responses are scored in two or more nominal categories. *Psychometrika, 37,* 29–51.

Bock, R. D. (1997). The nominal categories model. In W. J. van der Linden & R. K. Hambleton (Eds.), *Handbook of modern item response theory* (pp. 33–49). New York, NY: Springer.

Böckenholt, U. (2006). Thurstonian-based analysis: Past, present and future utilities. *Psychometrika, 71,* 615–629.

Böckenholt, U., & Tsai, R.-C. (2006). Random-effects models for preference data. In C. R. Rao & S. Sinharay (Eds.), *Handbook of statistics, Volume 26. Psychometrics* (pp. 447–468). Amsterdam, The Netherlands: Elsevier.

Bollen, K. A. (1989). *Structural equations with latent variables.* New York, NY: Wiley.

Bolsinova, M., & Tijmstra, J. (2018). Improving precision of ability estimation: Getting more from response times. *British Journal of Mathematical and Statistical Psychology, 71,* 13–38.

Bond, T. G., & Fox, C. M. (2015). *Applying the Rasch model: Fundamental measurement in the human sciences* (3rd edition). New York, NY: Routledge.

Boomsma, A., Van Duijn, M. A. J., & Snijders, T. A. B. (Eds.) (2001). *Essays on item response theory.* New York, NY: Springer.

Boring, E. G. (1923). Intelligence as the tests test it. *New Republic, 35,* 35–37.

Borsboom, D. (2005). *Measuring the mind. Conceptual issues in contemporary psychometrics.* Cambridge, UK: Cambridge University Press.

Borsboom, D. (2017). A network theory of mental disorders. *World Psychiatry, 16,* 5–13.

Borsboom, D., & Cramer, A. O. J. (2013). Network analysis: An integrative approach to the structure of psychopathology. *Annual Review of Clinical Psychology, 9,* 91–121.

Borsboom, D., & Mellenbergh, G. J. (2004). Why psychometrics is not pathological. *Theory & Psychology, 14,* 105–120.

Borsboom, D., Mellenbergh, G. J., & Van Heerden, J. (2004). The concept of validity. *Psychological Review, 111,* 1061–1071.

Bouwmeester, S., & Sijtsma, K. (2004). Measuring the ability of transitive reasoning, using product and strategy information. *Psychometrika, 69,* 123–146.

Bouwmeester, S., & Sijtsma, K. (2007). Latent class modeling of phases in the development of transitive reasoning. *Multivariate Behavioral Research, 42,* 457–480.

Bouwmeester, S., Sijtsma, K., & Vermunt, J. K. (2004). Latent class regression analysis for describing cognitive developmental phenomena: An application to transitive reasoning. *European Journal of Developmental Psychology, 1,* 67–86.

Bouwmeester, S., Vermunt, J. K., & Sijtsma, K. (2007). Development and individual differences in transitive reasoning: A fuzzy trace theory approach. *Developmental Review, 27,* 41–74.

Box, G. E. P. (1976). Science and statistics. *Journal of the American Statistical Association, 71,* 791–799.

Bradley, R. A., & Terry, M. E. (1952). Rank analysis of incomplete block designs: I. The method of paired comparisons. *Biometrika, 39,* 324–345.

Brainerd, C. J., & Kingma, J. (1984). Do children have to remember to reason? A fuzzy-trace theory of transitivity development. *Developmental Review, 4,* 311–377.

Brandmaier, A. M., Prindle, J. J., McArdle, J. J., & Lindenberger, U. (2016). Theory-guided exploration with structural equation model forests. *Psychological Methods, 21,* 566–582.

Brennan, R. L. (2001). *Generalizability theory.* New York, NY: Springer.

Bringmann, L. F., Vissers, N., Wichers, M., Geschwind, N., Kuppens, P., et al. (2013). A network approach to psychopathology: New insights into clinical longitudinal data. *PLoS ONE, 8*(4): e60188. doi:10.1371/journal.pone.0060188

Brogden, H. E. (1946). Variation in test validity with variation in the distribution of item difficulties, number of items, and degree of their intercorrelation. *Psychometrika, 11,* 197–214.

Brown, W. (1910). Some experimental results in the correlation of mental abilities. *British Journal of Psychology, 3,* 296–322.

Brusco, M. J., Köhn, H.-F., & Steinley, D. (2015). An exact method for partitioning dichotomous items within the framework of the monotone homogeneity model. *Psychometrika, 80,* 949–967.

Camilli, G. (1994). Origin of the scaling constant d = 1.7 in item response theory. *Journal of Educational Statistics, 19,* 293–295.

Carter, N. T., Dalal, D. K., Boyce, A. S., O'Connell, M. S., Kung, M.-C., & Delgado, K. M. (2014). Uncovering curvilinear relationships between conscientiousness and job performance: How theoretically appropriate measurement makes an empirical difference. *Journal of Applied Psychology, 99,* 564–586.

Casella, G., & Berger, R. L. (1990). *Statistical inference.* Belmont, CA: Duxbury Press.

Cattelan, M. (2012). Models for paired comparison data: A review with emphasis on dependent data. *Statistical Science, 27,* 412–433.

Cattell, J. McK. (1890). Mental tests and measurements. *Mind, 15,* 373–381.

Cattell, R. B. (1966). The scree test for the number of factors. *Multivariate Behavioral Research, 1,* 245–276.

Chen, J., de la Torre, J., & Zhang, Z. (2013). Relative and absolute fit evaluation in cognitive diagnosis modeling. *Journal of Educational Measurement, 50,* 123–140.

Chen, W. H., & Thissen, D. (1997). Local dependence indices for item pairs using item response theory. *Journal of Educational and Behavioral Statistics, 22,* 265–289.

Cheung, M. W.-L., & Jak, S. (2016). Analyzing big data in psychology: A split/analyze/meta-analyze approach. *Frontiers in Psychology, Quantitative Psychology and Measurement, 7*:738. doi:10.3389/fpsyg.2016.00738

Chiu, C.-Y., & Douglas, J. A. (2013). A nonparametric approach to cognitive diagnosis by proximity to ideal response patterns. *Journal of Classification, 30*, 225–250. doi:10.1007/s00357-013-9132-9

Chiu, C.-Y., Douglas, J. A., & Li, X. (2009). Cluster analysis for cognitive diagnosis: Theory and applications. *Psychometrika, 74*, 633–665.

Cho, E., & Kim, S. (2015). Cronbach's coefficient alpha: Well known but poorly understood. *Organizational Research Methods, 18*, 207–230.

Chung, M., & Johnson, M. S. (2017). An MCMC algorithm for estimating the Reduced RUM. Unpublished manuscript. Downloaded from https://arxiv.org/abs/1710.08412

Coombs, C. H. (1964). *A theory of data*. Ann Arbor, MI: Mathesis Press.

Cortina, J. M. (1993). What is coefficient alpha? An examination of theory and application. *Journal of Applied Psychology, 78*, 98–104.

Cover, T. M., & Thomas, J. A. (2006). *Elements of information theory* (2nd edition). Hoboken, NJ: Wiley.

Cramer, A. O. J. (2013). *The glue of (ab)normal mental life: Networks of interacting thoughts, feelings and behaviors*. PhD dissertation, University of Amsterdam, Amsterdam, The Netherlands.

Cressie, N., & Read, T. R. C. (1984). Multinomial goodness-of-fit tests. *Journal of the Royal Statistical Society, Series B, 46*, 440–464.

Cronbach, L. J. (1949). *Essentials of psychological testing*. New York, NY: Harper & Brothers, Publishers.

Cronbach, L. J. (1951). Coefficient alpha and the internal structure of tests. *Psychometrika, 16*, 297–334.

Cronbach, L. J., Gleser, G. C., Nanda, H., & Rajaratnam, N. (1972). *The dependability of behavioral measurements: Theory of generalizability for scores and profiles*. New York, NY: Wiley.

Cronbach, L. J., & Meehl, P. E. (1955). Construct validity in psychological tests. *Psychological Bulletin, 52*, 281–302.

Cronbach, L. J., & Warrington, W. G. (1952). Efficiency of multiple-choice tests as a function of spread of item difficulties. *Psychometrika, 17*, 127–147.

Croon, M. (1990). Latent class analysis with ordered latent classes. *British Journal of Mathematical and Statistical Psychology, 43*, 171–192. doi:10.1111/j.2044-8317.1990.tb00934.x

Croon, M. A. (1991). Investigating Mokken scalability of dichotomous items by means of ordinal latent class analysis. *British Journal of Mathematical and Statistical Psychology, 44*, 315–331. doi:10.1111/j.2044-8317.1991.tb00964.x

Croon, M. A. (2002). Using predicted latent scores in general latent structure models. In G. A. Marcoulides & I. Moustaki (Eds.), *Latent variable and latent structure models* (pp. 195–224). Mahwah, NJ: Erlbaum.

Culpepper, S. A. (2019). Estimating the cognitive diagnosis Q-matrix with expert knowledge: Application to the fraction-subtraction dataset. *Psychometrika, 84*, 333–357.

Dalege, J., Borsboom, D., van Harreveld, F., & van der Maas, H. L. J. (2017). Network analysis on attitudes: A brief tutorial. *Social Psychological and Personality Science, 8*, 528–537. doi:10.1177/1948550617709827

Davenport, E. C., Davison, M. L., Liou, P.-Y., & Love, Q. U. (2015). Reliability, dimensionality, and internal consistency as defined by Cronbach: Distinct albeit related concepts. *Educational Measurement: Issues and Practice, 34,* 4–9.

Dayton, C. M., & Macready, G. B. (1976). A probabilistic model for validation of behavioral hierarchies. *Psychometrika, 41,* 189–204.

Dayton, C. M., & Macready, G. B. (1980). A scaling model with response errors and intrinsically unscalable respondents. *Psychometrika, 45,* 343–356.

De Boeck, P., & Wilson, M. (Eds.) (2004). *Explanatory item response models. A generalized linear and nonlinear approach.* New York, NY: Springer.

De Boeck, P., Wilson, M., & Acton, G. S. (2005). A conceptual and psychometric framework for distinguishing categories and dimensions. *Psychological Review, 112,* 129–158.

De Groot, A. D. (1961). *Methodologie (Methodology).* Den Haag, The Netherlands: Mouton.

de la Torre, J. (2008). An empirically based method of Q-matrix validation for the DINA model: Development and applications. *Journal of Educational Measurement, 45,* 343–362.

de la Torre, J. (2009). DINA model and parameter estimation: A didactic. *Journal of Educational and Behavioral Statistics, 34,* 115–130.

de la Torre, J. (2011). The generalized DINA model framework. *Psychometrika, 76,* 179–199.

de la Torre, J., & Chiu, C.-Y. (2016). A general method of empirical Q-matrix validation. *Psychometrika, 81,* 253–273.

de la Torre, J., & Douglas, J. (2004). Higher-order latent trait models for cognitive diagnosis. *Psychometrika, 69,* 333–353.

de la Torre, J., & Lee, Y.-S. (2013). Evaluating the Wald test for item-level comparison of saturated and reduced models in cognitive diagnosis. *Journal of Educational Measurement, 50,* 355–373.

de la Torre, J., & Minchen, N. D. (2019). The G-DINA model framework. In M. von Davier & Y.-S. Lee (Eds.), *Handbook of diagnostic classification models. Models and model extensions, applications, software packages* (pp. 155–169). New York, NY: Springer.

de la Torre, J., Stark, S., & Chernyshenko, O. S. (2006). Markov chain Monte Carlo estimation of item parameters for the generalized graded unfolding model. *Applied Psychological Measurement, 30,* 216–232.

de la Torre, J., Van der Ark, L. A., & Rossi, G. (2018). Analysis of clinical data from cognitive diagnosis modelling framework. *Measurement and Evaluation in Counseling and Development, 51,* 281–296.

De Raad, B., & Perugini, M. (Eds.) (2002). *Big Five assessment.* Seattle, WA: Hogrefe & Huber Publishers.

Dempster, A. P., Laird, N. M., & Rubin, D. B. (1977). Maximum likelihood from incomplete data via the EM algorithm. *Journal of the Royal Statistical Society, Series B (Methodological), 39,* 1–38.

Denollet, J. (2005). DS14: Standard assessment of negative affectivity, social inhibition, and Type D personality. *Psychosomatic Medicine, 67,* 89–97.

DeSarbo, W. S., & Hoffman, D. L. (1986). Simple and weighted unfolding threshold models for the spatial representation of binary choice data. *Applied Psychological Measurement, 10,* 247–264.

DiBello, L. V., Roussos, L. A., & Stout, W. (2007). Review of cognitively diagnostic assessment and a summary of psychometric models. In C. R. Rao & S. Sinharay (Eds.), *Handbook of statistics, Volume 26. Psychometrics* (pp. 979–1030). Amsterdam, The Netherlands: Elsevier.

Dolan, C. V. (1994). Factor analysis of variables with 2, 3, 5 and 7 response categories: A comparison of categorical variable estimators using simulated data. *British Journal of Mathematical and Statistical Psychology, 47*, 309–326.

Domingos, P. (2015). *The master algorithm. How the quest for the ultimate learning machine will remake our world.* London, UK: Penguin Books.

Douglas, J. (1997). Joint consistency of nonparametric item characteristic curve and ability estimation. *Psychometrika, 62*, 7–28.

Douglas, J. A. (2001). Asymptotic identifiability of nonparametric item response models. *Psychometrika, 66*, 531–540.

Douglas, J., & Cohen, A. (2001). Nonparametric item response function estimation for assessing parametric model fit. *Applied Psychological Measurement, 25*, 234–243.

Dziak, J. J., Coffman, D. L., Lanza, S. T., & Li, R. (2017). *Sensitivity and specificity of information criteria.* Unpublished manuscript. Downloaded from https://peerj.com/preprints/1103.pdf

Edgeworth, F. Y. (1888). The statistics of examinations. *Journal of the Royal Statistical Society, 51*, 599–635.

Eggen, T. J. H. M., & Verhelst, N. D. (2011). Item calibration in incomplete testing designs. *Psicológica, 32*, 107–132.

Eliason, S. R. (1993). *Maximum likelihood estimation. Logic and practice.* Thousand Oaks, CA: Sage.

Ellis, J. L. (2014). An inequality for correlations in unidimensional monotone latent variable models for binary variables. *Psychometrika, 79*, 303–316.

Ellis, J. L., & Junker, B. W. (1997). Tail-measurability in monotone latent variable models. *Psychometrika, 62*, 495–523.

Ellis, J. L., & Van den Wollenberg, A. L. (1993). Local homogeneity in latent trait models. A characterization of the homogeneous monotone IRT model. *Psychometrika, 58*, 417–429.

Embretson, S. E. (1997). Multicomponent response models. In W. J. van der Linden & R. K. Hambleton (Eds.), *Handbook of modern item response theory* (pp. 305–321). New York, NY: Springer.

Embretson, S. E. (2016). Multicomponent models. In W. J. van der Linden (Ed.), *Handbook of item response theory. Volume One. Models* (pp. 225–242). Boca Raton, FL: Chapman & Hall/CRC.

Embretson, S. E., & Gorin, J. (2001). Improving construct validity with cognitive psychology principles. *Journal of Educational Measurement, 38*, 343–368.

Emons, W. H. M., Sijtsma, K., & Meijer, R. R. (2004). Testing hypotheses about the person-response function in person-fit analysis. *Multivariate Behavioral Research, 39*, 1–35.

Emons, W. H. M., Sijtsma, K., & Meijer, R. R. (2007). On the consistency of individual classification using short scales. *Psychological Methods, 12*, 105–120.

Epskamp, S., Borsboom, D., & Fried, E. I. (2018). Estimating psychological networks and their accuracy: A tutorial paper. *Behavioral Research Methods, 50*, 195–212.

Epskamp, S., Maris, G., Waldorp, L. J., & Borsboom, D. (2018). Network psychometrics. In P. Irwing, T. Booth, & D. J. Hughes (Eds.), *The Wiley handbook of psychometric testing. A multidisciplinary reference on survey, scale and test development. Volume II* (pp. 953–986). Hoboken, NJ: Wiley.

Epskamp, S., Waldorp, L. J., Mõttus, R., & Borsboom, D. (2018). The Gaussian graphical model in cross-sectional and time-series data. *Multivariate Behavioral Research, 53*, 453–480. doi:10.1080/00273171.2018.1454823

Falk, C. F., & Savalei, V. (2011). The relationship between unstandardized and standardized alpha, true reliability, and the underlying measurement model. *Journal of Personality Assessment, 93*, 445–453.

Fan, X., & Thompson, B. (2001). Confidence intervals for effect sizes, confidence intervals about score reliability coefficients, please: An EPM guidelines editorial. *Educational and Psychological Measurement, 61*, 517–531.

Feldt, L. S. (1965). The approximate sampling distribution of Kuder-Richardson reliability coefficient twenty. *Psychometrika, 30*, 357–370.

Feldt, L. S., Steffen, M., & Gupta, N. C. (1985). A comparison of five methods for estimating the standard error of measurement at specific score levels. *Applied Psychological Measurement, 9*, 351–361.

Feldt, L. S., Woodruff, D. J., & Salih, F. A. (1987). Statistical inference for coefficient alpha. *Applied Psychological Measurement, 11*, 93–103.

Ferguson, G. A. (1942). Item selection by the constant process. *Psychometrika, 7*, 19–29.

Finney, D. J. (1944). The application of probit analysis to the results of mental tests. *Psychometrika, 9*, 31–39.

Fischer, G. H. (1973). The linear logistic test model as an instrument in educational research. *Acta Psychologica, 37*, 359–374.

Fischer, G. H. (1974). *Einführung in die Theorie psychologischer Tests* [Introduction to the theory of psychological tests]. Bern, Switserland: Huber.

Fischer, G. H. (1995a). Derivations of the Rasch model. In G. H. Fischer & I. W. Molenaar (Eds.), *Rasch models. Foundations, recent developments, and applications* (pp. 15–38). New York, NY: Springer.

Fischer, G. H. (1995b). Linear logistic models for change. In G. H. Fischer & I. W. Molenaar (Eds.), *Rasch models. Foundations, recent developments, and applications* (pp. 157–180). New York, NY: Springer.

Fischer, G. H. (1997). Unidimensional linear logistic Rasch models. In W. J. van der Linden & R. K. Hambleton (Eds.), *Handbook of modern item response theory* (pp. 225–243). New York, NY: Springer.

Fischer, G. H., & Formann, A. K. (1982). Some applications of logistic latent trait models with linear constraints on the parameters. *Applied Psychological Measurement, 6*, 397–416.

Fischer, G. H., & Molenaar, I. W. (Eds.) (1995). *Rasch models. Foundations, recent developments, and applications.* New York, NY: Springer.

Fischer, G. H., & Ponocny, I. (1994). An extension of the partial credit model with an application to the measurement of change. *Psychometrika, 59*, 177–192.

Forero, C. G., Maydeu-Olivares, A., & Gallardo-Pujol, D. (2009). Factor analysis with ordinal indicators: A Monte Carlo study comparing DWLS and ULS estimation. *Structural Equation Modeling, 16*, 625–641.

Formann, A. K. (2003). Modeling data from water-level tasks: A test theoretical analysis. *Perceptual Motor Skills, 96*, 1153–1172.

Fox, J. (1997). *Applied regression analysis, linear models, and related models.* Thousand Oaks, CA: Sage.

Fox, J. (2000). *Nonparametric simple regression. Smoothing scatterplots.* Thousand Oaks, CA: Sage.

Fox, J.-P., Klein Entink, R., & van der Linden, W. (2007). Modeling of responses and response times with the package cirt. *Journal of Statistical Software, 20*(7), 1–14.

Furr, R. M., & Bacharach, V. R. (2013). *Psychometrics. An introduction.* Thousand Oaks, CA: Sage.

Galindo Garre, F., & Vermunt, J. K. (2006). Avoiding boundary estimates in latent class analysis by Bayesian posterior mode estimation. *Behaviormetrika, 33,* 43–59.

Garrido, L. E., Abad, F. J., & Ponsoda, V. (2013). A new look at Horn's parallel analysis with ordinal variables. *Psychological Methods, 18,* 454–474.

Glas, C. A. W. (1988a). The derivation of some tests for the Rasch model from the multinomial distribution. *Psychometrika, 53,* 525–546.

Glas, C. A. W. (1988b). The Rasch model and multistage testing. *Journal of Educational Statistics, 13,* 45–52.

Glas, C. A. W. (1999). Modification indices for the 2-PL and the nominal response model. *Psychometrika, 64,* 273–294.

Glas, C. A. W. (2010). Testing fit to IRT models for polytomously scored items. In M. L. Nering & R. Ostini (Eds.), *Handbook of polytomous item response models* (pp. 185–208). New York, NY: Routledge.

Glas, C. A. W. (2016). Frequentist model-fit tests. In W. J. van der Linden (Ed.), *Handbook of item response theory. Volume Two. Statistical tools* (pp. 343–362). Boca Raton, FL: Chapman & Hall/CRC.

Glas, C. A. W., & Van der Linden, W. J. (2010). Marginal likelihood inference for a model for item responses and response times. *British Journal of Mathematical and Statistical Psychology, 63,* 603–626.

Glas, C. A. W., & Verhelst, N. D. (1989). Extensions of the partial credit model. *Psychometrika, 53,* 525–546.

Glas, C. A. W., & Verhelst, N. D. (1995a). Testing the Rasch model. In G. H. Fischer & I. W. Molenaar (Eds.), *Rasch models. Foundations, recent developments, and applications* (pp. 69–95). New York, NY: Springer.

Glas, C. A. W., & Verhelst, N. D. (1995b). Tests of fit for polytomous Rasch models. In G. H. Fischer & I. W. Molenaar (Eds.), *Rasch models. Foundations, recent developments, and applications* (pp. 325–352). New York, NY: Springer.

Goldstein, H. (1980). Dimensionality, bias, independence and measurement scale problems in latent trait test score models. *British Journal of Mathematical and Statistical Psychology, 33,* 234–246.

Goodman, L. A. (1975). A new model for scaling response patterns: An application of the quasi-independence concept. *Journal of the American Statistical Association, 70,* 755–768.

Goodman, L. A. (2002). Latent class analysis: The empirical study of latent types, latent variables, and latent structures. In J. A. Hagenaars & A. L. McCutcheon (Eds.), *Applied latent class analysis* (pp. 3–55). Cambridge, UK: Cambridge University Press.

Gorsuch, R. L. (1983). *Factor analysis* (2nd edition). Hillsdale, NJ: Erlbaum.

Grayson, D. A. (1988). Two-group classification in latent trait theory: Scores with monotone likelihood ratio. *Psychometrika, 53,* 383–392.

Green, P. J., & Silverman, B. W. (1994). *Nonparametric regression and generalized linear models.* London, UK: Chapman & Hall.

Green, S. A., & Yang, Y. (2009). Commentary on coefficient alpha: A cautionary tale. *Psychometrika, 74,* 121–135.

Green, S. B., Lissitz, R. W., & Mulaik, S. A. (1977). Limitations of coefficient alpha as an index of test unidimensionality. *Educational and Psychological Measurement, 37*, 827–838.

Green, S. B., & Yang, Y. (2015). Evaluation of dimensionality in the assessment of internal consistency reliability: Coefficient alpha and omega coefficients. *Educational Measurement: Issues and Practice, 34*, 14–20.

Gu, Z., Emons, W. H. M., & Sijtsma, K. (2018). Review of issues about classical change scores: A multilevel modeling perspective on some enduring beliefs. *Psychometrika, 83*, 674–695.

Guilford, J. P. (1936, 1954). *Psychometric methods.* New York, NY: McGraw-Hill.

Gulliksen, H. (1950). *Theory of mental tests.* New York, NY: Wiley.

Gustafsson, J. E. (1980). Testing and obtaining fit of data to the Rasch model. *British Journal of Mathematical and Statistical Psychology, 33*, 205–233.

Guttman, L. (1944). A basis for scaling qualitative data. *American Sociological Review, 9*, 139–150.

Guttman, L. (1945). A basis for analyzing test-retest reliability. *Psychometrika, 10*, 255–282.

Guttman, L. (1950). The basis for scalogram analysis. In S. A. Stouffer, L. Guttman, E. A. Suchman, P. F. Lazarsfeld, S. A. Star, & J. A. Clausen (Eds.), *Measurement and prediction* (pp. 60–90). Princeton, NJ: Princeton University Press.

Haberman, S. J. (1978). *Analysis of qualitative data. Volume 1: Introductory topics.* New York, NY: Academic Press.

Habing, B. (2001). Nonparametric regression and the parametric bootstrap for local dependence assessment. *Applied Psychological Measurement, 25*, 221–233.

Haertel, E. H. (1989). Using restricted latent class models to map the skill structure of achievement items. *Journal of Educational Measurement, 26*, 301–321.

Hagenaars, J. A. (1990). *Categorical longitudinal data; loglinear panel, trend and cohort analysis.* Newbury Park, CA: Sage.

Hagenaars, J. A., & McCutcheon, A. L. (Eds.) (2002). *Applied latent class analysis.* Cambridge, UK: Cambridge University Press.

Hambleton, R. K., & Swaminathan, H. (1985). *Item response theory. Principles and applications.* Boston, MA: Kluwer Nijhoff Publishing.

Hambleton, R. K., Swaminathan, H., & Rogers, H. J. (1991). *Fundamentals of item response theory.* Newbury Park, CA: Sage.

Hand, D. J. (2013). Data, not dogma: Big data, open data, and the opportunities ahead. In A. Tucker, F. Höppner, A. Siebes, & S. Swift (Eds.), *Advances in Intelligent Data Analysis XII. In Proceedings of the 12th International Symposium IDA 2013 London GB* (LNCS 8207; pp. 1–12). Berlin Heidelberg, Germany: Springer.

Hansen, M., Cai, L., Monroe, S., & Li, Z. (2016). Limited-information goodness-of-fit testing of diagnostic classification item response models. *British Journal of Mathematical and Statistical Psychology, 69*, 225–252.

Harmann, H. H. (1976). *Modern factor analysis* (3rd edition revised). Chicago, IL, and London, UK: The University of Chicago Press.

Hartz, S. M. (2002). *A Bayesian framework for the unified model for assessing cognitive abilities: Blending theory with practicality.* Unpublished doctoral dissertation, University of Illinois, Champaign, IL, USA.

Harvill, L. M. (1991). An NCME instructional module on standard error of measurement. *Educational Measurement: Issues and Practice, 10*(2), 33–41.

Hayashi, K., & Kamata, A. (2005). A note on the estimator of the alpha coefficient for standardized variables under normality. *Psychometrika, 70,* 579–586.

Heinen, T. (1993). *Discrete latent variable models.* Tilburg, The Netherlands: Tilburg University Press.

Heinen, T. (1996). *Latent class and discrete latent trait models. Similarities and differences.* Thousand Oaks, CA: Sage.

Heiser, W. J. (1981). *Unfolding analysis of proximity data.* PhD dissertation, University of Leiden, The Netherlands.

Hemker, B. T. (1996). *Unidimensional IRT models for polytomous items, with results for Mokken scale analysis.* PhD dissertation, Utrecht University, Utrecht, The Netherlands.

Hemker, B. T., Sijtsma, K, & Molenaar, I. W. (1995). Selection of unidimensional scales from a multidimensional item bank in the polytomous Mokken IRT model. *Applied Psychological Measurement, 19,* 337–352.

Hemker, B. T., Sijtsma, K., Molenaar, I. W., & Junker, B. W. (1996). Polytomous IRT models and monotone likelihood ratio of the total score. *Psychometrika, 61,* 679–693.

Hemker, B. T., Sijtsma, K., Molenaar, I. W., & Junker, B. W. (1997). Stochastic ordering using the latent trait and the sum score in polytomous IRT models. *Psychometrika, 62,* 331–347.

Hemker, B. T., Van der Ark, L. A., & Sijtsma, K. (2001). On measurement properties of continuation ratio models. *Psychometrika, 66,* 487–506.

Henning, G. (1989). Meanings and implications of the principle of local independence. *Language Testing, 6,* 95–108.

Henry, K. L., & Muthén, B. (2010). Multilevel latent class analysis: An application of adolescent smoking typologies with individual and contextual predictors. *Structural Equation Modeling, 17,* 193–215.

Henson, R. A., Templin, J. L., & Willse, J. T. (2009). Defining a family of cognitive diagnosis models using log-linear models with latent variables. *Psychometrika, 74,* 191–210.

Hofman, A. D., Visser, I., Jansen, B. R. J., & Van der Maas, H. L. J. (2015). The balance-scale task revisited: A comparison of statistical models for rule-based and information-integration theories of proportional reasoning. *PLoS ONE, 10*(10): e0136449. doi:10.1371/journal.pone.0136449

Hogarty, K. Y., Hines, C. V., Kromrey, J. D., Ferron, J. M., & Mumford, K. R. (2005). The quality of factor solutions in exploratory factor analysis: The influence of sample size, communality, and overdetermination. *Educational and Psychological Measurement, 65,* 202–226.

Hoijtink, H. (1990). A latent trait model for dichotomous choice data. *Psychometrika, 55,* 641–655.

Hoijtink, H. (1991). The measurement of latent traits by proximity items. *Applied Psychological Measurement, 15,* 153–169.

Hoijtink, H., & Boomsma, A. (1995). On person parameter estimation in the dichotomous Rasch model. In G. H. Fischer & I. W. Molenaar (Eds.), *Rasch models. Foundations, recent developments, and applications* (pp. 53–68). New York, NY: Springer.

Hoijtink, H., & Molenaar, I. W. (1997). A multidimensional item response model: Constrained latent class analysis using Gibbs sampler and posterior predictive checks. *Psychometrika, 62,* 171–189.

Holland, P. W. (1990). On the sampling theory foundations of item response theory models. *Psychometrika, 55,* 577–601.

Holland, P. W., & Rosenbaum, P. R. (1986). Conditional association and unidimensionality in monotone latent variable models. *The Annals of Statistics, 14,* 1523–1543.

Holland, P. W., & Wainer, H. (Eds.) (1993). *Differential item functioning.* Hillsdale, NJ: Erlbaum.

Holt, J. A., & Macready, G. B. (1989). A simulation study of the difference chi-square statistic for comparing latent class models under violation of regularity conditions. *Applied Psychological Measurement, 13,* 221–231.

Holzinger, K. J. (1944). A simple method of factor analysis. *Psychometrika, 9,* 257–262.

Hooker, G., Finkelman, M., & Schwartzman, A. (2009). Paradoxical results in multidimensional item response theory. *Psychometrika, 74,* 419–442.

Horn, J. L. (1965). A rationale and test for the number of factors in factor analysis. *Psychometrika, 30,* 179–185.

Hoyt, C. (1941). Test reliability estimated by analysis of variance. *Psychometrika, 6,* 153–160.

Hu, L., & Bentler, P. M. (1999). Cutoff criteria for fit indexes in covariance structure analysis: Conventional criteria versus new alternatives. *Structural Equation Modeling, 6,* 1–55.

Hubert, M., & Vandervieren, E. (2008). An adjusted boxplot for skewed distributions. *Computational Statistics and Data Analysis, 52,* 5186–5201.

Hulin, C. L., Drasgow, F., & Parsons, C. K. (1983). *Item response theory.* Homewood, IL: Dow-Jones-Irwin.

Humphry, S. M. (2013). A middle path between abandoning measurement and measurement theory. *Theory & Psychology, 23,* 770–785.

Huynh, H. (1994). A new proof for monotone likelihood ratio for the sum of independent Bernoulli random variables. *Psychometrika, 59,* 77–79.

IBM Corp. (2016). *IBM SPSS Statistics for Windows, Version 25.0.* Armonk, NY: IBM Corp.

Ioannidis, J. P. A. (2005). Why most published research findings are false. *PLoS Medicine, 2*(8): e124.

Ip, E. H. (2010). Empirically indistinguishable multidimensional IRT and locally dependent unidimensional item response models. *British Journal of Mathematical and Statistical Psychology, 63,* 395–416.

Irtel, H. (1995). An extension of the concept of specific objectivity. *Psychometrika, 60,* 115–118.

Ising, E. (1925). Beitrag zur Theorie des Ferromagnetismus. *Zeitschrift für Physik, 31,* 253–258.

Jackson, P. H., & Agunwamba, C. C. (1977). Lower bounds for the reliability of the total score on a test composed of non-homogeneous items: I. Algebraic lower bounds. *Psychometrika, 42,* 567–578.

Jacobson, N. S., & Truax, P. (1991). Clinical significance: A statistical approach to defining meaningful change in psychotherapy research. *Journal of Consulting and Clinical Psychology, 59,* 12–19.

Jansen, B. R. J., & Van der Maas, H. L. J. (1997). Statistical test of the rule assessment methodology by latent class analysis. *Developmental Review, 17,* 321–357.

Jansen, B. R. J., & Van der Maas, H. L. J. (2002). The development of children's rule use on the balance scale task. *Journal of Experimental Child Psychology, 81,* 383–416.

Jansen, M. G. H. (1997). Rasch model for speed tests and some extensions with applications to incomplete designs. *Journal of Educational and Behavioral Statistics, 22*, 125–140.

Jansen, P. G. W. (1984). Relationships between the Thurstone, Coombs, and Rasch approaches to item scaling. *Applied Psychological Measurement, 8*, 373–383.

Janssen, R. (2016). Linear logistic models. In W. J. van der Linden (Ed.), *Handbook of item response theory. Volume One. Models* (pp. 211–224). Boca Raton, FL: Chapman & Hall/CRC.

Janssen, R., & De Boeck, P. (1997). Psychometric modeling of componentially designed synonym tasks. *Applied Psychological Measurement, 21*, 37–50.

Johnson, M. S. (2006). Nonparametric estimation of item and respondent locations from unfolding-type items. *Psychometrika, 71*, 257–279.

Johnson, M. S., & Junker, B. W. (2003). Using data augmentation and Markov Chain Monte Carlo for the estimation of unfolding response models. *Journal of Educational and Behavioral Statistics, 28*, 195–230.

Johnson, M. S., & Junker, B. W. (2005). Attitude scaling. In B. S. Everitt & D. C. Howell (Eds.), *Encyclopedia of statistics in behavioral science. Volume 1* (pp. 102–110). Hoboken, NJ: Wiley.

Johnson, N. L., & Kotz, S. (1969). *Distributions in statistics—discrete distributions.* New York, NY: Wiley.

Jordan, P., & Spiess, M. (2012). Generalizations of paradoxical results in multidimensional item response theory. *Psychometrika, 77*, 127–152.

Jordan, P., & Spiess, M. (2018). A new explanation and proof of the paradoxical scoring results in multidimensional item response models. *Psychometrika, 83*, 831–846.

Jöreskog, K. G. (1971). Statistical analysis of sets of congeneric tests. *Psychometrika, 36*, 109–133.

Junker, B. W. (1991). Essential independence and likelihood-based ability estimation for polytomous items. *Psychometrika, 56*, 255–278.

Junker, B. W. (1993). Conditional association, essential independence and monotone unidimensional item response models. *The Annals of Statistics, 21*, 1359–1378.

Junker, B. W. (1996). *Examining monotonicity in polytomous item response data.* Paper presented at the Annual Meeting of the Psychometric Society, June 27–30, 1996, Banff, AB, Canada.

Junker, B. W. (1998). Some remarks on Scheiblechner's treatment of ISOP models. *Psychometrika, 63*, 73–85.

Junker, B. W., & Ellis, J. L. (1997). A characterization of monotone unidimensional latent variable models. *The Annals of Statistics, 25*, 1327–1343.

Junker, B. W., & Sijtsma, K. (2000). Latent and manifest monotonicity in item response models. *Applied Psychological Measurement, 24*, 65–81.

Junker, B. W., & Sijtsma, K. (2001a). Nonparametric item response theory in action: An overview of the special issue. *Applied Psychological Measurement, 25*, 211–220.

Junker, B. W., & Sijtsma, K. (2001b). Cognitive assessment models with few assumptions, and connections with nonparametric item response theory. *Applied Psychological Measurement, 25*, 258–272.

Kagan, J. (2005). A time for specificity. *Journal of Personality Assessment, 85*, 125–127.

Kaiser, H. F. (1960). The application of electronic computers to factor analysis. *Educational and Psychological Measurement, 20*, 141–151.

Karabatsos, G., & Sheu, C.-F. (2004). Order-constrained Bayes inference for dichotomous models of unidimensional nonparametric IRT. *Applied Psychological Measurement, 28,* 110–125.

Kelley, K., & Cheng, Y. (2012). Estimation of and confidence interval formation for reliability coefficients of homogeneous measurement instruments. *Methodology: European Journal of Research Methods for the Behavioral and Social Sciences, 8,* 39–50.

Kendall, M. G., & Babington Smith, B. (1939). The problem of m rankings. *The Annals of Mathematical Statistics, 10,* 275–287.

Kim, S.-H. (2002). *A continuation ratio model for ordered category items.* Paper presented at the Annual Meeting of the Psychometric Society, Chapel Hill, NC. Downloaded from https://files.eric.ed.gov/fulltext/ED475828.pdf.

Kindermann, R., & Snell, J. L. (1980). *Markov random fields and their applications* (series Contemporary Mathematics; v. 1). Providence, RI: American Mathematical Society.

Kistner, E. O., & Muller, K. E. (2004). Exact distributions of intraclass correlation and Cronbach's alpha with Gaussian data and general covariance. *Psychometrika, 69,* 459–474.

Klinkenberg, S., Straatemeier, M., & Van der Maas, H. L. J. (2011). Computer adaptive practice of Maths ability using a new item response model for on the fly ability and difficulty estimation. *Computers & Education, 57,* 1813–1824.

Köhn, H.-F., & Chiu, C.-Y. (2016). A proof of the duality of the DINA model and the DINO model. *Journal of Classification, 33,* 171–184.

Kolen, M. J., & Brennan, R. L. (2014). *Test equating, scaling, and linking. Methods and practices.* New York, NY: Springer.

Krantz, D. H., Luce, R. D., Suppes, P., & Tversky, A. (1971). *Foundations of measurement. Volume I. Additive and polynomial representations.* London, UK: Academic Press.

Kristof, W. (1974). Estimation of reliability and true score variance from a split of the test into three arbitrary parts. *Psychometrika, 39,* 491–499.

Kroonenberg, P. M., & Ten Berge, J. M. F. (1987). Cross-validation of the WISC-R factorial structure using three-mode principal components analysis and perfect congruence analysis. *Applied Psychological Measurement, 11,* 195–201.

Kruyen, P. M., Emons, W. H. M., & Sijtsma, K. (2012). Test length and decision quality in personnel selection: When is short too short? *International Journal of Testing, 12,* 321–344.

Kruyen, P. M., Emons, W. H. M., & Sijtsma, K. (2013). On the shortcomings of shortened tests: A literature review. *International Journal of Testing, 13,* 223–248.

Kuder, G. F., & Richardson, M. W. (1937). The theory of estimation of test reliability. *Psychometrika, 2,* 151–160.

Kuijpers, R. E., Van der Ark, L. A., & Croon, M. A. (2013). Testing hypotheses involving Cronbach's alpha using marginal models. *British Journal of Mathematical and Statistical Psychology, 66,* 503–520.

Kunina-Habenicht, O., Rupp, A. A., & Wilhelm, O. (2012). The impact of model misspecification on parameter estimation and item-fit assessment in log-linear diagnostic classification models. *Journal of Educational Measurement, 49,* 59–81.

Langeheine, R., Pannekoek, J., & Van de Pol, F. (1996). Bootstrapping goodness-of-fit measures in categorical data analysis. *Sociological Methods & Research, 24,* 492–516.

Larsen, F. B., Pedersen, M. H., Friis, K., Glümer, C., & Lasgaard, M. (2017). A latent class analysis of multimorbidity and the relationship to socio-demographic factors and health-related quality of life. A national population-based study of 162,283 Danish adults. *PLoS ONE, 12*(1): e0169426. doi:10.1371/journal.pone.0169426

Lawley, D. N. (1943). On problems connected with item selection and test construction. *Proceedings of the Royal Society of Edinburgh, 61,* 73–287.

Lazarsfeld, P. F. (1950). The logical and mathematical foundation of latent structure analysis. In S. A. Stouffer, L. Guttman, E. A. Suchman, P. F. Lazarsfeld, S. A. Star, & J. A. Clausen (Eds.), *Measurement and prediction* (pp. 362–412). Princeton, NJ: Princeton University Press.

Lee, Y.-S., & Luna-Bazaldua, D. A. (2019). How to conduct a study with diagnostic models. In M. von Davier & Y.-S. Lee (Eds.), *Handbook of diagnostic classification models. Models and model extensions, applications, software packages* (pp. 525–545). New York, NY: Springer.

Li, H., Rosenthal, R., & Rubin, D. B. (1996). Reliability of measurement in psychology: From Spearman-Brown to maximal reliability. *Psychological Methods, 1,* 98–107.

Ligtvoet, R., Van der Ark, L. A., Bergsma, W. P., & Sijtsma, K. (2011). Polytomous latent scales for the investigation of the ordering of items. *Psychometrika, 76,* 200–216.

Ligtvoet, R., Van der Ark, L. A., Te Marvelde, J. M., & Sijtsma, K. (2010). Investigating an invariant item ordering for polytomously scored items. *Educational and Psychological Measurement, 70,* 578–595.

Ligtvoet, R., & Vermunt, J. K. (2012). Latent class models for testing monotonicity and invariant item ordering for polytomous items. *British Journal of Mathematical and Statistical Psychology, 65,* 237–250.

Lissitz, R. W. (2009). *The concept of validity. Revisions, new directions, and applications.* Charlotte, NC: Information Age Publishing, Inc.

Liu, C.-W., & Chalmers, R. P. (2018). Fitting item response unfolding models to Likert-scale data using mirt in R. *PLoS ONE, 13*(5): e0196292. doi:10.1371/journal.pone.0196292

Liu, Y., Tian, W., & Xin, T. (2016). An application of M_2 statistic to evaluate the fit of cognitive diagnostic models. *Journal of Educational and Behavioral Statistics, 41,* 3–26.

Loevinger, J. (1948). The technique of homogeneous tests compared with some aspects of 'scale analysis' and factor analysis. *Psychological Bulletin, 48,* 507–530.

Loevinger, J. (1954). The attenuation paradox in test theory. *Psychological Bulletin, 51,* 493–504.

Lord, F. M. (1952). A theory of test scores. *Psychometric Monograph No. 7,* Psychometric Society.

Lord, F. M. (1980). *Applications of item response theory to practical testing problems.* Hillsdale, NJ: Erlbaum.

Lord, F. M. (1983). Unbiased estimators of ability parameters, of their variance, and of their parallel-forms reliability. *Psychometrika, 48,* 233–245.

Lord, F. M., & Novick, M. R. (1968). *Statistical theories of mental test scores.* Reading, MA: Addison-Wesley.

Luce, R. D. (1959). *Individual choice behavior.* New York, NY: Wiley.

Lumsden, J. (1976). Test theory. *Annual Review of Psychology, 27,* 251–280.

Ma, W., & de la Torre, J. (2018). *GDINA: The generalized DINA model framework. R package version 2.1.* [Computer software] Downloaded from https://CRAN.R-project.org/package=GDINA

Ma, W., Iaconangelo, C., & de la Torre, J. (2016). Model similarity, model selection, and attribute classification. *Applied Psychological Measurement, 40*, 200–217.

MacCallum, R. C., Roznowski, M., & Necowitz, L. B. (1992). Model modifications in covariance structure analysis: The problem of capitalization on chance. *Psychological Bulletin, 111*, 490–504.

Macready, G. B., & Dayton, C. M. (1977). The use of probabilistic models in the assessment of mastery. *Journal of Educational Statistics, 2*, 99–120.

Magidson, J., & Vermunt, J. K. (2001). Latent class factor and cluster models, bi-plots, and related graphical displays. *Sociological Methodology, 31*, 223–264.

Magidson, J., & Vermunt, J. K. (2004). Latent class models. In D. Kaplan (Ed.), *The Sage handbook of quantitative methodology for the social sciences* (pp. 175–198). Thousand Oaks, CA: Sage.

Marianti, S., Fox, J. P., Avetisyan, M., Veldkamp, B. P., & Tijmstra, J. (2014). Testing for aberrant behavior in response time modeling. *Journal of Educational and Behavioral Statistics, 39*, 426–451.

Maris, E. (1993). Additive and multiplicative models for gamma distributed variables, and their application as psychometric models for response times. *Psychometrika, 58*, 445–469.

Maris, E. (1999). Estimating multiple classification latent class models. *Psychometrika, 64*, 187–212.

Markus, K. A., & Borsboom, D. (2013). *Frontiers of test validity theory: Measurement, causation, and meaning.* New York, NY: Routledge.

Marsh, L. C., & Cormier, D. R. (2002). *Spline regression models.* Thousand Oaks, CA: Sage.

Marsman, M., Borsboom, D., Kruis, J., Epskamp, S., Van Bork, R., Waldorp, L. J., Van der Maas, H. L. J., & Maris, G. (2018). An introduction to network psychometrics: Relating Ising network models to item response theory models. *Multivariate Behavioral Research, 53*, 15–35. doi:10.1080/00273171.2017.1379379

Marsman, M., Maris, G., Bechger, T., & Glas, C. (2015). Bayesian inference for low-rank Ising networks. *Scientific Reports, 5*: 9050. doi:10.1038/srep09050

Masicampo, E. J., & Lalande, D. R. (2012). A peculiar preference of p values just below .05. *The Quarterly Journal of Experimental Psychology, 65*, 2271–2279.

Masters, G. N. (1982). A Rasch model for partial credit scoring. *Psychometrika, 47*, 149–174.

Masters, G. N. (1988). Measurement models for ordered response categories. In R. Langeheine & J. Rost (Eds.), *Latent trait and latent class models* (pp. 11–29). New York, NY: Plenum Press.

Masters, G. N. (2016). Partial credit model. In W. J. van der Linden (Ed.), *Handbook of item response theory. Volume One. Models* (pp. 109–126). Boca Raton, FL: Chapman & Hall/CRC.

Masters, G. N., & Wright, B. D. (1984). The essential process in a family of measurement models. *Psychometrika, 49*, 529–544.

Maxwell, A. E. (1959). Maximum likelihood estimates of item parameters using the logistic function. *Psychometrika, 24*, 221–227.

Maydeu-Olivares, A. (2001). Limited information estimation and testing of Thurstonian models for paired comparison data under multiple judgment sampling. *Psychometrika, 66*, 209–228.

Maydeu-Olivares, A. (2013). Goodness-of-fit assessment of item response theory models. *Measurement: Interdisciplinary Research and Perspectives, 11*, 71–101.

Maydeu-Olivares, A., Coffman, D. L., & Hartmann, W. M. (2007). Asymptotically distribution-free (ADF) interval estimation of coefficient alpha. *Psychological Methods, 12,* 157–176.

Maydeu-Olivares, A., & Joe, H. (2006). Limited information goodness-of-fit testing in multidimensional contingency tables. *Psychometrika, 71,* 713–732.

McCutcheon, A. L. (2002). Basic concepts and procedures in single- and multiple-group latent class analysis. In J. A. Hagenaars & A. L. McCutcheon (Eds.), *Applied latent class analysis* (pp. 56–85). Cambridge, UK: Cambridge University Press.

McDonald, R. P. (1981). The dimensionality of tests and items. *British Journal of Mathematical and Statistical Psychology, 34,* 100–117.

McDonald, R. P. (1999). *Test theory: A unified treatment.* Mahwah, NJ: Erlbaum.

McLachlan, G. J. (1987). On bootstrapping the likelihood ratio test statistic for the number of components in a normal mixture. *Journal of the Royal Statistical Society: Series C (Applied Statistics), 36,* 318–324.

Meijer, R. R., & Baneke, J. J. (2004). Analyzing psychopathology items: A case for non-parametric item response theory modeling. *Psychological Methods, 9,* 354–368.

Meijer, R. R., & Sijtsma, K. (2001). Methodology review: Evaluating person fit. *Applied Psychological Measurement, 25,* 107–135.

Meijer, R. R., Sijtsma, K., & Molenaar, I. W. (1995). Reliability estimation for single dichotomous items based on Mokken's IRT model. *Applied Psychological Measurement, 19,* 323–335.

Mellenbergh, G. J. (1994). A unidimensional latent trait model for continuous item responses. *Multivariate Behavioral Research, 29,* 223–236.

Mellenbergh, G. J. (1995). Conceptual notes on models for discrete polytomous item responses. *Applied Psychological Measurement, 19,* 91–100.

Mellenbergh, G. J. (1996). Measurement precision in test score and item response models. *Psychological Methods, 1,* 293–299.

Mellenbergh, G. J. (1998). Het één-factor model voor continue en metrische responsen (The one-factor model for continuous and metric responses). In W. P. van den Brink & G. J. Mellenbergh (Eds.), *Testleer en testconstructie* (pp. 155–186). Amsterdam, The Netherlands: Boom.

Michell, J. (1999). *Measurement in psychology: A critical history of a methodological concept.* Cambridge, UK: Cambridge University Press.

Michell, J. (2000). Normal science, pathological science and psychometrics. *Theory & Psychology, 10,* 639–667.

Michell, J. (2014). The Rasch paradox, conjoint measurement, and psychometrics: Response to Humphry and Sijtsma. *Theory & Psychology, 24,* 111–123.

Millon, T., Millon, C., Davis, R., & Grossman, S. (2009). *MCMI–III Manual* (4th edition). Minneapolis, MN: Pearson Assessments.

Millsap, R. E. (2011). *Statistical approaches to measurement invariance.* New York, NY: Routledge.

Mislevy, R. J. (2018). *Sociocognitive foundations of educational measurement.* New York, NY: Routledge.

Mislevy, R. J., & Sheenan, K. M. (1989). The role of collateral information about examinees in item parameter estimation. *Psychometrika, 54,* 661–679.

Mokken, R. J. (1971). *A theory and procedure of scale analysis.* The Hague: Mouton/Berlin: De Gruyter.

Mokken, R. J. (1997). Nonparametric models for dichotomous responses. In W. J. van der Linden & R. K. Hambleton (Eds.), *Handbook of modern item response theory* (pp. 351–367). New York, NY: Springer.

Mokken, R. J., & Lewis, C. (1982). A nonparametric approach to the analysis of dichotomous item responses. *Applied Psychological Measurement, 6,* 417–430.

Mokken, R. J., Lewis, C., & Sijtsma, K. (1986). Rejoinder to "The Mokken scale: A critical discussion". *Applied Psychological Measurement, 10,* 279–285.

Molenaar, D., Tuerlinckx, F., & Van der Maas, H. L. J. (2015). A bivariate generalized linear item response theory modeling framework to the analysis of responses and response times. *Multivariate Behavioral Research, 50,* 56–74.

Molenaar, D., & Visser, I. (2017). Cognitive and psychometric modelling of responses and response times. *British Journal of Mathematical and Statistical Psychology, 70,* 185–186.

Molenaar, I. W. (1970). Approximations to the poisson, binomial, and hypergeometric distribution functions. *Mathematical Centre Tracts, No. 31.* Amsterdam, The Netherlands: Mathematisch Centrum.

Molenaar, I. W. (1983). Some improved diagnostics for failure of the Rasch model. *Psychometrika, 48,* 49–72.

Molenaar, I. W. (1995). Estimation of item parameters. In G. H. Fischer & I. W. Molenaar (Eds.), *Rasch models. Foundations, recent developments, and applications* (pp. 39–51). New York, NY: Springer.

Molenaar, I. W. (1997). Nonparametric models for polytomous items. In W. J. van der Linden & R. K. Hambleton (Eds.), *Handbook of modern item response theory* (pp. 369–380). New York, NY: Springer.

Molenaar, I. W., & Sijtsma, K. (1988). Mokken's approach to reliability estimation extended to multicategory items. *Kwantitatieve Methoden, 9*(28), 115–126.

Molenaar, I. W., & Sijtsma, K. (2000). *MSP5 for Windows. A program for Mokken scale analysis for polytomous items.* Groningen, The Netherlands: iecProGAMMA.

Mollenkopf, W. G. (1949). Variation on the standard error of measurement. *Psychometrika, 14,* 189–229.

Mosteller, F. (1951a). Remarks on the method of paired comparisons: I. The least squares solution assuming equal standard deviations and equal correlations. *Psychometrika, 16,* 3–9.

Mosteller, F. (1951b). Remarks on the method of paired comparisons: III. A test of significance for paired comparisons when equal standard deviations and equal correlations are assumed. *Psychometrika, 16,* 207–218.

Mulaik, S. A. (1972). *The foundations of factor analysis.* New York, NY: McGraw-Hill.

Muraki, E. (1992). A generalized partial credit model: Application of an EM algorithm. *Applied Psychological Measurement, 16,* 159–176.

Muthén, B. O., & Kaplan, D. (1985). A comparison of some methodologies for the factor analysis of non-normal Likert variables. *British Journal of Mathematical and Statistical Psychology, 38,* 171–189.

Nering, M. L., & Ostini, R. (Eds.) (2011). Handbook of polytomous item response theory models. New York, NY: Routledge.

Neyman, J., & Scott, E. L. (1948). Consistent estimates based on partially consistent observations. *Econometrica, 16,* 1–32.

Nichols, P. D., Chipman, S. F., & Brennan, R. L. (Eds.) (1995). *Cognitively diagnostic assessment* (pp. 327–359). Hillsdale, NJ: Erlbaum.

Novick, M. R. (1966). The axioms and principal results of classical test theory. *Journal of Mathematical Psychology, 3,* 1–18.

Novick, M. R., & Lewis, C. (1967). Coefficient alpha and the reliability of composite measurements. *Psychometrika, 32,* 1–13.

Nunnally, J. C. (1978). *Psychometric theory.* New York, NY: McGraw-Hill.

Nyholt, D. R., Gillespie, N. G., Heath, A. C., Merikangas, K. R., Duffy, D. L., & Martin, N. G. (2004). Latent class and genetic analysis does not support migraine with aura and migraine without aura as separate entities. *Genetic Epidemiology, 26,* 231–244.

Oosterwijk, P. R., Van der Ark, L. A., & Sijtsma, K. (2016). Numerical differences between Guttman's reliability coefficients and the GLB. In L. A. van der Ark, D. M. Bolt, W.-C. Wang, J. A. Douglas, & M. Wiberg (Eds.), *Quantitative psychology research: The 80th Annual Meeting of the Psychometric Society, Beijing, China, 2015* (pp. 155–172). New York, NY: Springer.

Oosterwijk, P. R., Van der Ark, L. A., & Sijtsma, K. (2017). Overestimation of reliability by Guttman's λ_4, λ_5, and λ_6 and the greatest lower bound. In L. A. van der Ark, S. Culpepper, J. A. Douglas, W.-C. Wang, & M. Wiberg (Eds.), *Quantitative psychology research: The 81th Annual Meeting of the Psychometric Society 2016, Asheville, NC* (pp. 159–172). New York, NY: Springer.

Oosterwijk, P. R., Van der Ark, L. A., & Sijtsma, K. (2019). Using confidence intervals for assessing reliability of real tests. *Assessment, 26,* 1207–1216.

Orlando, M., & Thissen, D. (2000). Likelihood-based item-fit indices for dichotomous item response theory models. *Applied Psychological Measurement, 24,* 50–64.

Padilla, M. A., Divers, J., & Newton, M. (2012). Coefficient alpha bootstrap confidence interval under nonnormality. *Applied Psychological Measurement, 36,* 331–348.

Piaget, J. (1947). *La psychologie de l'intelligence* [The psychology of intelligence]. Paris, France: Collin.

Piaget, J., Inhelder, B., & Szeminska, A. (1948). *La géométrie spontanée de l'enfant* [The child's conception of geometry]. Paris, France: Presses Universitaires de France.

Polak, M. G. (2011). *Item analysis of single-peaked response data. The psychometric evaluation of bipolar measurement scales.* PhD dissertation, Leiden University, Leiden, The Netherlands.

Popper, K. (1935, 2002). *The logic of scientific discovery.* Abingdon, UK: Routledge.

Post, W. J. (1992). *Nonparametric unfolding models. A latent structure approach.* Leiden, The Netherlands: DSWO Press.

Proctor, C. H. (1970). A probabilistic formulation and statistical analysis of Guttman scaling. *Psychometrika, 35,* 73–78.

Raijmakers, M. E. J., Jansen, B. R. J., & Van der Maas, H. L. J. (2004). Rules and development in triad classification task performance. *Developmental Review, 24,* 289–321.

Ramsay, J. O. (1991). Kernel smoothing approaches to nonparametric item characteristic curve estimation. *Psychometrika, 56,* 611–630.

Ramsay, J. O. (1997). A functional approach to modeling test data. In W. J. van der Linden & R. K. Hambleton (Eds.), *Handbook of modern item response theory* (pp. 381–394). New York, NY: Springer.

Ramsay, J. O. (2000). *TestGraf. A program for the graphical analysis of multiple choice test and questionnaire data.* Montreal, QC: McGill University.

Ramsay, J. O., & Silverman, B. W. (1997). *Functional data analysis.* New York, NY: Springer.

Rao, C. R., & Sinharay, S. (Eds.) (2007). *Handbook of statistics, Volume 26. Psychometrics.* Amsterdam, The Netherlands: Elsevier.

Rasch, G. (1960). *Probabilistic models for some intelligence and attainment tests.* Copenhagen, Denmark: Nielsen & Lydiche.

Rasch, G. (1968). An individualistic approach to item analysis. In P. F. Lazarsfeld & N. W. Henry (Eds.), *Latent structure analysis* (pp. 89–107). Boston, MA: Houghton Mifflin.

Ratcliff, R. (1978). A theory of memory retrieval. *Psychological Review, 85,* 59–108.

Raykov, T. (1997). Estimation of composite reliability for congeneric measures. *Applied Psychological Measurement, 21,* 173–184.

Raykov, T. (2001). Bias of coefficient α for fixed congeneric measures with correlated errors. *Applied Psychological Measurement, 25,* 69–76.

Raykov, T. (2008). Alpha if item deleted: A note on loss of criterion validity in scale development if maximizing coefficient alpha. *British Journal of Mathematical and Statistical Psychology, 61,* 275–285.

Raykov, T., & Marcoulides, G. A. (2011). *Introduction to Psychometric Theory.* New York, NY: Taylor & Francis.

Raykov, T., & Shrout, P. E. (2002). Reliability of scales with general structure: Point and interval estimation using a structural equation modeling approach. *Structural Equation Modeling, 9,* 195–212.

Reckase, M. D. (1997). A linear logistic multidimensional model for dichotomous item response data. In W. J. van der Linden & R. K. Hambleton (Eds.), *Handbook of modern item response theory* (pp. 271–286). New York, NY: Springer.

Reckase, M. D. (2016). Logistic multidimensional models. In W. J. van der Linden (Ed.), *Handbook of item response theory. Volume One. Models* (pp. 189–209). Boca Raton, FL: Chapman & Hall/CRC.

Revelle, W., & Condon, D. (2019). Reliability from α to ω: A tutorial. *Psychological Assessment, 32,* 1395–1411.

Revelle, W., & Zinbarg, R. E. (2009). Coefficients alpha, beta, omega, and the GLB: Comments on Sijtsma. *Psychometrika, 74,* 145–154.

Rhemtulla, M., Brosseau-Liard, P. É., & Savalei, V. (2012). When can categorical variables be treated as continuous? A comparison of robust continuous and categorical SEM estimation methods under suboptimal conditions. *Psychological Methods, 17,* 354–373.

Richardson, M. W. (1936). The relation between the difficulty and the differential validity of a test. *Psychometrika, 1,* 33–49.

Rindskopf, D. (1983). A general framework for using latent class analysis to test hierarchical and nonhierarchical learning models. *Psychometrika, 48,* 85–97.

Roberts, J. S. (2016). Generalized graded unfolding model. In W. J. van der Linden (Ed.), *Handbook of item response theory. Volume One. Models* (pp. 369–390). Boca Raton, FL: Chapman & Hall/CRC.

Roberts, J. S., & Laughlin, J. E. (1996). A unidimensional item response model for unfolding responses from a graded disagree-agree response scale. *Applied Psychological Measurement, 20,* 231–255.

Robitzsch, A. (2018). *Package "sirt", Supplementary item response theory models.* Downloaded from https://cran.r-project.org/web/packages/sirt/sirt.pdf

Rosenbaum, P. R. (1984). Testing the conditional independence and monotonicity assumptions of item response theory. *Psychometrika, 49,* 425–435.

Rosenbaum, P. R. (1987a). Probability inequalities for latent scales. *British Journal of Mathematical and Statistical Psychology, 40,* 157–168.

Rosenbaum, P. R. (1987b). Comparing item characteristic curves. *Psychometrika, 52,* 217–233.

Rosenthal, R. L. (1969). Interpersonal expectations: Effects of the experimenter's hypothesis. In R. Rosenthal & R. L. Rosnow (Eds.), *Artifacts in behavioral research* (pp. 181–277). New York, NY: Academic Press.

Rosenthal, R. L., & Rosnow, R. L. (1969). *Artifacts in behavioral research.* New York, NY: Academic Press.

Rosseel, Y. (2012). lavaan: An R package for structural equation modeling. *Journal of Statistical Software, 48*(2), 1–36.

Rossi, G., Elklit, A., & Simonsen, E. (2010). Empirical evidence for a four factor framework of personality disorder organization: Multigroup confirmatory factor analysis of the Millon Clinical Multiaxial Inventory – III personality disorders scales across Belgian and Danish data samples. *Journal of Personality Disorders, 24,* 128–150. doi:10.1521/pedi.2010.24.1.128

Rossi, G., Sloore, H., & Derksen, J. (2008). The adaptation of the MCMI-III in two non-English-speaking countries: State of the art of the Dutch language version. In T. Millon & C. Bloom (Eds.), *The Millon inventories: A practitioner's guide to personalized clinical assessment* (2nd edition) (pp. 369–386). New York, NY: The Guilford Press.

Rost, J., & Langeheine, R. (1997). *Applications of latent trait and latent class models in the social sciences* (pp. 118–126). Münster, Germany: Waxmann.

Roussos, L. A., DiBello, L. V., Stout, W., Hartz, S. M., Henson, R. A., & Templin, J. L. (2007). The fusion model skills diagnosis system. In In J. A. Leighton & M. J. Gierl (Eds.), *Cognitive diagnostic assessment for education. Theory and applications* (pp. 275–318). Cambridge, UK: Cambridge University Press.

Roussos, L. A., Templin, J. L., & Henson, R. A. (2007). Skills diagnosis using IRT-based latent class models. *Journal of Educational Measurement, 44,* 293–311.

Rupp, A. A., Templin, J. L., & Henson, R. A. (2010). *Diagnostic measurement. Theory, methods, and applications.* New York, NY: The Guilford Press.

Ruscio, J, Haslam, N., & Ruscio, A. M. (2006). *Introduction to the taxometric method: A practical guide.* Mahwah, NJ: Erlbaum.

Samejima, F. (1969). Estimation of latent trait ability using a response pattern of graded scores. *Psychometric Monograph, No. 17,* Psychometric Society.

Samejima, F. (1972). A general model for free-response data. *Psychometric Monograph, No. 18,* Psychometric Society.

Samejima, F. (1995). Acceleration model in the heterogeneous case of the general graded response model. *Psychometrika, 60,* 549–572.

Samejima, F. (1997). Graded response model. In W. J. van der Linden & R. K. Hambleton (Eds.), *Handbook of modern item response theory* (pp. 85–100). New York, NY: Springer.

Samejima, F. (2016). Graded response models. In W. J. van der Linden (Ed.), *Handbook of item response theory. Volume One. Models* (pp. 95–107). Boca Raton, FL: Chapman & Hall/CRC.

Scheiblechner, H. (1972). Das Lernen und Lösen komplexer Denkaufgaben (Learning and solving complex thought problems). *Zeitschrift für experimentelle und angewandte Psychologie, 19,* 476–506.

Scheiblechner, H. (1995). Isotonic ordinal probabilistic models (ISOP). *Psychometrika, 60,* 281–304.

Schmitt, N. (1996). Uses and abuses of coefficient alpha. *Psychological Assessment, 8,* 350–353.

Schulman, R. S., & Haden, R. L. (1975). A test theory model for ordinal measurement. *Psychometrika, 40,* 455–472.

Schwarz, G. E. (1978). Estimating the dimension of a model. *The Annals of Statistics, 6,* 461–464. doi:10.2307/2958889

Sclove, L. S. (1987). Application of model-selection criteria to some problems in multivariate analysis. *Psychometrika, 52,* 333–343.

Sedere, M. U., & Feldt, L. S. (1977). The sampling distributions of the Kristof reliability coefficient, the Feldt coefficient, and Guttman's lambda-2. *Journal of Educational Measurement, 14,* 53–62.

Sen, S., & Bradshaw, L. (2017). Comparison of relative fit indices for diagnostic model selection. *Applied Psychological Measurement, 41,* 422–438.

Siegel, S. (1956). *Nonparametric statistics for the behavioral sciences.* Tokyo: McGraw-Hill.

Siegler, R. S. (1976). Three aspects of cognitive development. *Cognitive Psychology, 8,* 481–520.

Siegler, R. S. (1981). Developmental sequences within and between concepts. *Monographs of the Society for Research in Child Development, 46*(2, Serial No. 189).

Shannon, C. E. (1948). A mathematical theory of communication. *The Bell System Technical Journal, 27,* 379–423, 623–656 (reprinted version with corrections).

Shavelson, R. J., & Webb, N. M. (1991). *Generalizability theory. A primer.* Thousand Oaks, CA: Sage.

Sijtsma, K. (1983). Rasch-homogeniteit empirisch onderzocht (Rasch homogeneity empirically examined). *Tijdschrift voor Onderwijsresearch, 8,* 104–121.

Sijtsma, K. (2009). On the use, the misuse, and the very limited usefulness of Cronbach's alpha. *Psychometrika, 74,* 107–120.

Sijtsma, K. (2012). Psychological measurement between physics and statistics. *Theory & Psychology, 22,* 786–809.

Sijtsma, K. (2015). Delimiting coefficient alpha from internal consistency and unidimensionality. *Educational Measurement: Issues and Practices, 34*(4), 10–13.

Sijtsma, K., & Emons, W. H. M. (2011). Advice on total-score reliability issues in psychosomatic measurement. *Journal of Psychosomatic Research, 70,* 565–572.

Sijtsma, K., & Emons, W. H. M. (2013). Separating models, ideas, and data to avoid a paradox: Rejoinder to Humphry. *Theory & Psychology, 23,* 786–796.

Sijtsma, K., & Hemker, B. T. (1998). Nonparametric polytomous IRT models for invariant item ordering, with results for parametric models. *Psychometrika, 63,* 183–200.

Sijtsma, K., & Hemker, B. T. (2000). A taxonomy of IRT models for ordering persons and items using simple sum scores. *Journal of Educational and Behavioral Statistics, 25,* 391–415.

Sijtsma, K., & Junker, B. W. (1996). A survey of theory and methods of invariant item ordering. *British Journal of Mathematical and Statistical Psychology, 49,* 79–105.

Sijtsma, K., & Junker, B. W. (2006). Item response theory: Past performance, present developments, and future expectations. *Behaviormetrika, 33,* 75–102.

Sijtsma, K., & Meijer, R. R. (1992). A method for investigating intersection of item response functions in Mokken's nonparametric IRT model. *Applied Psychological Measurement, 16,* 149–157.

Sijtsma, K., & Meijer, R. R. (2007). Nonparametric item response theory and related topics. In C. R. Rao & S. Sinharay (Eds.), *Handbook of statistics, Volume 26: Psychometrics* (pp. 719–746). Amsterdam, The Netherlands: Elsevier.

Sijtsma, K., Meijer, R. R., & Van der Ark, L. A. (2011). Mokken scale analysis as time goes by: An update for scaling practitioners. *Personality and Individual Differences, 50*, 31–37.

Sijtsma, K., & Molenaar, I. W. (1987). Reliability of test scores in nonparametric item response theory. *Psychometrika, 52*, 79–97.

Sijtsma, K., & Molenaar, I. W. (2002). *Introduction to nonparametric item response theory.* Thousand Oaks, CA: Sage.

Sijtsma, K., & Molenaar, I. W. (2016). Mokken models. In W. J. van der Linden (Ed.), *Handbook of item response theory. Volume One. Models* (pp. 303–321). Boca Raton, FL: Chapman & Hall/CRC.

Sijtsma, K., & Van der Ark, L. A. (2001). Progress in NIRT analysis of polytomous item scores: Dilemma's and practical solutions. In A. Boomsma, M. A. J. van Duijn, & T. A. B. Snijders (Eds.), *Essays on item response theory* (pp. 297–318). New York, NY: Springer.

Sijtsma, K., & Van der Ark, L. A. (2015). Conceptions of reliability revisited and practical recommendations. *Nursing Research, 64*, 128–136.

Sijtsma, K., & Van der Ark, L. A. (2017). A tutorial on how to do a Mokken scale analysis on your test and questionnaire data. *British Journal of Mathematical and Statistical Psychology, 70*, 137–158.

Sixtl, F. (1973). Probabilistic unfolding. *Psychometrika, 38*, 235–248.

Smits, I. A. M., Timmerman, M. E., & Meijer, R. R. (2012). Exploratory Mokken scale analysis as a dimensionality assessment tool: Why scalability does not imply unidimensionality. *Applied Psychological Measurement, 36*, 516–539.

Sorrel, M. A., Abad, F. J., Olea, J., de la Torre, J., & Barrada, J. R. (2017). Inferential item-fit evaluation in cognitive diagnosis modeling. *Applied Psychological Measurement, 41*, 614–631.

Sorrel, M. A., Olea, J., Abad, F. J., de la Torre, J., Aguado, D., & Lievens, F. (2016). Validity and reliability of situational judgement test scores: A new approach based on cognitive diagnosis models. *Organizational Research Methods, 19*, 506–532. doi:10.1177/1094428116630065

Spearman, C. (1904). 'General intelligence,' objectively determined and measured. *American Journal of Psychology, 15*, 201–293.

Spearman, C. (1907). Demonstration of formulæ for true measurement of correlation. *American Journal of Psychology, 18*, 161–169.

Spearman, C. (1910). Correlation calculated from faulty data. *British Journal of Psychology, 3*, 271–295.

Stark, S., Chernyshenko, O. S., Drasgow, F., & Williams, B. A. (2006). Examining assumptions about item responding in personality assessment: Should ideal point methods be considered for scale development and scoring? *Journal of Applied Psychology, 91*, 25–39.

Stevens, S. S. (1946). On the theory of scales of measurement. *Science, 103*, 677–680.

Stigler, S. M. (1986). *The history of statistics. The measurement of uncertainty before 1900.* Cambridge, MA: Harvard University Press.

Stout, W. F. (1987). A nonparametric approach for assessing latent trait unidimensionality. *Psychometrika, 52*, 589–617.

Stout, W. F. (1990). A new item response theory modeling approach with applications to unidimensionality assessment and ability estimation. *Psychometrika, 55*, 293–325.

Stout, W. F. (2002). Psychometrics: From practice to theory and back. *Psychometrika, 67*, 485–518.

Stout, W. F., Habing, B., Douglas, J., Kim, H., Roussos, L., & Zhang, J. (1996). Conditional covariance based nonparametric multidimensionality assessment. *Applied Psychological Measurement, 20,* 331–354.

Straat, J. H., Van der Ark, L. A., & Sijtsma, K. (2013). Comparing optimization algorithms for item selection in Mokken scale analysis. *Journal of Classification, 30,* 75–99.

Straat, J. H., Van der Ark, L. A., & Sijtsma, K. (2014). Minimum sample size requirements for Mokken scale analysis. *Educational and Psychological Measurement, 74,* 809–822.

Straat, J. H., Van der Ark, L. A., & Sijtsma, K. (2016). Using conditional association to identify locally independent item sets. *Methodology: European Journal of Research Methods for the Behavioral and Social Sciences, 12,* 117–123.

Stuive, I. (2007). *A comparison of confirmatory factor analysis methods. Oblique multiple group method versus confirmatory common factor method.* PhD dissertation, University of Groningen, Groningen, The Netherlands.

Suppes, P., & Zanotti, M. (1981). When are probabilistic explanations possible? *Synthese, 48,* 191–199.

Tekle, F. B., Gudicha, D. W., & Vermunt, J. K. (2016). Power analysis for the bootstrap likelihood ratio test for the number of classes in latent class models. *Advances in Data Analysis and Classification, 10,* 209–224.

Templin, J. L., & Henson, R. A. (2006). Measurement of psychological disorders using cognitive diagnosis models. *Psychological Methods, 11,* 287–305.

Ten Berge, J. M. F., Snijders, T. A. B., & Zegers, F. E. (1981). Computational aspects of the greatest lower bound to the reliability and constrained minimum trace factor analysis. *Psychometrika, 46,* 201–213.

Ten Berge, J. M. F., & Sočan, G. (2004). The greatest lower bound to the reliability of a test and the hypothesis of unidimensionality. *Psychometrika, 69,* 613–625.

Ten Berge, J. M. F., & Zegers, F. E. (1978). A series of lower bounds to the reliability of a test. *Psychometrika, 43,* 575–579.

Thissen, D. (1983). Timed testing: An approach using item response theory. In D. J. Weiss (Ed.), *New horizons in testing: Latent trait test theory and computerized adaptive testing* (pp. 179–203). New York, NY: Academic Press.

Thissen, D., & Cai, L. (2016). Nominal categories models. In W. J. van der Linden (Ed.), *Handbook of item response theory. Volume 1. Models* (pp. 51–73). Boca Raton, FL: Chapman & Hall/CRC.

Thissen, D., & Steinberg, L. (1984). A response model for multiple choice items. *Psychometrika, 49,* 501–519.

Thissen, D., & Steinberg, L. (1986). A taxonomy of item response models. *Psychometrika, 51,* 567–577.

Thissen, D., & Steinberg, L. (1997). A response model for multiple-choice items. In W. J. van der Linden & R. K. Hambleton (Eds.), *Handbook of modern item response theory* (pp. 51–65). New York, NY: Springer.

Thomson, G. H. (1938, 1951). *The factorial analysis of human ability.* Boston, MA: Houghton Mifflin.

Thorndike, R. L. (1951). Reliability. In E. F. Lindquist (Ed.), *Educational measurement* (pp. 560–620). Washington, DC: American Council on Education.

Thurstone, L. L. (1927a). A law of comparative judgment. *Psychological Review, 34,* 273–286.

Thurstone, L. L. (1927b). Psychophysical analysis. *The American Journal of Psychology, 38,* 368–389.

Thurstone, L. L. (1928). Attitudes can be measured. *The American Journal of Sociology, 33*, 529–554.

Thurstone, L. L. (1931). Multiple factor analysis. *Psychological Review, 38*, 406–427.

Thurstone, L. L. (1935). *The vectors of mind.* Chicago, IL: University of Chicago Press.

Thurstone, L. L. (1937). Ability, motivation, and speed. *Psychometrika, 2*, 249–254.

Thurstone, L. L., & Chave, E. J. (1929). *The Measurement of Attitude.* Chicago, IL: University of Chicago Press.

Tijmstra, J., Hessen, D. J., Van der Heijden, P. G. M., & Sijtsma, K. (2011). Invariant ordering of item-total regressions. *Psychometrika, 76*, 217–227.

Tijmstra, J., Hessen, D. J., Van der Heijden, P. G. M., & Sijtsma, K. (2013). Testing manifest monotonicity using order-constrained statistical inference. *Psychometrika, 78*, 83–97.

Tijmstra, J., Hoijtink, H., & Sijtsma, K. (2015). Evaluating manifest monotonicity using Bayes factors. *Psychometrika, 80*, 880–896.

Timmerman, M. E., & Lorenzo-Seva, U. (2011). Dimensionality assessment of ordered polytomous items with parallel analysis. *Psychological Methods, 16*, 209–220.

Trabasso, T. (1977). The role of memory as a system in making transitive inferences. In R. V. Kail, J. W. Hagen, & J. M. Belmont (Eds.), *Perspectives on the development of memory and cognition* (pp. 333–366). Hillsdale, NJ: Erlbaum.

Tsai, R.-C. (2000). Remarks on the identifiability of Thurstonian ranking models: Case V, Case III, or neither? *Psychometrika, 65*, 233–240.

Tsai, R.-C. (2003). Remarks on the identifiability of Thurstonian paired comparison models under multiple judgment. *Psychometrika, 68*, 361–372.

Tsukida, K., & Gupta, M. R. (2011). How to analyze paired comparison data. *UWEE Technical Report Number UWEETR-2011-0004.* Downloaded from https://aps.dtic.mil/dtic/tr/fulltext/u2/a543806.pdf

Tucker, L. R (1946). Maximum validity of a test with equivalent items. *Psychometrika, 11*, 1–13.

Tuerlinkcx, F., & De Boeck, P. (2005). Two interpretations of the discrimination parameter. *Psychometrika, 70*, 629–650.

Tuerlinckx, F., Molenaar, D., & Van der Maas, H. L. J. (2016). Diffusion-based response-time models. In W. J. van der Linden (Ed.), *Handbook of item response theory. Volume One. Models* (pp. 283–300). Boca Raton, FL: Chapman & Hall/CRC.

Tutz, G. (1990). Sequential item response models with an ordered response. *British Journal of Mathematical and Statistical Psychology, 43*, 39–55.

Tutz, G. (2016). Sequential models for ordered responses. In W. J. van der Linden, (Ed.), *Handbook of item response theory. Volume One. Models* (pp. 139–151). Boca Raton, FL: Chapman & Hall/CRC.

Ünlü, A. (2007). Nonparametric item response theory axioms and properties under nonlinearity and their exemplification with knowledge space theory. *Journal of Mathematical Psychology, 51*, 383–400.

Van Abswoude, A. A. H., Van der Ark, L. A., & Sijtsma, K. (2004). A comparative study of test data dimensionality assessment procedures under nonparametric IRT models. *Applied Psychological Measurement, 28*, 3–24.

Van Abswoude, A. A. H., Vermunt, J. K., Hemker, B. T., & Van der Ark, L. A. (2004). Mokken scale analysis using hierarchical clustering procedures. *Applied Psychological Measurement, 28*, 332–354.

Van Blokland-Vogelesang, R. (1991). *Unfolding and group consensus ranking for individual preferences.* University of Leiden, Leiden, The Netherlands: DSWO Press.

Van Bork, R., Grasman, R. P. P. P., & Waldorp, L. J. (2018a). Unidimensional factor models imply weaker partial correlations than zero-order correlations. *Psychometrika, 83,* 443–452.

Van Bork, R., Van Borkulo, C. D., Waldorp, L. J., Cramer, A. O. J., & Borsboom, D. (2018b). Network models for clinical psychology. In J. T. Wixted & E. J. Wagenmakers (Eds.), *Stevens' handbook of experimental psychology and cognitive neuroscience* (4th edition). *Volume 5. Methodology* (pp. 693–727). Hoboken, NJ: Wiley.

Van Borkulo, C. D., Borsboom, D., Epskamp, S., Blanken, T. F., Boschloo, L., Schoevers, R. A., & Waldorp, L. J. (2014). A new method for constructing networks from binary data. *Scientific Reports, 4:* 5918. doi:10.1038/srep05918

Van den Wollenberg, A. L. (1979). *The Rasch model and time-limit tests.* PhD dissertation, Radboud University Nijmegen, Nijmegen, The Netherlands: Stichting Studentenpers Nijmegen.

Van den Wollenberg, A. L. (1982a). Two new test statistics for the Rasch model. *Psychometrika, 47,* 123–139.

Van den Wollenberg, A. L. (1982b). A simple and effective method to test the dimensionality axiom of the Rasch model. *Applied Psychological Measurement, 6,* 83–91.

Van der Ark, L. A. (2001). Relationships and properties of polytomous item response theory models. *Applied Psychological Measurement, 25,* 273–282.

Van der Ark, L. A. (2005). Practical consequences of stochastic ordering of the latent trait under various polytomous IRT models. *Psychometrika, 70,* 283–304.

Van der Ark, L. A. (2007). Mokken scale analysis in R. *Journal of Statistical Software, 20,* 1–19.

Van der Ark, L. A. (2010). Computation of the Molenaar Sijtsma statistic. In A. Fink, B. Lausen, W. Seidel, & A. Ultsch (Eds.), *Advances in data analysis, data handling and business intelligence* (pp. 775–784). Berlin, Germany: Springer.

Van der Ark, L. A. (2012). New developments in Mokken scale analysis in R. *Journal of Statistical Software, 48,* 1–27.

Van der Ark, L. A. (2014). Visualizing uncertainty of estimated item response functions in nonparametric item response theory. In R. E. Millsap, L. A. van der Ark, D. Bolt, & C. M. Woods (Eds.), *New developments in quantitative psychology* (pp. 59–68). New York, NY: Springer.

Van der Ark, L. A., & Bergsma, W. P. (2010). A note on stochastic ordering of the latent trait using the sum of polytomous item scores. *Psychometrika, 75,* 272–279.

Van der Ark, L. A., Croon, M. A., & Sijtsma, K. (2008). Mokken scale analysis for dichotomous items using marginal models. *Psychometrika, 73,* 183–208.

Van der Ark, L. A., Hemker, B. T., & Sijtsma, K. (2002). Hierarchically related nonparametric IRT models, and practical data analysis methods. In G. Marcoulides & I. Moustaki (Eds.), *Latent variable and latent structure modeling* (pp. 40–62). Mahwah, NJ: Erlbaum.

Van der Ark, L. A., Rossi, G., & Sijtsma, K. (2019). Nonparametric item response theory and Mokken scale analysis, with relations to latent class models and cognitive diagnostic models. In M. von Davier & Y.-S. Lee (Eds.), *Handbook of diagnostic classification models. Models and model extensions, applications, software packages* (pp. 21–45). New York, NY: Springer.

Van der Ark, L. A., Van der Palm, D. W., & Sijtsma, K. (2011). A latent class approach to estimating test-score reliability. *Applied Psychological Measurement, 35,* 380–392.

Van der Linden, W. J. (2006). A lognormal model for response times on test items. *Journal of Educational and Behavioral Statistics, 31,* 181–204.

Van der Linden, W. J. (2007). A hierarchical framework for modeling speed and accuracy on test items. *Psychometrika, 72*, 287–308.

Van der Linden, W. J. (2009). Conceptual issues in response-time modeling. *Journal of Educational Measurement, 46*, 247–272.

Van der Linden, W. J. (2012). On compensation in multidimensional response modeling. *Psychometrika, 77*, 21–30.

Van der Linden, W. J. (Ed.) (2016a). *Handbook of item response theory. Volume One. Models*. Boca Raton, FL: Chapman & Hall/CRC.

Van der Linden, W. J. (2016b). Lognormal response-time model. In W. J. van der Linden, (Ed.), *Handbook of item response theory. Volume One. Models* (pp. 261–282). Boca Raton, FL: Chapman & Hall/CRC.

Van der Linden, W. J., & Barrett, M. D. (2016). Parameter linking. In W. J. van der Linden (Ed.), *Handbook of item response theory. Volume Three. Applications* (pp. 21–45). Boca Raton, FL: Chapman & Hall/CRC.

Van der Linden, W. J., & Glas, C. A. W. (Eds.) (2010). *Elements of adaptive testing*. New York, NY: Springer.

Van der Linden, W. J., & Hambleton, R. K. (1997). Item response theory: Brief history, common models, and extensions. In W. J. van der Linden & R. K. Hambleton (Eds.), *Handbook of modern item response theory* (pp. 1–28). New York, NY: Springer.

Van der Linden, W. J., Scrams, D. J., & Schnipke, D. L. (1999). Using response-time constraints to control for differential speededness in computerized adaptive testing. *Applied Psychological Measurement, 23*, 195–210.

Van der Linden, W. J., & Van Krimpen-Stoop, E. M. (2003). Using response times to detect aberrant responses in computerized adaptive testing. *Psychometrika, 68*, 251–265.

Van der Maas, H. L. J., Dolan, C. V., Grasman, R. P. P. P., Wicherts, J. M., Huizenga, H. M., & Raijmakers, M. E. J. (2006). A dynamic model of general intelligence: The positive manifold of intelligence by mutualism. *Psychological Review, 113*, 842–861.

Van der Maas, H. L. J., & Jansen, B. R. J. (2003). What response times tell of children's behavior on the balance scale task. *Journal of Experimental Child Psychology, 85*, 141–177.

Van der Maas, H. L. J., Kan, K.-J., & Borsboom, D. (2014). Intelligence is what the intelligence test measures. Seriously. *Journal of Intelligence, 2*, 12–15.

Van der Maas, H. L. J., Kan, K.-J., Marsman, M., & Stevenson, C. E. (2017). Network models for cognitive development and intelligence. *Journal of Intelligence, 5(2)*, 16; doi:10.3390/jintelligence5020016.

Van der Maas, H. L. J., Molenaar, D., Maris, G., Kievit, R. A., & Borsboom, D. (2011). Cognitive psychology meets psychometric theory: On the relation between process models for decision making and latent variable models for individual differences. *Psychological Review, 118*, 339–356.

Van der Palm, D. W., Van der Ark, L. A., & Sijtsma, K. (2014). A flexible latent class approach to estimating test-score reliability. *Journal of Educational Measurement, 51*, 339–357.

Van Engelenburg, G. (1997). *On psychometric models for polytomous items with ordered categories within the framework of item response theory*. Unpublished doctoral dissertation, University of Amsterdam, Amsterdam, The Netherlands.

Van Ginkel, J. R., Sijtsma, K., Van der Ark, L. A., & Vermunt, J. K. (2010). Incidence of missing item scores in personality measurement, and simple item-score imputation. *Methodology: European Journal of Research Methods for the Behavioral and Social Sciences, 6*, 17–30.

Van Maanen, L., Been, P. H., & Sijtsma, K. (1989). Problem solving strategies and the Linear Logistic Test Model. In E. E. Ch. I. Roskam (Ed.), *Mathematical psychology in progress* (pp. 267–287). New York, NY/Berlin: Springer.

Van Onna, M. J. H. (2002). Bayesian estimation and model selection in ordered latent class models for polytomous items. *Psychometrika, 67*, 519–538.

Van Rees, C. J., Vermunt, J. K., & Verboord, M. (1999). Cultural classifications under discussion. Latent class analysis of highbrow and lowbrow reading. *Poetics: Journal of Empirical Research on Literature, the Media and the Arts, 26*, 349–366.

Van Rijn, P. W., Eggen, T. J. H. M., Hemker, B. T., Sanders, P. F. (2002). Evaluation of selection procedures for computerized adaptive testing with polytomous items. *Applied Psychological Measurement, 26*, 393–411.

Van Rijn, P. W., & Rijmen, F. (2012). A note on explaining away and paradoxical results in multidimensional item response theory. *ETS Report Research Series*. Princeton, NJ: Educational Testing Service.

Van Schuur, W. H. (1984). *Structure in political beliefs. A new model for stochastic unfolding with application to European party activists*. PhD dissertation, University of Groningen, Groningen, The Netherlands.

Van Schuur, W. H. (1988). Stochastic unfolding. In W. E. Saris & I. N. Gallhofer (Eds.), *Sociometric research, Volume I: Data collection and scaling* (pp. 137–158). London, UK: Macmillan.

Van Schuur, W. H. (2003). Mokken scale analysis: Between the Guttman scale and parametric item response theory. *Political Analysis, 11*, 139–163.

Van Schuur, W. H. (2011). *Ordinal item response theory. Mokken scale analysis*. Thousand Oaks, CA: Sage.

Van Zyl, J. M., Neudecker, H., & Nel, D. G. (2000). On the distribution of the maximum likelihood estimator of Cronbach's alpha. *Psychometrika, 65*, 271–280.

Verhelst, N. (1998). *Estimating the reliability of a test from a single test administration*. Unpublished report No. 98-2. Arnhem, The Netherlands: Cito.

Verhelst, N. D., & Glas, C. A. W. (1993). A dynamic generalization of the Rasch model. *Psychometrika, 58*, 395–415.

Verhelst, N. D., & Glas, C. A. W. (1995). The one parameter logistic model. In G. H. Fischer & I. W. Molenaar (Eds.), *Rasch models. Foundations, recent developments, and applications* (pp. 215–237). New York, NY: Springer.

Verhelst, N. D., Glas, C. A. W., & De Vries, H. H. (1997). A steps model to analyze partial credit. In W. J. van der Linden & R. K. Hambleton (Eds.), *Handbook of modern item response theory* (pp. 123–138). New York, NY: Springer.

Verhelst, N. D., Glas, C. A. W., & Van der Sluis, A. (1984). Estimation problems in the Rasch model: The basic symmetric functions. *Computational Statistics Quarterly, 1*, 245–262.

Verhelst, N. D., & Verstralen, H. H. F. M. (1993). A stochastic unfolding model derived from the partial credit model. *Kwantitatieve Methoden, (42)*, 73–92.

Verhelst, N. D., Verstralen, H. H. F. M., & Jansen, M. G. H. (1997). A logistic model for time-limit tests. In W. J. van der Linden & R. K. Hambleton (Eds.), *Handbook of modern item response theory* (pp. 169–185). New York, NY: Springer.

Vermunt, J. K. (1997). *LEM: A general program for the analysis of categorical data. User's manual.* [Computer software] Downloaded from http://cwis.kub.nl/~fsw:1/mto/

Vermunt, J. K. (2001). The use of restricted latent class models for defining and testing nonparametric and parametric item response theory models. *Applied Psychological Measurement, 25,* 283–294.

Vermunt, J. K. (2003). Multilevel latent class models. *Sociological Methodology, 33,* 213–239.

Vermunt, J. K. (2010). Latent class modeling with covariates: Two improved three-step approaches. *Political Analysis, 18,* 450–469.

Vermunt, J. K., & Magidson, J. (2002). Latent class cluster analysis. In J. A. Hagenaars & A. L. McCutcheon (Eds.), *Applied latent class analysis* (pp. 89–106). Cambridge, UK: Cambridge University Press.

Vermunt, J. K., & Magidson, J. (2004). Latent class analysis. In M. S. Lewis-Beck, A. Bryman, & T. F. Liao (Eds.), *The Sage Encyclopedia of Social Sciences Research Methods* (pp. 549–553). Thousand Oaks, CA: Sage Publications.

Von Davier, M. (2008). A general diagnostic model applied to language testing data. *British Journal of Mathematical and Statistical Psychology, 61,* 287–307.

Von Davier, M. (2010). Hierarchical mixtures of diagnostic models. *Psychological Test and Assessment Modeling, 52,* 8–28.

Von Davier, M. (2016). Rasch model. In W. J. van der Linden (Ed.), *Handbook of item response theory. Volume One. Models* (pp. 31–48). Boca Raton, FL: Chapman & Hall/CRC.

Von Davier, M., & Carstensen, C. H. (Eds.) (2007). *Multivariate and mixture distribution Rasch models.* New York, NY: Springer.

Von Davier, A., Holland, P. W., & Thayer, D. T. (2004). *The kernel method of test equating.* New York, NY: Springer.

Von Davier, M., & Lee, Y.-S. (Eds.) (2019). *Handbook of diagnostic classification models. Models and model extensions, applications, software packages.* New York, NY: Springer.

Von Davier, M., Naemi, B., & Roberts, R. D. (2012). Factorial versus typological models: A comparison of methods for personality data. *Measurement: Interdisciplinary Research and Perspectives, 10,* 185–208.

Von Davier, M., & Rost, J. (2016). Logistic mixture-distribution response models. In W. J. van der Linden (Ed.), *Handbook of item response theory. Volume One. Models* (pp. 393–406). Boca Raton, FL: Chapman & Hall/CRC.

Vrieze, S. I. (2012). Model selection and psychological theory: A discussion of the differences between the Akaike Information Criterion (AIC) and the Bayesian Information Criterion (BIC). *Psychological Methods, 17,* 228–243. doi:10.1037%2Fa0027127

Wagenmakers, E. J., Van der Maas, H. L. J., & Grasman, R. P. P. P. (2007). An EZ-diffusion model for response time and accuracy. *Psychonomic Bulletin & Review, 14,* 3–22.

Wagenmakers, E. J., Wetzels, R., Borsboom, D., Van der Maas, H. L. J., & Kievit, R. A. (2012). An agenda for purely confirmatory research. *Perspectives on Psychological Science, 7,* 632–638.

Wainer, H. (Ed.) (2000). *Computerized adaptive testing: A primer* (2nd edition). Hillsdale, NJ: Erlbaum.

Wainer, H., Bradlow, E. T., & Wang, X. (2007). *Testlet response theory and its applications.* New York, NY: Cambridge University Press.

Wainer, H., & Braun, H. I. (Eds.) (1988). *Test validity*. Hillsdale, NJ: Erlbaum.

Wainer, H., & Thissen, D. (2001). True score theory: The traditional method. In D. Thissen & H. Wainer (Eds.), *Test scoring* (pp. 23–72). Mahwah, NJ: Erlbaum.

Walker, D. A. (1931). Answer pattern and score scatter in tests and examinations. *British Journal of Psychology, 22*, 73–86.

Waller, N. G., & Meehl, P. E. (1998). *Multivariate taxometric procedures: Distinguishing types from continua*. Thousand Oaks, CA: Sage.

Warm, T. A. (1989). Weighted likelihood estimation of ability in item response models. *Psychometrika, 54*, 427–450.

Wasserman, L. A. (2004). *All of statistics. A concise course in statistical inference*. New York, NY: Springer.

Whitely, S. E. (1980). Multicomponent latent trait models for ability tests. *Psychometrika, 45*, 479–494.

Wood, R. (1978). Fitting the Rasch model—a heady tale. *British Journal of Mathematical and Statistical Psychology, 31*, 27–32.

Woodhouse, B., & Jackson, P. H. (1977). Lower bounds for the reliability of the total score on a test composed of non-homogeneous items: II. A search procedure to locate the greatest lower bound. *Psychometrika, 42*, 579–591.

Woodward, J. A., & Bentler, P. M. (1978). A statistical lower-bound to population reliability. *Psychological Bulletin, 85*, 1323–1326.

Wright, B. D., & Stone, M. H. (1979). *Best test design. Rasch measurement*. Chicago, IL: Mesa Press.

Yang, X., & Embretson, S. E. (2007). Construct validity and cognitive diagnostic assessment. In J. A. Leighton & M. J. Gierl (Eds.), *Cognitive diagnostic assessment for education. Theory and applications* (pp. 119–145). Cambridge, UK: Cambridge University Press.

Yen, W. (1981). Using simulation results to choose a latent trait model. *Applied Psychological Measurement, 5*, 245–262. doi:10.1177/014662168100500212

Yoder, J. G. (2005). Christiaan Huygens, book on the pendulum clock (1673). In I. Grattan-Guinness (Ed.), *Landmark writings in Western mathematics* (pp. 33–45). Amsterdam, the Netherlands: Elsevier.

Zegers, F. E., & Ten Berge, J. M. F. (1982). Necessary and sufficient conditions for parallelism of tests in classical test theory. *Tijdschrift voor Onderwijsresearch, 7*, 76–79.

Zermelo, E. (1929). Die Berechnung der Turnier-Ergebnisse als ein Maximumproblem der Wahrscheinlichkeitsrechnung. *Mathematische Zeitschrift, 29*, 436–460.

Zhang, J., & Stout, W. F. (1999a). The theoretical DETECT index of dimensionality and its application to approximate simple structure. *Psychometrika, 64*, 213–249.

Zhang, J., & Stout, W. F. (1999b). Conditional covariance structure of generalized compensatory multidimensional items. *Psychometrika, 64*, 129–152.

Zinbarg, R. E., Revelle, W., Yovel, I., & Li, W. (2005). Cronbach's α, Revelle's β, and McDonald's ω_H: Their relations with each other and two alternative conceptualizations of reliability. *Psychometrika, 70*, 123–133.

Zijlmans, E. A. O., Van der Ark, L. A., Tijmstra, J., & Sijtsma, K. (2018). Methods for estimating item-score reliability. *Applied Psychological Measurement, 42*, 553–570.

Zijlstra, W. P., Van der Ark, L. A., & Sijtsma, K. (2007). Outlier detection in test and questionnaire data. *Multivariate Behavioral Research, 42*, 531–555.

Zwinderman, A. H. (1991). A generalized Rasch model for manifest predictors. *Psychometrika, 56*, 589–600.

Zwinderman, A. H. (1997). Response models with manifest predictors. In W. J. van der Linden & R. K. Hambleton (Eds.), *Handbook of modern item response theory* (pp. 245–256). New York, NY: Springer.

Zwitser, R. J., & Maris, G. (2016). Ordering individuals with sum scores: The introduction of the nonparametric Rasch model. *Psychometrika, 81*, 39–59.

Index

Printed in the United States
by Baker & Taylor Publisher Services